Expression Systems

METHODS EXPRESS

The **METHODS EXPRESS** series

Series editor: B. David Hames

Faculty of Biological Sciences, University of Leeds, Leeds LS2 9JT, UK

Animal Cell Culture

Bioinformatics

Biosensors

Cell Imaging

DNA Microarrays

Expression Systems

Genomics

G-PCR

Immunohistochemistry

In Situ Hybridization

PCR

Plant Cell Culture

Protein Arrays

Proteomics

Whole Genome Amplification

METHODS EXPRESS

Expression Systems

METHODS EXPRESS

edited by **M.R. Dyson**
*The Wellcome Trust Sanger Institute,
Cambridge, UK*

and **Y. Durocher**
*Animal Cell Technology Group,
Biotechnology Research Institute,
Montreal, Canada*

© Scion Publishing Ltd, 2007

First published 2007

All rights reserved. No part of this book may be reproduced or transmitted, in any form or by any means, without permission.

A CIP catalogue record for this book is available from the British Library.

ISBN: 978 1 904842 43 9 (paperback)
ISBN: 978 1 904842 45 3 (hardback)

Scion Publishing Limited
Bloxham Mill, Barford Road, Bloxham, Oxfordshire OX15 4FF
www.scionpublishing.com

Important Note from the Publisher

The information contained within this book was obtained by Scion Publishing Limited from sources believed by us to be reliable. However, while every effort has been made to ensure its accuracy, no responsibility for loss or injury whatsoever occasioned to any person acting or refraining from action as a result of information contained herein can be accepted by the authors or publishers.

Typeset by Phoenix Photosetting, Chatham, Kent, UK
Printed by Ajanta Offset and Packagings Ltd, Delhi, India

Cover image: The cover illustrates a montage of maltose binding protein fusions to murine immunoglobulin superfamily (IgSF) receptor intracellular domains expressed in *E. coli* and purified by immobilized metal affinity chromatography. Proteins were analyzed using a LabChip® 90 automated electrophoresis system (Caliper Life Sciences Inc, Mountain View, CA 94043, USA) which employs a microfluidic chip to separate proteins based on molecular weight, stain and destain prior to fluorescence detection. The image is a gel view in pseudo color where the bands are colored as a heat map spectrum ranging from red (most intense) to green–blue (low intensity). The data for the image were generated by R.L. Perera and M.R. Dyson (The Wellcome Trust Sanger Institute, Cambridge, UK).

Contents

Contributors xi
Preface xiv
Abbreviations xv

Color section xix

Chapter 1. Expression strategy
Michael R. Dyson
1. Introduction 1
2. Methods and approaches 3
   2.1 Case study 1: mouse Rab3a 4
   2.2 Case study 2: human epidermal growth factor receptor 8
   2.3 Expression of previously unexpressed proteins 9
3. Troubleshooting 10
4. References 10

Chapter 2. Protein expression in *Escherichia coli*
Rosalind Kim
1. Introduction 13
2. Methods and approaches 14
   2.1 Ligation-independent cloning 14
   2.2 Small-scale expression 19
   2.3 Cell paste growth 22
   2.4 Clear lysate preparation 23
   2.5 Purification of His_6-tagged protein 24
   2.6 Tobacco etch virus protease digestion of fusion protein 25
3. Troubleshooting 27
4. References 28

Chapter 3. Expression engineering of synthetic antibodies using ribosome display
Lydia M. Contreras Martínez and Matthew P. DeLisa
1. Introduction — 29
2. Methods and approaches — 31
   - 2.1 Principles of ribosome display — 31
   - 2.2 Experimental overview of ribosome display — 32
   - 2.3 ScFv gene sequence and control elements — 33
   - 2.4 Construction of an scFv ribosome display vector (pscFvDisplay) — 34
   - 2.5 Amplification and purification of the DNA target — 37
   - 2.6 *In vitro* transcription of a purified DNA library — 39
   - 2.7 *In vitro* translation of a mRNA library — 40
   - 2.8 Affinity selection of ribosome complexes and mRNA isolation — 44
   - 2.9 Reverse transcriptase PCR — 44
   - 2.10 *In vivo* analysis of selected scFvs — 49
3. Troubleshooting — 49
4. References — 50

Chapter 4. Refolding proteins from inclusion bodies
Renaud Vincentelli
1. Introduction — 53
2. Methods and approaches — 54
   - 2.1 Preparation and purification of inclusion bodies — 57
   - 2.2 Production — 59
   - 2.3 Solubilization of inclusion bodies — 60
   - 2.4 Refolding of solubilized proteins — 60
   - 2.5 Analysis of refolded protein — 63
   - 2.6 Conclusion — 64
3. References — 64

Chapter 5. Selection of protein variants with improved expression using green fluorescent protein-derived folding and solubility reporters
Stéphanie Cabantous and Geoffrey S. Waldo
1. Introduction — 67
2. Methods and approaches — 68
   - 2.1 GFP insertion technology — 68
   - 2.2 Principles of GFP insertion — 69
   - 2.3 Methodology — 71
   - 2.4 Selecting optima with improved solubility — 72
3. Troubleshooting — 85
4. References — 86

Chapter 6. Protein expression in the wheat-germ cell-free system
Tatsuya Sawasaki and Yaeta Endo

1.	Introduction	87
2.	Methods and approaches	88
	2.1 Principles of the new wheat-germ cell-free system	88
	2.2 Preparation of the extract from wheat embryos	89
	2.3 Development of the 5'UTR of mRNA to enhance translation	92
	2.4 Expression vector pEU	92
	2.5 PCR-directed generation of DNA template	93
	2.6 Preparation of mRNA	95
	2.7 Translation	97
	2.8 Adapting the CFCF reaction for transcription and translation in one tube	100
	2.9 Applications based on the cell-free system	102
3.	Troubleshooting	106
4.	References	107

Chapter 7. *Saccharomyces cerevisiae*: a microbial eukaryotic expression system
Christine Lang

1.	Introduction	109
2.	Methods and approaches	110
	2.1 Vectors and promoters	110
	2.2 Strains	112
	2.3 Analysis of expression	113
2.4 Recommended protocols		114
3.	Troubleshooting	120
4.	References	121

Chapter 8. Expression of proteins in *Pichia pastoris*
Geoff P. Lin-Cereghino, Wilson Leung, and Joan Lin-Cereghino

1.	Introduction	123
	1.1 Background	124
	1.2 Choosing a plasmid	124
	1.3 Choosing a host strain	127
2.	Methods and approaches	130
	2.1 Transformation	130
	2.2 Screening of transformants	133
	2.3 Small-scale expression	135
	2.4 Optimization	141
	2.5 Small-scale purification	141
	2.6 Considerations for scaling up expression and purification	142
3.	Troubleshooting	143
4.	References	144

Chapter 9. Improved baculovirus expression vectors
Richard B. Hitchman, Robert D. Possee, and Linda A. King

1. Introduction — 147
2. Methods and approaches — 147
   - 2.1 Principles of *flash*BAC — 147
   - 2.2 Insect cell culture requirements — 149
   - 2.3 Maintaining insect cells in monolayer culture — 153
   - 2.4 Construction of recombinant baculoviruses using *flash*BAC — 155
   - 2.5 Automated production of recombinant baculoviruses — 162
3. Troubleshooting — 167
4. References — 167

Chapter 10. Transient transfection of insect cells for rapid expression screening and protein production
Kathryn H. Loomis, Courtney R. Rockwell, Heather D. Sternard, Keith W. Yaeger, and Robert E. Novy Jr

1. Introduction — 169
2. Methods and approaches — 170
   - 2.1 Media and insect cells — 170
   - 2.2 Transient expression vectors — 171
   - 2.3 Transfection reagent — 171
   - 2.4 Insect cell lysis — 172
   - 2.5 IMAC purification of His•Tag fusion proteins — 172
   - 2.6 Recommended protocols — 172
   - 2.7 Examples of results — 177
3. Troubleshooting — 180
4. References — 182

Chapter 11. Generation of stable CHO cell lines for protein expression
Zhijian Lu, Haley Laken, Jimin Zhang, Xiaotian Zhong, and Richard Zollner

1. Introduction — 183
   - 1.1 Cell lineage — 184
   - 1.2 Selection markers — 185
2. Methods and approaches — 185
   - 2.1 Cell line maintenance — 185
   - 2.2 Cryopreservation of CHO cells — 188
   - 2.3 Cell line construction — 190
   - 2.4 Scaling up of CHO cells for recombinant protein production — 196
   - 2.5 Specialized application: host engineering — 198
3. References — 200

Chapter 12. Transient expression in HEK293-EBNA1 cells
Roseanne Tom, Louis Bisson, and Yves Durocher
1. Introduction 203
2. Methods and approaches 206
   2.1 Cell culture 206
   2.2 Plasmid DNA preparation 209
   2.3 Preparation of PEI 210
   2.4 Transfection of 293E and 293-6E cells 212
   2.5 Purification of His-tagged r-proteins 218
   2.6. Results 220
3. Troubleshooting 221
4. References 223

Chapter 13. Nisin- and subtilin-controlled gene expression systems for Gram-positive bacteria
Oscar P. Kuipers and Jan Kok
1. Introduction 225
2. Methods and approaches 228
   2.1 The NICE system 228
   2.2 The SURE system 235
   2.3 Future applications 238
3. Troubleshooting 238
4. References 239

Chapter 14. Protein production using lentiviral vectors
Rénald Gilbert, Sophie Broussau, and Bernard Massie
1. Introduction 241
   1.1 Properties of lentiviral vectors 241
   1.2 Essential components of LVs 241
   1.3 Strategies to control protein expression from LVs 243
2. Methods and approaches 244
   2.1 Production of LVs 244
   2.2 Concentration of LVs 250
   2.3 LV titration 252
   2.4 Cell marking and protein production using LVs 255
3. Troubleshooting 258
4. References 259

Chapter 15. Expression in mammalian cells using BacMam viruses
Hsiao-Ping Lee and Yu-Chen Hu
1. Introduction 261
2. Methods and approaches 262
   2.1 Construction and production of BacMam virus 262

2.2 Transduction of mammalian cells using culture medium as the
      surrounding solution   262
  2.3 Improved protocol for BacMam transduction   265
  2.4 Protein production in a BelloCell-500 bioreactor   267
  2.5 Determination of baculovirus transducing ability in
      mammalian cells   271
3. Troubleshooting   274
4. References   274

Appendix 1
List of suppliers   277

Index   281

Contributors

Bisson, Louis Animal Cell Technology Group, Biotechnology Research Institute, 6100 Royalmount Avenue, Montreal, QC, Canada H4P 2R2.
E-mail: louis.bisson@cnrc-nrc.gc.ca

Broussau, Sophie Genomics & Gene Therapy Vectors, Biotechnology Research Institute, NRC, 6100 Royalmount Avenue, Montreal, QC, Canada H4P 2R2.
E-mail: Sophie.Brousau@cnrc-nrc.gc.ca

Cabantous, Stéphanie Bioscience Division, MS-M888, Los Alamos National Laboratory, PO Box 1663, Los Alamos, NM 87545, USA.
E-mail: scaban@lanl.gov

Contreras Martínez, Lydia M. 304 Olin Hall, Cornell University, Ithaca, NY 14853, USA. E-mail: lmc67@cornell.edu

DeLisa, Matthew P. 254 Olin Hall, Cornell University, Ithaca, NY 14853, USA.
E-mail: md255@cornell.edu

Durocher, Yves Animal Cell Technology Group, Biotechnology Research Institute, 6100 Royalmount Avenue, Montreal, QC, Canada H4P 2R2.
E-mail: yves.durocher@nrc-cnrc.gc.ca

Dyson, Michael R. The Wellcome Trust Sanger Institute, Wellcome Trust Genome Campus, Hinxton, Cambridge, CB10 1SA, UK.
E-mail: mrd@sanger.ac.uk

Endo, Yaeta Cell-Free Science and Technology Research Center, Ehime University, Matsuyama 790-8577, Japan. E-mail: yendo@eng.ehime-u.ac.jp

Gilbert, Rénald Genomics & Gene Therapy Vectors, Biotechnology Research Institute, NRC, 6100 Royalmount Avenue, Montreal, QC, Canada H4P 2R2.
E-mail: renald.gilbert@cnrc-nrc.gc.ca

Hitchman, Richard B. School of Biological and Molecular Sciences, Oxford Brookes University, Headington Campus, Gipsy Lane, Oxford, OX3 0BP, UK.
E-mail: rhitchman@brookes.ac.uk

Hu, Yu-Chen Department of Chemical Engineering, National Tsing Hua University, Hsinchu, Taiwan 300. E-mail: yuchen@che.nthu.edu.tw

Kim, Rosalind Lawrence Berkeley National Laboratory, Physical Biosciences Division, One Cyclotron Road, Mail Stop: Calvin Lab – 350, Berkeley, CA 94720, USA. Current address: Department of Chemistry, University of California, Berkeley, CA 94729, USA. E-mail: r_kim@lbl.gov

King, Linda A. School of Biological and Molecular Sciences, Oxford Brookes University, Headington Campus, Gipsy Lane, Oxford, OX3 0BP, UK. E-mail: laking@brookes.ac.uk

Kok, Jan Department of Molecular Genetics, Groningen Biomolecular Sciences and Biotechnology Institute, University of Groningen, Kerklaan 30, 9751 NN Haren, The Netherlands. E-mail: jan.kok@rug.nl

Kuipers, Oscar P. Department of Molecular Genetics, Groningen Biomolecular Sciences and Biotechnology Institute, University of Groningen, Kerklaan 30, 9751 NN Haren, The Netherlands. E-mail: o.p.kuipers@rug.nl

Laken, Haley Wyeth Biotech, 1 Butt Road, Andover, MA 01810, USA.

Lang, Christine Institut für Biotechnologie, Technische Universität Berlin, FG Mikrobiologie und Genetik, Gustav-Meyer-Allee 25, D-13355 Berlin, Germany. E-mail: christine.lang@tu-berlin.de

Lee, Hsiao-Ping Department of Chemical Engineering, National Tsing Hua University, Hsinchu, Taiwan 300. E-mail: d913648@oz.nthu.edu.tw

Leung, Wilson Department of Biological Sciences, University of the Pacific, 1050 W. Brookside Road Stockton, CA 95211, USA. E-mail: w_leung@pacific.edu

Lin-Cereghino, Geoff P. Department of Biological Sciences, University of the Pacific, 1050 W. Brookside Road Stockton, CA 95211, USA. E-mail: glincere@pacific.edu

Lin-Cereghino, Joan Department of Biological Sciences, University of the Pacific, 1050 W. Brookside Road Stockton, CA 95211, USA. E-mail: jlincere@pacific.edu

Loomis, Kathryn H. EMD Biosciences, Inc., 441 Charmany Drive, Madison, WI 53719, USA. E-mail: kathryn.loomis@emdbiosciences.com

Lu, Zhijian Wyeth Research, 87 CambridgePark Drive, Cambridge, MA 02140, USA. E-mail: zlu@wyeth.com

Massie, Bernard Biotechnology Research Institute, 6100 Royalmount Avenue, Montreal, QC, Canada H4P 2R2. E-mail: bernard.massie@nrc.gc.ca

Novy, Robert E. Jr. EMD Biosciences, Inc., 441 Charmany Drive, Madison, WI 53719, USA. E-mail: bob.novy@emdbiosciences.com

Possee, Robert D. CEH Oxford, Mansfield Road, Oxford, OX1 3SR, UK. E-mail: rdpo@ceh.ac.uk

Rockwell, Courtney R. EMD Biosciences, Inc., 441 Charmany Drive, Madison, WI 53719, USA. E-mail: courtney.rockwell@emdbiosciences.com

Sawasaki, Tatsuya Cell-Free Science and Technology Research Center, Ehime University, Matsuyama 790-8577, Japan. E-mail: sawasaki@eng.ehime-u.ac.jp

Sternard, Heather D. EMD Biosciences, Inc., 441 Charmany Drive, Madison, WI 53719, USA. E-mail: heather.sternard@emdbiosciences.com

Tom, Roseanne Animal Cell Technology Group, Biotechnology Research Institute, 6100 Royalmount Avenue, Montreal, QC, Canada H4P 2R2. E-mail: roseanne.tom@cnrc-nrc.gc.ca

Vincentelli, Renaud AFMB-UMR CNRS 6098 Case 932, 163 avenue de Luminy, 13288 Marseille Cedex 09, France. E-mail: renaud.vincentelli@afmb.univ-mrs.fr

Waldo, Geoffrey S. Bioscience Division, MS-M888, Los Alamos National Laboratory, PO Box 1663, Los Alamos, NM 87545, USA. E-mail: waldo@lanl.gov

Yaeger, Keith W. EMD Biosciences, Inc., 441 Charmany Drive, Madison, WI 53719, USA. E-mail: keith.yaeger@emdbiosciences.com

Zhang, Jimin Wyeth Research, 87 CambridgePark Drive, Cambridge, MA 02140, USA. E-mail: jzhang@wyeth.com

Zhong, Xiaotian Wyeth Research, 87 CambridgePark Drive, Cambridge, MA 02140, USA. E-mail: xzhong@wyeth.com

Zollner, Richard Wyeth Research, 87 CambridgePark Drive, Cambridge, MA 02140, USA. E-mail: rzollner@wyeth.com

Preface

Protein expression and purification used to be the preserve of the biochemist working to elucidate the function or enzymatic activity of a protein or the molecular/structural biologist aiming to solve the structure of a protein, perhaps in association with its binding partner. However, as science becomes increasingly inter-disciplinary in nature and the best journals demand a combined approach to describing gene function, so protein expression is now being performed by a broader cross-section of the scientific community in both academia and industry. This demands laboratory manuals that clearly describe and provide state-of-the-art protocols for protein expression, similar to the *Molecular Cloning: A Laboratory Manual* (Cold Spring Harbor Laboratory Press) series, which taught molecular biology to a generation of scientists. The protocols should be written for both the novice and more-experienced researchers looking to attempt a new expression system. With this single volume, covering the major expression hosts, it is hoped that an informed choice can be made at the beginning of an expression project, taking into account the protein target, the time involved, the ultimate use of the expressed protein, and the laboratory equipment required. Although not exhaustive, the main expression systems are covered from prokaryotes to lower and higher eukaryotes and this should enable the majority of proteins to be expressed successfully.

We would like to thank David Hames (Leeds University) for useful discussions, Jonathan Ray and Clare Boomer (Scion Publishing Ltd) for production of this volume, and Jane Hoyle for copy-editing.

Michael R. Dyson & Yves Durocher
June 2007

Abbreviations

aa	amino acid(s)
AcMNPV	*Autographa californica* multiple nucleopolyhedrovirus
Amp	ampicillin
AOX	alcohol oxidase
AP	alkaline phosphatase
ATCC	American Type Culture Collection
BAC	bacterial artificial chromosome
BSA	bovine serum albumin
BSGC	Berkeley Structural Genomics Center
BV	budded virus
cAMP	cyclic AMP
CFCF	continuous-flow cell-free
CHO	Chinese hamster ovary
CMV	cytomegalovirus
cPPT	central polypurine track
cSIN-LV	conditional self-inactivating lentiviral vector
C_t	cycle threshold
cTA	cumate transactivator
DHFR	dihydrofolate reductase
DMEM	Dulbecco's modified Eagle's medium
DMSO	dimethyl sulfoxide
DNase	deoxyribonuclease I
DSMZ	German Collection of Microorganisms and Cell Cultures
DTT	dithiothreitol
EGFP	enhanced green fluorescent protein
EGFR	epidermal growth factor receptor
ELISA	enzyme-linked immunosorbent assay
FBS	fetal bovin serum
FLD	formaldehyde dehydrogenase
FMN	flavin mononucleotide
GAP	glyceraldehyde 3-phosphate dehydrogenase
GFP	green fluorescent protein
GlcNAc-TI	UDP-*N*-acetylglucosamine : α-3-D-mannoside β-1,2-*N*-acetylglucosaminyltransferase I
GPI	glycosylphosphatidylinositol

GST	glutathione *S*-transferase
HBsAg	hepatitis B virus surface antigen
HDV	hepatitis delta virus
HEK	human embryonic kidney
HIV	human immunodeficiency virus
hMSC	human mesenchymal stem cell
HRP	horseradish peroxidase
IB	inclusion body
IMAC	immobilized metal affinity chromatography
IPTG	isopropyl β-D-thiogalactopyranoside
LB	Luria–Bertani
LCH	*Lens culinaris* lectin
LC-SFM	low-calcium, serum-free medium
L-HDAg	large hepatitis virus delta antigen
LIC	ligation-independent cloning
L-PHA	*Phaseolus vulgaris* L lectin
LTR	long terminal repeat
LV	lentiviral vector
m.o.i.	multiplicity of infection
MAD	multi-wavelength anomalous diffraction
MBP	maltose-binding protein
MTX	methotrexate
NCBI	National Center for Biotechnology Information
NICE	nisin-controlled expression
NMR	nuclear magnetic resonance
ORF	open reading frame
p.f.u.	plaque-forming units
p.i.	post-infection
p.t.	post-transfection
PAGE	polyacrylamide gel electrophoresis
PBS	phosphate-buffered saline
PBS-CMF	phosphate-buffered saline, calcium- and magnesium-free
PCR	polymerase chain reaction
PDB	Protein Data Bank
PDI	protein disulfide isomerase
PEI	polyethylenimine
PMSF	phenylmethylsulfonyl fluoride
PVDF	polyvinylidene difluoride
QPCR	quantitative polymerase chain reaction
RIP	ribosome-inactivating protein
Rluc	*Renilla* luciferase
r-protein	recombinant protein
RRE	Rev-responsive element
RSV	Rous sarcoma virus
RT	reverse transcriptase
RT-PCR	reverse transcriptase PCR

scFvs	single-chain Fv antibody fragments
SD	Shine–Dalgarno
SDS	sodium dodecyl sulfate
SIN-LV	self-inactivating lentiviral vector
SMART	Simple Modular Architecture Research Tool
SURE	subtilin-regulated expression
TEV	tobacco etch virus
tmRNA	transfer messenger RNA
TU	transducing unit
UTR	untranslated region
VLP	virus-like particle
VSV-G	vesicular stomatitis virus glyco protein

Color Section

Chapter 1. Expression strategy

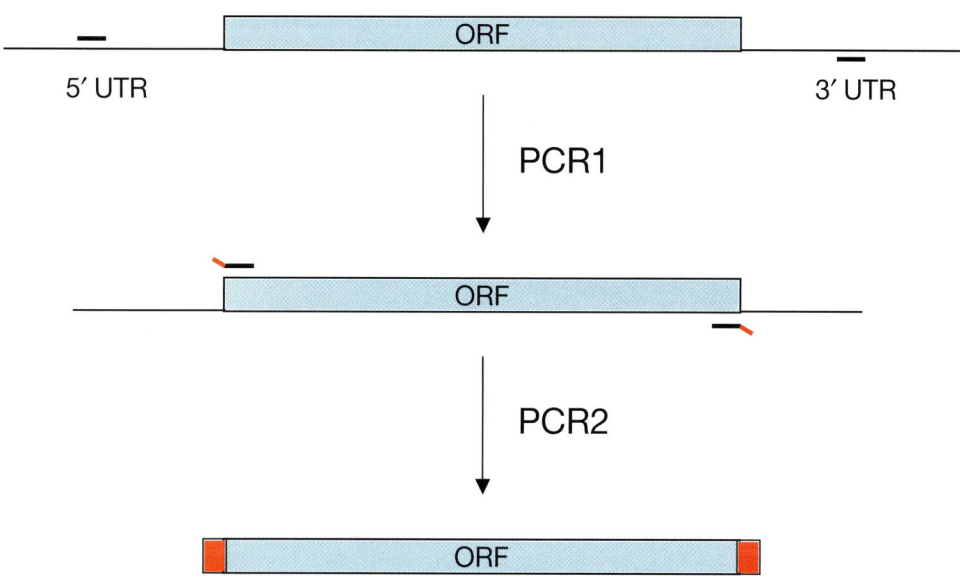

Figure 2. Nested PCR strategy (see page 6).
PCR1 amplifies from a mixed cDNA pool with primers either placed in the UTRs or 5' and 3' of the DNA encoding an individual protein domain. PCR2 amplifies the DNA encoding the protein to be expressed with tags (red) to aid restriction- or recombination-based cloning into an expression vector.

Figure 3. Domain topology of human EGFR (1210 aa) (see page 9).
Marked protein features and Pfam domains include: A, signal peptide (aa 1–24); B, receptor L domain (aa 57–168 and 361–481); C, furin-like cysteine rich region (aa 184–338); D, transmembrane region (aa 646–668); E, low-compexity regions (aa 651–666, 1002–1015 and 1025–1046); and F, protein tyrosine kinase domain (aa 712–968).

Chapter 5. Selection of protein variants with improved expression using green fluorescent protein-derived folding and solubility reporters

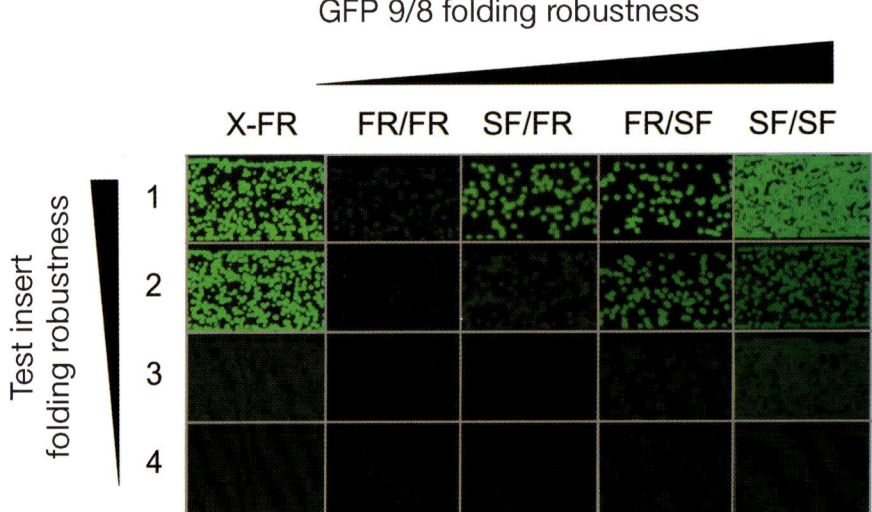

Figure 2 (c). GFP insertion vector topology and reporter stringency (see page 70).
Reporter stringency for X-FR (conventional C-terminal folding reporter GFP) and the four GFP insertion vectors. The reporter abbreviation above each column designates the identity of the GFP variant from which the flanking fragment is derived. Each row corresponds to one of four *Pyrobaculum aerophilum* test proteins with different solubility levels (#1 is fully soluble, #2 and #3 are partially soluble, and #4 is totally insoluble). The least-stringent vectors (SF/SF) are able to detect most proteins. As the number of SF mutations decreases, they detect proteins with increased solubility, as for the most stringent FR/FR.

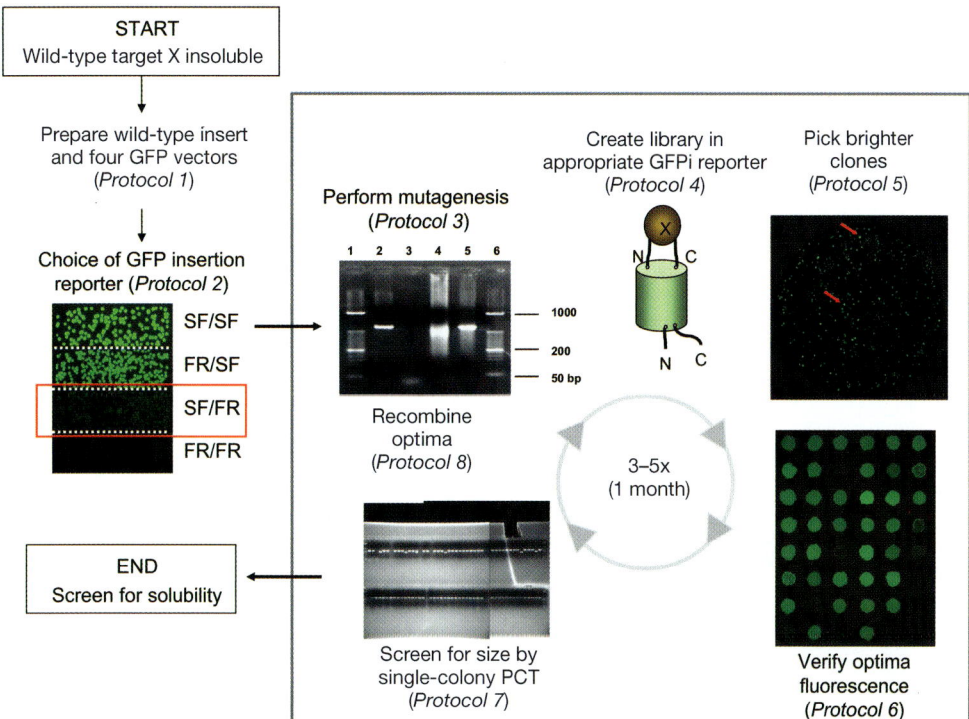

Figure 3. Directed evolution of protein folding (see page 71).
The most suitable GFP insertion vector is identified and then libraries of target genes encoding insoluble targets are subjected to random mutagenesis by DNA shuffling: a pool of homologous genes (*lane 2*) is fragmented into small ~20 bp fragments with DNase I (*lane 3*), which are reassembled by primerless PCR into full-length genes (*lane 4*) and reamplified with vector-specific primers to yield the full-length gene (*lane 5*). The purified mutant library is digested by restriction endonuclease, sized by preparative gel electrophoresis, and cloned into the appropriate GFP folding reporter (*Protocol 4*). Variants with improved folding robustness are associated with brighter fluorescent colonies (see image of *in vivo* induction plate, *Protocol 5*), whose fluorescence is verified by spotting individual clones onto nitrocellulose membrane (*Protocol 6*). Optima clones are screened for insert length by single-colony PCR and analytical agarose gel electrophoresis (*Protocol 7*). Full-length clones with improved folding are recombined for another cycle of evolution (*Protocol 8*).

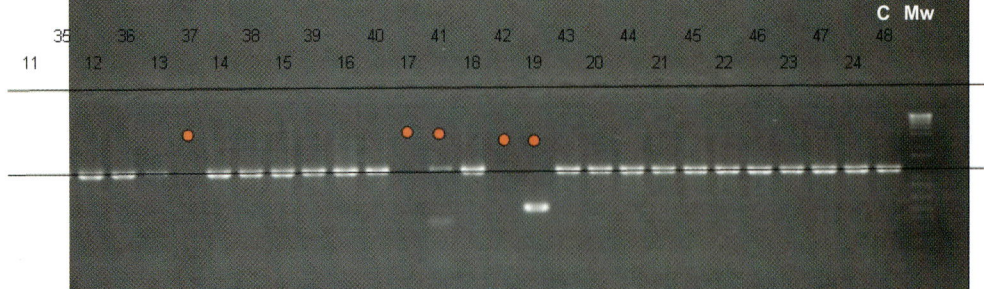

Figure 4. Analytical gel of single-colony PCR screen (see page 73).
PCR products for optima clones and the positive control (C, lane 48). Red dots indicate clones to discard (truncated or missing sites).

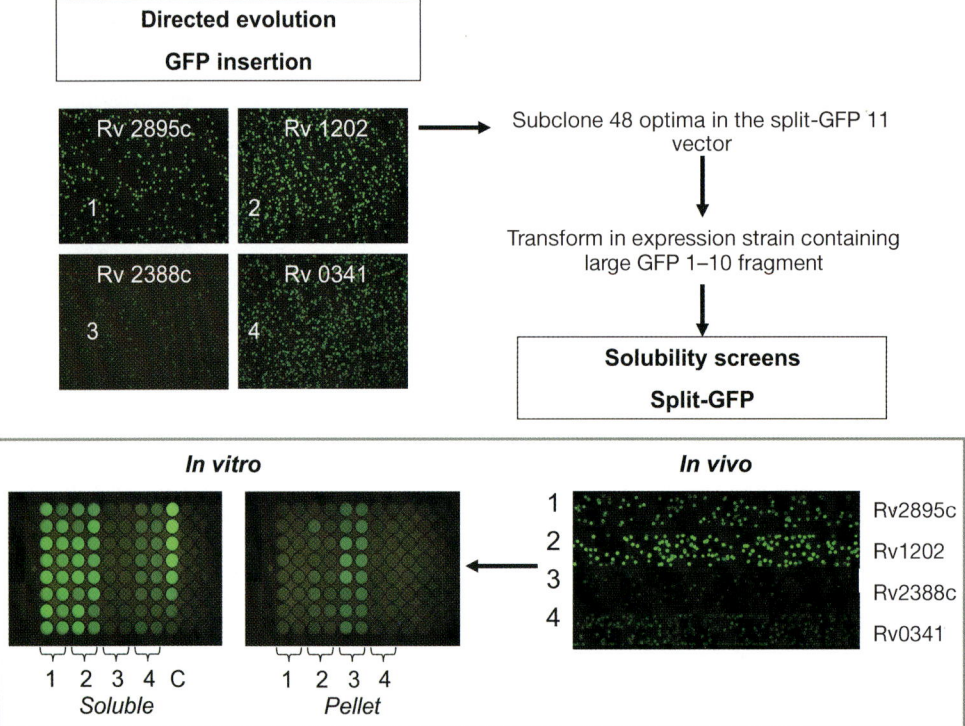

Figure 5. Assessing solubility on variants with improved folding (see page 84).
At the fourth round of directed evolution in the GFP folding reporter vector, the library of *M. tuberculosis* protein variants display stable levels of fluorescence (top left). Forty-eight optima are selected from the brightest colonies and the variants are subcloned into the split-GFP 11 solubility reporter vector and transformed into an *E. coli* strain containing the large, complementing GFP 1–10 fragment. The four *M. tuberculosis* targets are screened for solubility *in vivo* (bottom right) and the 16 brightest clones are grown and induced in 96-well plates. Soluble and pellet fractions are assayed *in vitro* by adding the large, complementing GFP fragment assay reagent for precise quantification. Protein #3 (Rv2388c) is an example of slow evolution that requires a change in strategy (faint in both GFP insertion and split-GFP reporters). Proteins 1, 2, and 4 display most fluorescence in the soluble fraction (bottom left).

Chapter 6. Protein expression in the wheat-germ cell-free system

Figure 2. Wheat embryos (see page 90).
Extensive washing of wheat embryos eliminates endosperm contaminants, visible as a white color on the unwashed embryos, producing clean, washed embryos.

Chapter 9. Improved baculovirus expression vectors

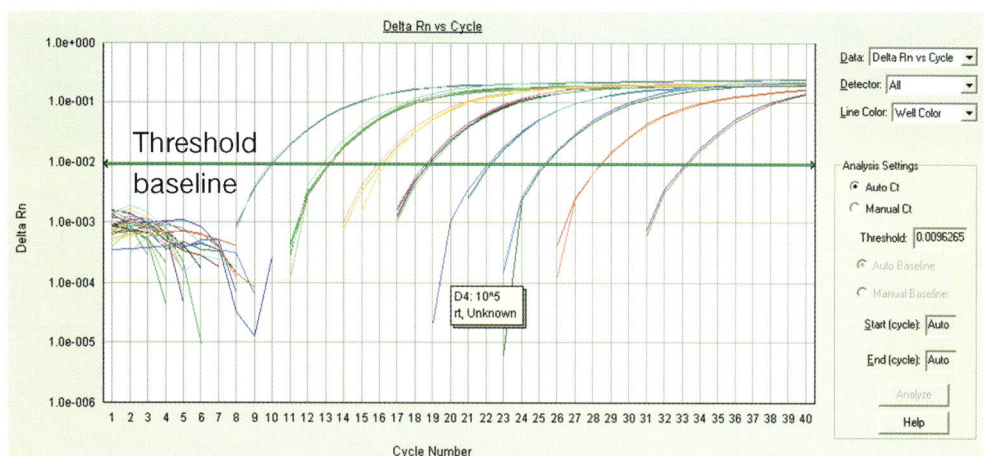

Figure 4. Typical amplification plot generated using the Applied Biosystems Sequence Detection System 7500, showing the threshold baseline (see page 165).
A clear distinction between different virus titers can be seen from the plot as represented by cycle number.

CHAPTER 1
Expression strategy
Michael R. Dyson

1. INTRODUCTION

A pre-requisite for performing many biochemical studies is the expression and purification of proteins. To confirm experimentally the functional role the protein plays *in vivo* is crucial to enrich our knowledge of gene function, which too often relies on theories derived from sequence alignment or whole-organism genetic studies. Once a protein has been successfully expressed and purified, this opens the door to several new and exciting experiments including testing for enzyme activity, interaction studies with other proteins or nucleic acids, antibody generation, and ultimately structure determination. Modern molecular biology would be virtually impossible to perform without the enzymes to amplify, cut, and ligate DNA, and the DNA polymerases used to sequence genomes. Similarly, the biotechnology industry also depends heavily on protein expression and purification. For example the protein pharmaceutical 'blockbuster' erythropoietin (Epogen; Amgen), used to treat anemia, has current sales of over $1 billion annually, and to answer the current clinical needs for recombinant antibodies, the pharmaceutical industry has now expanded its bioreactor production capacity to 200 000 liters (1). Also, it is important to produce active enzymes both for drug screening programs, such as the development of human immunodeficiency virus inhibitors, and for use in the diagnostic, food, and domestic product industries.

Over 90% of all proteins whose structures have been determined and coordinates deposited in the Brookhaven Protein Data Bank (PDB) (http://www.pdb.org[1.1]) have been expressed in *Escherichia coli*. Chapter 2 describes the latest approaches used in a successful structural genomics laboratory for *E. coli* expression. Attempted expression of some proteins in *E. coli* can result in protein misfolding and insoluble inclusion-body formation. This can also be a problem when a protein is expressed in *E. coli* with a solubility-enhancing fusion, such as maltose binding protein, but the protein crashes out of solution after proteolytic cleavage. These situations can be rectified by denaturing the protein, usually with chaotropic agents, and refolding by dilution. Although it can be time-consuming to find the correct refolding conditions, this has been employed successfully in solving the structure of the human T-cell receptor (2) and also for

Expression Systems: *Methods Express* (M.R. Dyson and Y. Durocher, eds)
© Scion Publishing Limited, 2007

several industrial processes including the production of human interferon and granulocyte colony-stimulating factor. Chapter 4 guides the researcher through a new protein refolding screening method that can be automated. Proteins may be unstable or prone to misfolding when overexpressed alone because naturally they exist as part of large multi-component complexes *in vivo*. The powerful methods of combinatorial biology can be employed in such cases by making a library of variants and then selecting for those that have the best properties. Two new methods are described for doing this using either green fluorescent protein (GFP) (Chapter 5) or ribosome display (Chapter 3). The protocols contained in Chapter 3 have been adapted specifically for the high-yield synthesis and affinity maturation of active antibody fragments and other proteins that contain disulfide bonds. Chapter 5 describes an elegant GFP selection method that overcomes many of the problems associated with C-terminal solubility reporter fusions. An alternative prokaryotic host is the Gram-positive *Lactococcus lactis* and this has proved to be very useful, when coupled with the nisin-controlled expression (NICE) system, for producing food-grade proteins and also for overexpression of toxic genes such as integral membrane proteins (Chapter 13).

The remaining chapters cover expression in eukaryotic systems, and these are most suitable for expression of eukaryotic proteins because they possess the most appropriate chaperones to aid protein folding, contain homologous binding partners that may help to stabilize proteins that naturally exist as part of a multi-component complex, and also have the capability to perform the full range of post-translational modifications. Until recently, it was only possible to synthesize small amounts of labeled protein in eukaryotic *in vitro*-translation systems, such as rabbit reticulocytes, for binding studies. However, new developments in the wheat-germ cell-free system has meant that it is now possible to synthesize preparative amounts of protein and this is described in detail in Chapter 6. This is very useful for the expression of toxic proteins, which would result in cell death if overexpressed *in vivo*. Many laboratories are hesitant to start eukaryotic expression because of the cost of purchasing specialized equipment. However, the lower eukaryotic organisms *Saccharomyces cerevisiae* and *Pichia pastoris* can easily be introduced into a standard molecular biology laboratory without the need to purchase additional equipment. The latest techniques for successful expression in these systems are presented in Chapters 7 and 8. Insect cells have proved to be an invaluable expression host for several eukaryotic proteins including kinases and the P450 drug metabolizing enzymes. Usually this is done using baculovirus expression vectors, and raising large baculovirus stocks can be a very good way of achieving scalable expression. Recently, new developments have improved both the time taken from expression vector to expressed protein and also the quality of product in the case of secreted or membrane protein expression (Chapter 9). Transient transfection of insect cells is also possible and this is described in detail in Chapter 10. Unlike mammalian cells, insect cells do not require CO_2 incubators, adding to their ease of use. However, for many expression projects, there is no substitute for choosing a mammalian expression system and this is particularly true for many human secreted proteins that require authentic glycosylation or other post-translational modifications. For example, it is estimated that 60–70% of all recombinant protein pharmaceuticals and 100% of the 18 current licensed therapeutic antibodies are produced in mammalian cells. The remaining chapters concentrate on various aspects of mammalian expression. If the aim is to develop an expression system for large-scale protein expression, such as the kilogram production of protein pharmaceuticals, it is worth spending the time (usually greater than 6 months) developing a high-yielding stable cell line and Chapter 11 describes methods for doing this in the Chinese hamster ovary (CHO) cell line. The use of lentiviral vectors offers a more rapid and powerful method of making stable pools of cells expressing high levels of recombinant proteins, even though it requires access to a level 2 biosafety facility (Chapter 14). Perhaps more convenient to the researcher is the use of the transient transfection system, which allows rapid, high-level protein expression that is also scalable (Chapter 12). Finally, Chapter 15 describes the interesting method of using BacMam viruses to transduce mammalian cells with high frequency for both transient and stable expression.

The aim of this chapter is to give some guidance as to the best expression host to choose and whether to express a full-length protein or a shorter version. This is highly dependent on the protein target to be expressed and three examples are described that illustrate the need for several expression systems to be available to the researcher.

2. METHODS AND APPROACHES

Before starting a new protein expression project, it is important to ask the following questions:

1. What is the target protein and what properties does it possess? Properties that should be examined include:
   - Molecular weight
   - The presence of signal peptides, transmembrane domains, multiple disulfide bridges, and post-translational modification sites.
   - Domain structure.
   - The presence of low-complexity or disordered regions.
2. From what organism does the protein derive?
3. What is the natural cellular location (nuclear, cytoplasmic, organelle, or extracellular) of the protein?
4. Has it or a homologous protein been expressed previously?
5. What is the protein to be used for and how much is required?

These questions are important to answer at the beginning of a new project as the answers will guide the choice of expression system and whether to truncate a multi-domain-containing protein to express the domain of interest. Most of these questions can now be answered by navigating the Internet to search the literature, and in the case studies listed below, the relevant web sites and search engines will be listed. Section 2.3 suggests some approaches to tackle proteins that have not been expressed previously. It should be noted that there are usually several ways to solve a protein expression problem and the examples below serve more for illustrative purposes.

2.1 Case study 1: mouse Rab3a

Rab3a is a small GTPase important for vesicle transport (3, 4). It is a mouse protein and so one would predict that a eukaryotic expression system, such as yeast, insect, or mammalian cells, would be the most appropriate host. However, the protein is relatively small with the polypeptide chain being 220 amino acids (aa) long and having a mass of 25 kDa. Studies have revealed that expression of eukaryotic proteins in *E. coli* is most likely to succeed when the proteins have a low molecular weight and a low number of contiguous hydrophobic amino acids and low-complexity regions (5), so Rab3a is a good candidate protein for *E. coli* expression. Indeed, a search of the literature (http://www.pubmed.gov[1.2]) reveals that the protein has been expressed in *E. coli* with a thrombin-cleavable N-terminal hexahistidine affinity purification tag, and has been purified, crystallized, and its structure solved by X-ray diffraction (6). The construct that gave crystals that diffracted at the highest resolution contained aa 19–217, although the aa 1–217 construct also expressed well. This information is valuable to decide whether to express the full-length protein or a truncated version. An alternative search engine to ask whether the protein has been expressed previously and its structure solved is the OCA browser (http://oca.ebi.ac.uk/oca-bin/ocamain[1.3]), which searches the PDB database and lists the expression host, vector, and the amino acid sequence of the polypeptide expressed. The C-terminal residues CAC of Rab3a are post-translationally modified *in vivo* by geranylgeranylation for targeting to the membrane of a specific cell organelle. *E. coli* does not possess the post-translational machinery to perform this modification, so for soluble protein expression it is better to remove this recognition sequence. For proteins that have not been expressed previously or are not annotated, code has been written to predict C-terminal protein prenylation (7) and this is available via a web interface (http://mendel.imp.ac.at/sat/PrePS/[1.4]).

An excellent source of integrated information on proteins is the UniProt (Universal Protein Resource) database (8) and this can be searched both using a protein name text (http://www.ebi.uniprot.org/uniprot-srv/index.do[1.5]) and with the amino acid sequence (http://www.ebi.ac.uk/blast2/index.html?UniProt[1.6]). The UNIPROT accession code for mouse Rab3a is P63011 and the entry lists a wealth of information including alternative gene names, function, whether the protein is part of a larger complex, interacting partners, protein features including sites of post-translational modification and polypeptide length, molecular weight, and sequence. There are also links to sites that give more background knowledge of protein function and one of these is the mouse genome informatics site (http://www.informatics.jax.org/[1.7]), which has distilled RNA and protein expression knowledge (9), gene knockout data and also links to the National Center for Biotechnology Information (NCBI) (http://www.ncbi.nlm.nih.gov/Database/[1.8]) Entrez Gene database (http://www.ncbi.nlm.nih.gov/entrez/query.fcgi?db=gene[1.9]), listing literature references for gene function. There are also links to several servers that list protein domain information, and perhaps the most useful of these is the

Figure 1. Domain topology of mouse Rab3a (220 aa) depicted with the Pfam Ras domain at aa 24–185.

Protein family or Pfam database (http://www.sanger.ac.uk/Software/Pfam/[1.10]) as it gives not only the coordinates of protein domains but also several other protein features including signal peptides, transmembrane domains, and low-complexity regions (10). The domain coordinates are calculated based on multiple alignment using hidden Markov models and, although very good at identifying a core consensus, are less good at predicting the precise structural domain boundaries. To design a construct to express a stable domain, it is important to base this on the structural boundaries. For example, the Pfam coordinates for the Ras domain of mouse Rab3a are aa 24–185 (see *Fig. 1*), whereas the construct that was expressed stably for structure determination was aa 19–217 and residues visualized in the structure were aa 16–192. It is likely that attempted expression based on the Pfam domain coordinates would result in an unstable protein and low expression yield as this would truncate the first β-sheet and last α-helix.

Having gathered the answers to as many of the questions listed in section 2 as possible, the next stage is to obtain a DNA sample of the protein to be expressed, design oligonucleotide primers to amplify the target region, amplify the coding sequence using the polymerase chain reaction (PCR) and clone into a suitable expression vector. One can obtain a clone directly from a laboratory that has actually worked with the protein or search the public clone collection resources; the clone can then be ordered through various distributors: in Europe this is GeneService Ltd (http://www.geneservice.co.uk/home/[1.11]) or the RZPD (http://www.rzpd.de/[1.12]), and in Asia is DNAFORM (http://www.dnaform.jp/index_e.html[1.13]); the USA distributors are listed at http://image.llnl.gov/image/html/idistributors.shtml[1.14]. The main public clone collections are the Japanese Fantom2 and later Fantom3 collections (11), which can be both text or accession code searched (http://fantom2.gsc.riken.jp/db/search/[1.15] and http://fantom3.gsc.riken.jp/db/search/[1.16]) and BLAST searched with the target DNA sequence (http://fantom2.gsc.riken.go.jp/blast/[1.17] and http://fantom3.gsc.riken.jp/blast/[1.18]), and the Mammalian Gene Collection (MGC) and Integrated Molecular Analysis of Genomes and their Expression (IMAGE) Consortium's clones (12, 13), which can be text searched at http://mgc.nci.nih.gov/[1.19] and sequence BLAST searched at http://mgc.nci.nih.gov/Reagents/MGCBlast[1.20]. The mouse Fantom2 collection has now been included in a book where one can hole-punch to isolate the clone of interest (14). Another public clone collection originating from the Kazusa DNA Research Institute specializes in isolating long cDNAs (15). The human KIAA clones are available through the Human Unidentified Gene-Encoded (HUGE) Large Proteins database (http://www.kazusa.or.jp/huge/index.html[1.21]) and the mouse KIAA clones are available from the Rodent Unidentified Gene-Encoded (ROUGE) Large Proteins database (http://www.kazusa.or.jp/rouge/index.html[1.22]). The

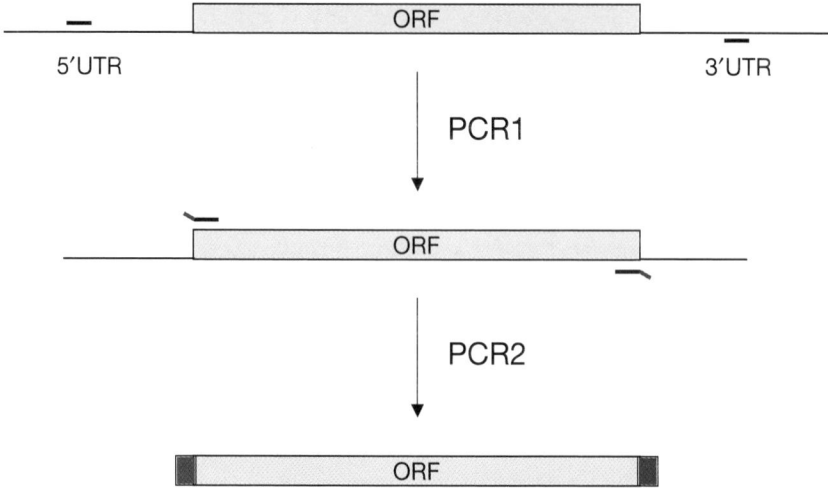

Figure 2. Nested PCR strategy (see page xix for color version).
PCR1 amplifies from a mixed cDNA pool with primers either placed in the UTRs or 5′ and 3′ of the DNA encoding an individual protein domain. PCR2 amplifies the DNA encoding the protein to be expressed with tags (red) to aid restriction- or recombination-based cloning into an expression vector.

ORFeome Collaboration (OC) was started in 2005 to integrate the various clone collections described above and to sequence confirmed clones from other laboratories (16) including the Wellcome Trust Sanger Institute (http://www.sanger.ac.uk/HGP/Chr22/ORFcloning/[1.23]), the Harvard Institute of Proteomics (http://www.hip.harvard.edu/[1.24]), and the Dana Farber Cancer Institute (http://ccsb.dfci.harvard.edu/home.html[1.25]). These full-length cDNA clones will be cloned into GATEWAY compatible cloning vectors (17) and will be a very useful resource for the future (http://www.orfeomecollaboration.org/html/index.shtml[1.26]). In addition to the public clone resources, there are various commercial suppliers of clones and these include (not exclusively) Invitrogen, Open Biosystems, and GeneCopoeia. The collections listed here cover human, mouse, and *Caenorhabditis elegans*, but specialized sites can usually be found to cover other organisms.

If the target clone cannot be found in the clone collections, it must be either isolated from genomic (if prokaryotic or the gene does not contain introns) or cDNA, or chemically synthesized (18, 19). Chemical synthesis allows one to design the gene for optimal expression in the chosen host including codon optimization, elimination of internal restriction enzyme sites, and the removal of RNA secondary structure, which can reduce translation rates (20, 21). However, whole-gene synthesis can be expensive when working on several expression projects. A method for open reading frame (ORF) cloning from cDNA is detailed in *Protocol 1* and is based on the method of Collins *et al.* (22). The specificity required to amplify a single cDNA from a complex mixture is achieved by a nested PCR where the first PCR is performed with primers that sit 5′ and 3′ of the target to be cloned. If the target is a full-length gene, then the primers for this first PCR must sit in the 5′ and 3′ untranslated regions (UTRs) of the cDNA (see *Fig. 2*, also available in the color section). If this reaction is analyzed by agarose gel electrophoresis, then a clean single product is rarely seen, but after the second PCR with ORF-specific primers, a single clean product is often observed. The primers for the second PCR contain 5′ tags to enable either restriction enzyme or recombination-based cloning into an expression vector (23) and the second PCR protocol can be equally applied to amplify ORFs from cloned plasmid DNA. For primer design, the DNA sequence must first be downloaded and this can be done by either following the links from the UniProt entry or downloading the Refseq sequence via the Entrez Gene database (24). There are several commercial software packages that can be used to aid primer design and DNA sequence confirmation including ACCELRYS GENE or DS GENE (Accelrys), MACVECTOR (MacVector Inc.) and LASERGENE (DNASTAR). VECTOR NTI (Invitrogen) is now freely available to academic researchers and there are also several freeware software programs available (http://molbiol-tools.ca/molecular_biology_freeware.htm[1,27]). Primers should ideally be designed to have a melting temperature close to 60°C, be between 18 and 25 nucleotides (not including the 5′ tag), and have a GC content of between 40 and 60%.

Protocol 1

ORF amplification from cDNA using PCR

Equipment and Reagents
- KOD hot-start DNA polymerase[a], 2 mM dNTPs, 25 mM MgCl$_2$, and 10× buffer, supplied with kit (Novagen)
- Oligonucleotide primers for nested PCR
- QUICK-Clone II human universal cDNA or QUICK-Clone mouse cDNA (Clontech). The original 10 µl (1 ng/µl) stock is diluted by the addition of 190 µl of 10 mM Tris/HCl (pH 8) and stored at −70°C in 20 µl aliquots
- Thermocycler

Method
1. Prepare the PCR 1 mix as follows:
   - 2 µl of 10× KOD buffer
   - 2 µl of 2 mM dNTPs
   - 0.8 µl of 25 mM MgCl$_2$
   - 1 µl of 50 pg/µl QUICK-Clone cDNA[b]
   - 1 µl of 5 µM forward primer
   - 1 µl of 5 µM reverse primer
   - 0.5 µl of 0.5 units/µl KOD hot-start DNA polymerase
   - 11.7 µl of deionized water such as Millipore analytical grade
2. Place the PCR mix in a thermocycler with the following program:

Number of cycles	Program
1	2 min at 94°C
30	15 s at 94°C, 30 s at 54°C, 2 min[c] at 68°C
1	4 min[d] at 68°C
1	Hold at 4°C

3. Add 2 µl of PCR 1 mix to 98 µl of 10 mM Tris/HCl (pH 8)
4. Prepare the PCR 2 mix using the internal primers to amplify the ORF to be expressed as detailed below:
   - 2 µl of 10× KOD buffer
   - 2 µl of 2 mM dNTPs
   - 0.8 µl of 25 mM MgCl$_2$
   - 2 µl of a 1:50 dilution of the completed PCR from step 2
   - 1 µl of 5 µM forward primer
   - 1 µl of 5 µM reverse primer
   - 0.5 µl of 0.5 units/µl KOD hot-start DNA polymerase
   - 10.7 µl of deionized water such as Millipore analytical grade
5. Place the PCR mix in a thermocycler and run the same program as in step 2.
6. Analyze the PCR products by agarose gel electrophoresis for the expected insert size. If a single band is observed, this can be used directly for cloning into the expression plasmid (23).
7. The DNA sequence of the clone should be confirmed prior to starting any expression experiments.

Notes

[a] It is important to use a proof-reading DNA polymerase, to minimize the possibility of the PCR introducing mutations. KOD DNA polymerase has both good fidelity and processivity and the hot-start version contains two monoclonal antibodies that inhibit nonspecific amplification due to mis-priming events that can occur during set-up and primer degradation at room temperature due to the inhibition of the exonuclease activity (25).
[b] The QUICK-Clone cDNA could be substituted by any good-quality cDNA source. If PCR amplifying from a clone, start at step 4 and substitute the 1:50 dilution of completed PCR from step 2 with 10 ng of clone plasmid DNA.
[c] A 1 min extension time per kb is standard.
[d] This time is 2 min longer than the extension time.

2.2 Case study 2: human epidermal growth factor receptor

The human epidermal growth factor receptor (EGFR) is involved in cell growth and differentiation. Its abnormal activation is associated with human cancer and it is therefore a therapeutic target. Starting with UniProt entry P00533 and following the link to the Pfam database reveals the domain architecture of the 1210 aa (134 kDa) protein (see *Fig. 3*, also available in the color section). EGFR is an example of a type 1 integral membrane protein with an N-terminal signal peptide (aa 1–24), several annotated extracellular domains, a transmembrane domain (aa 646–668) and a cytoplasmic kinase domain (aa 712–968). A search of the PDB via the OCA query search engine shows that two regions of the protein have been expressed and their structures solved, and both were expressed using eukaryotic expression systems. The intracellular kinase domain (aa 672–998) was expressed using baculovirus (26) and a truncated extracellular domain (aa 1–501) was expressed using a CHO stable cell line (27). It is interesting to note the differences between the structural and Pfam domain boundaries, and again this illustrates that for good soluble expression of correctly folded protein one should always aspire to follow the structural domain boundaries. The choice of expression

A B C B D E F E E

Figure 3. Domain topology of human EGFR (1210 aa) (see page xix for color version).
Marked protein features and Pfam domains include: A, signal peptide (aa 1–24); B, receptor L domain (aa 57–168 and 361–481); C, furin-like cysteine rich region (aa 184–338); D, transmembrane region (aa 646–668); E, low-compexity regions (aa 651–666, 1002–1015 and 1025–1046); and F, protein tyrosine kinase domain (aa 712–968).

system for the intra- and extracellular regions of the EGFR shows that baculovirus is historically the first choice for producing active eukaryotic enzymes, whereas the requirement for authentic glycosylation makes the mammalian expression systems a good choice for production of the ectodomains of cell-surface receptors.

2.3 Expression of previously unexpressed proteins

Both of the case studies above were fortunate to have examples that were well annotated in the databases and had also been expressed previously. However, sometimes the researcher is presented with a protein that has not been expressed or had its structure solved by X-ray crystallography or nuclear magnetic resonance. In this case, it is possible to search the structural databases to ask whether a homologous protein has been studied and the structural domain boundaries can be inferred by multiple sequence alignment. For example, this can be performed by entering the amino acid sequence of the target protein into a BLASTP search and selecting to query the PDB database (http://www.ncbi.nlm.nih.gov/BLAST/[1.28]). The search results provide multiple alignments of the query sequence with any matches in the database. Alternatively, one can perform text searches of the literature via the PubMed search engine. If the protein is large (>50 kDa), one should consider using a eukaryotic expression system, which is more capable of expressing large, multi-domain-containing proteins, or one should truncate the protein to a smaller size for soluble expression in *E. coli*. For protein truncation, the domain architecture of the protein needs to be mapped onto the linear amino acid sequence. If the protein is novel and has not yet been annotated in UniProt, there are various search engines and algorithms available through web sites that allow one to do this. For example, the polypeptide sequence can be entered into the Pfam (10) search engine (http://www.sanger.ac.uk/Software/Pfam/search.shtml[1.29]) or the Simple Modular Architecture Research Tool (SMART) (28) (http://smart.embl-heidelberg.de/[1.30]). Different algorithms are employed by Pfam and SMART, so it is often worth comparing the results from these two sites. SMART is also good at mapping low-complexity regions onto proteins and these can be tested as interdomain linkers by truncating at these sites and attempting soluble expression (5). Low-complexity and intrinsically disordered regions can have an important functional role *in vivo*, acting as adapter molecules for binding to other proteins or as sites for post-translational modification such as phosphorylation. A Perl script has been written to predict disordered regions of proteins (29) and this can be mapped onto the linear sequence as a FoldIndex score (http://bip.weizmann.ac.il/fldbin/findex[1.31]). Prediction methods also exist to identify signal peptides (30), transmembrane domains (31), sites of glycosylphosphatidylinositol (GPI) lipid anchoring (32), prenylation (7), and N-terminal N-myristoylation (33), and can be accessed at: http://www.cbs.dtu.dk/services/SignalP/[1.32], http://www.cbs.dtu.dk/services/TMHMM/[1.33], http://mendel.imp.ac.at/gpi/gpi_server.html[1.34], http://mendel.imp.ac.at/sat/PrePS/index.html[1.4], and http://mendel.imp.ac.at/sat/myristate/index.html[1.35], respectively. Once the protein features have been mapped onto the target protein sequence, one is in a strong position to decide on the expression strategy including if and where to truncate a new target protein and also which expression system is the most appropriate.

3. TROUBLESHOOTING

- For *Protocol 1*, if several bands are observed, the target band should be gel purified before cloning.
- If no product is observed after attempted ORF amplification from a cDNA source, attempt with a tissue-specific cDNA source where there is known mRNA expression or use a random-primed cDNA (e.g. from Ambion), which can enrich for longer transcripts.
- It is essential to sequence the DNA fully to confirm the expression construct after cloning to check for mutations introduced by the PCR primers or by DNA polymerase. This can result in frame-shift mutations, the introduction of stop codons or destabilizing coding point mutations that will result in either reduced or zero product yield during expression.
- Some problems in expression can be solved by co-expression with a binding partner. Interacting proteins can be identified by a literature search or various public databases (34) such as the IntAct resource (35) (http://www.ebi.ac.uk/intact[1.36]) or the Reactome database (36) (http://www.reactome.org/[1.37]).

Acknowledgements

I would like to thank Yves Durocher (Biotechnology Research Institute, National Research Council Canada, Montreal, Canada) for helpful comments and Robert Finn (The Wellcome Trust Sanger Institute, Hinxton, Cambridge, UK) for assistance in the preparation of the Pfam figures.

4. REFERENCES

1. Farid SS (2006) *J. Chromatogr. B Analyt. Technol. Bimed. Life Sci.* **848**, 8–18.
2. Garboczi DN, Ghosh P, Utz U, Fan QR, Biddison WE & Wiley DC (1996) *Nature*, **384**, 134–141.
3. Stenmark H & Olkkonen V (2001) *Genome Biol.* **2**, reviews3007.3001.
4. Wennerberg K, Rossman KL & Der CJ (2005) *J. Cell Sci.* **118**, 843–846.

★★ 5. Dyson MR, Shadbolt SP, Vincent KJ, Perera RL & McCafferty J (2004) *BMC Biotechnol.*, **4**, 32. – *Highlights protein features that correlate with successful expression in* E. coli.
6. Ostermeier C & Brunger AT (1999) *Cell*, **96**, 363–374.
★ 7. Maurer-Stroh S & Eisenhaber F (2005) *Genome Biol.* **6**, R55. – *A protein prenylation prediction method.*
★★★ 8. The UniProt C (2007) *Nucleic Acids Res.* **35**, D193–D197. – *A review of the UniProt database.*
9. Smith CM, Finger JH, Hayamizu TF, *et al.* (2007) *Nucleic Acids Res.* **35**, D618–D623.
★★ 10. Finn RD, Mistry J, Schuster-Bockler B, *et al.* (2006) *Nucleic Acids Res.* **34**, D247–D251. – *An update of the Pfam database.*
11. Maeda N, Kasukawa T, Oyama R, *et al.* (2006) *PLoS Genet.* **2**, e62.
12. Lennon G, Auffray C, Polymeropoulos M & Soares MB (1996) *Genomics*, **33**, 151–152.
13. The MGC Project Team (2004) *Genome Res.* **14**, 2121–2127.
14. Kawai J & Hayashizaki Y (2003) *Genome Res.* **13**, 1488–1495.
15. Kikuno R, Nagase T, Nakayama M, *et al.* (2004) *Nucleic Acids Res.* **32**, D502–D504.
16. Lamesch P, Li N, Milstein S, *et al.* (2007) *Genomics*, **89**, 307–315.
★ 17. Hartley JL, Temple GF & Brasch MA (2000) *Genome Res.* **10**, 1788–1795. – *The original GATEWAY reference.*
★★ 18. Kodumal SJ, Patel KG, Reid R, Menzella HG, Welch M & Santi DV (2004) *Proc. Natl. Acad. Sci. U.S.A.* **101**, 15573–15578. – *Synthetic gene synthesis applied to a 32 kb gene cluster.*
★★ 19. Smith HO, Hutchison CA III, Pfannkoch C & Venter JC (2003) *Proc. Natl. Acad. Sci. U.S.A.* **100**, 15440–15445. – *A gene synthesis approach to ORF generation.*
20. Jayaraj S, Reid R & Santi DV (2005) *Nucleic Acids Res.* **33**, 3011–3016.
21. Villalobos A, Ness J, Gustafsson C, Minshull J & Govindarajan S (2006) *BMC Bioinformatics*, **7**, 285.
★ 22. Collins JE, Wright CL, Edwards CA, *et al.* (2004) *Genome Biol.* **5**, R84. – *The original method used to amplify ORFs from cDNA using nested PCR.*
23. Sambrook J & Russell D (2001) *Molecular Cloning: a Laboratory Manual*, 3rd edn. Cold Spring Harbor Laboratory Press, Cold Spring Harbor.
24. Maglott D, Ostell J, Pruitt KD & Tatusova T (2007) *Nucleic Acids Res.* **35**, D26–D31.
25. Mizuguchi H, Nakatsuji M, Fujiwara S, Takagi M & Imanaka T (1999) *J. Biochem. (Tokyo)* **126**, 762–768.
26. Stamos J, Sliwkowski MX & Eigenbrot C (2002) *J. Biol. Chem.* **277**, 46265–46272.
27. Garrett TPJ, McKern NM, Lou M, Elleman TC, Adams TE, Lovrecz GO, *et al.* (2002) *Cell*, **110**, 763–773.
28. Letunic I, Copley RR, Pils B, Pinkert S, Schultz J & Bork P (2006) *Nucleic Acids Res.* **34**, D257–D260.
★★ 29. Prilusky J, Felder CE, Zeev-Ben-Mordehai T, *et al.* (2005) *Bioinformatics*, **21**, 3435–3438. – *A method for predicting disordered regions within proteins.*
★ 30. Bendtsen JD, Nielsen H, von Heijne G & Brunak S (2004) *J. Mol. Biol.* **340**, 783–795. – *A method for predicting signal peptides.*
★ 31. Krogh A, Larsson B, von Heijne G & Sonnhammer ELL (2001) *J. Mol. Biol.* **305**, 567–580. – *A method for predicting transmembrane regions.*
★ 32. Eisenhaber B, Bork P & Eisenhaber F (1999) *J. Mol. Biol.* **292**, 741–758 – *A method for predicting GPI lipid anchoring.*
★ 33. Maurer-Stroh S, Eisenhaber B & Eisenhaber F (2002) *J. Mol. Biol.* **317**, 541–557 – *A method for predicting N-myristoylation.*
★ 34. Mathivanan S, Periaswamy B, Gandhi TKB, *et al.* (2006) *BMC Bioinformatics*, **7**, S19. – *A review of protein–protein interaction databases.*
35. Kerrien S, Alam-Faruque Y, Aranda B, *et al.* (2007) *Nucleic Acids Res.* **35**, D561–D565.
36. Joshi-Tope G, Gillespie M, Vastrik I, *et al.* (2005) *Nucleic Acids Res.* **33**, D428–D432.

CHAPTER 2
Protein expression in *Escherichia coli*
Rosalind Kim

1. INTRODUCTION

The field of structural genomics has led to the development of high-throughput and parallel approaches in the areas of cloning, expression, purification, and structure determination (1–3). The need for milligram quantities of numerous proteins for functional and structural studies requires an efficient and high-throughput pipeline. Due to the easy handling, low cost, and fast growth of *Escherichia coli*, this has become the expression system of choice for many structural genomic projects (4, 5).

In this chapter, we will describe the platform used at the Berkeley Structural Genomics Center (BSGC) from cloning to protein purification (http:// strgen.org/[2.1]). The ligation-independent cloning (LIC) method (6) is used with seven different *E. coli* expression vectors designed at BSGC. Our platform concentrates mainly on six-histidine (His_6) affinity-tagged protein constructs, but in some cases other fusions (i.e. maltose-binding protein (MBP) or glutathione *S*-transferase (GST)) are made to increase the solubility of the target. Parallel expression using a 96-well format is used for screening the highly soluble proteins (7). After selection of the best-behaving targets, large-scale protein expression and purification is performed. In most cases, the fusion tag is removed with the use of tobacco etch virus protease. These methods will be of broad interest in the fields of functional proteomics, structural biology, and structural genomics. We have adapted the cloning and the expression screening to the Biomek 2000 robot (Beckman Coulter), but the protocols can be adapted to any robot selected and can also be used in a smaller-scale set-up. This chapter will cover methods from cloning to purification of a His_6-tag fusion target including cleavage of the tag using tobacco etch virus protease.

2. METHODS AND APPROACHES

2.1 Ligation-independent cloning

LIC allows directional cloning of polymerase chain reaction (PCR) products without the use of restriction enzymes or DNA ligase (6). This approach takes advantage of the 3′→5′ exonuclease activity of T4 DNA polymerase for the generation of long, cohesive overhangs on both the vector and the DNA fragment to be cloned. The creation of highly specific, 12-nucleotide, single-stranded overhangs in the vector and the insert results in annealed products that are the desired molecules. At BSGC, we have constructed a series of LIC vectors (see *Table 1*). Novagen LIC vector kits are also available. The BSGC LIC vectors are based on pET21a (T7 promoter) with inserted sequences that generate the different fusions listed in *Table 1*. An insert prepared for this cloning protocol can be introduced into all of the BSGC LIC vectors (for design of the primers, refer to http://strgen.org/protocols/[2.2]). The necessary steps are:

1. Creation of PCR products with appropriate 5′ extensions into the primers. The PCR products are treated with LIC-qualified T4 DNA polymerase in the presence of the appropriate dNTP to generate the specific vector-compatible overhangs (see *Protocol 1*).
2. The LIC vector is prepared to have overhangs that are complementary to those in the insert (see *Protocol 2*).
3. The annealed LIC insert and vector is transformed into *E. coli* TOP10 competent cells where covalent bonds are formed at the junction of the vector–insert to yield a circular plasmid (see *Protocol 3*).
4. After verification of the size of the constructs, the plasmids are then transformed into the expression host (see *Protocol 4*).

Table 1. BSGC LIC vectors

All pB vectors are based on pET21a (ampicillin resistant) from Novagen. The pB3, pB4, pB5, pB6, and pB7 constructs express target proteins that, when cleaved with TEV protease, generate a target protein with six glycines at the N terminus. These have been designed to increase the efficiency of cleavage by TEV protease (D. Waugh, personal communication).

Vector Name	Tag
pB1	No tag
pB2	Noncleavable His_6 tag
pB3	His_6-TEV cleavage sequence
pB4	His_6-MBP-TEV cleavage sequence
pB5	His_6-GST-TEV cleavage sequence
pB6	His_6-thioredoxin-TEV cleavage sequence
pB7	His_6-NusA-TEV cleavage sequence

GST, glutathione *S*-transferase; His_6, six histidines; MBP, maltose binding protein; TEV, tobacco etch virus protease.

2.1.1 Preparation of insert

The design of the primers is based on the vector selected. When using the pB vectors, after PCR amplification of the target, the PCR product is treated with the 3'→5' exonuclease activity of T4 DNA polymerase in the presence of dTTP to degrade the DNA from the 3' ends up to the first dT residue.

Protocol 1

LIC insert preparation protocol

Equipment and Reagents
- Amplified target DNA
- 50 mM Tris/HCl (pH 8.0)
- T4 DNA polymerase (20 units/0.5 µl; New England Biolabs)
- 10× T4 reaction buffer (500 mM NaCl, 100 mM Tris/HCl, 100 mM $MgCl_2$, 10 mM dithiothreitol, pH 7.9, at 25°C; New England Biolabs)
- 100 mM dTTP (Promega)
- Bovine serum albumin (BSA) (10 mg/ml; New England Biolabs)
- Water (double deionized, autoclaved, room temperature)
- PCR Machine GeneAmp 9600 (PerkinElmer)

Method
1. Set the PCR machine to run one cycle of: 37°C for 30 min, 70°C for 20 min, and 4°C termination.
2. Bring 1 µg of amplified target DNA to 43.5 µl with 50 mM Tris/HCl (pH 8.0) in a PCR plate.
3. Add 5 µl of 10× T4 reaction buffer, 0.5 µl of 100 mM dTTP, and 0.5 µl of BSA (10 mg/ml) to the PCR plate. Mix the reaction mixture.
4. Add 0.5 µl (20 units) of T4 DNA polymerase.
5. Incubate the mixture in the PCR machine and cycle as in step 1 above.
6. Use 1 µl (20 ng of insert DNA) for each LIC reaction in *Protocol 3*.

2.1.2 Preparation of vector

The LIC vector is linearized and treated with T4 DNA polymerase to generate the complementary single-stranded tails in the presence of dATP to degrade the DNA from the 3' ends up to the first dA residue. It is then essential to gel purify the vector to eliminate high background due to plasmids with no inserts.

Protocol 2

LIC vector preparation protocol

Equipment and Reagents
- LIC vector, uncut (100 ng/μl)
- 10× NEB4 buffer (500 mM potassium acetate, 200 mM Tris/acetate, 100 mM magnesium acetate, 10 mM dithiothreitol, pH 7.9, at 25°C; New England Biolabs)
- *Sma*I restriction enzyme, 20 units/μl (New England Biolabs)
- T4 DNA polymerase (New England Biolabs)
- 100 mM dATP (Promega)
- BSA (10 mg/ml; New England Biolabs)
- Water (sterile, double deionized)
- Agarose (Research Organics)
- TAE buffer (0.04 M Tris/acetate, 0.01 M EDTA, pH 8)
- Horizon electrophoresis system (Gibco)
- 1 kb Plus DNA ladder (1 μg/μl; Invitrogen)
- PCR Machine GeneAmp 9600 (PerkinElmer)
- StrataPrep DNA gel extraction kit (Stratagene)

Method
1. Set the PCR machine to 25°C.
2. Mix 5 μg of BSGC LIC vector and 5 μl of 10× NEB4 buffer in the PCR tube and bring to a volume of 49 μl with water.
3. Add 1 μl (20 units) of *Sma*I to the PCR tube.
4. Incubate at 25°C for 1–2 h.
5. After incubation, run 1 μl of the reaction mix on a 1.2% agarose/TAE gel at 60 V for 1 h. Use undigested plasmid and 4 μl of the 1 kb Plus DNA ladder marker as standards. If the reaction is complete, proceed to the next step. If the reaction is incomplete, add another 1 μl of *Sma*I, repeat the incubation and rerun the reaction on an agarose gel.
6. Set up the PCR machine to run one cycle of: 37°C for 30 min, 70°C for 20 min, and 4°C termination.
7. Add 1 μl of 100 mM dATP and 0.5 μl of BSA (10 mg/ml) to the linearized vector and mix.
8. Add 1 μl (20 units) of T4 DNA polymerase to the PCR tube. Incubate in the PCR machine and cycle as above (step 6).
9. Run the T4 DNA polymerase-treated vector on a 1% agarose/TAE gel at 100 V for 45 min. Purify the vector following the StrataPrep DNA gel extraction kit protocol.
10. Use 0.5 μl (50 ng of plasmid) for each LIC reaction in *Protocol 3*.

2.1.3 Formation of hybrid products

Treated LIC vector and insert are incubated to allow the formation of hybrid products (circles containing single insert and vector molecules). These are then transformed into TOP10 *E. coli* competent cells.

Protocol 3

LIC reaction/transformation protocol

Equipment and Reagents
- Prepared LIC target insert (from *Protocol 1*)
- Prepared LIC vector (ampicillin (Amp) resistant) (from *Protocol 2*)
- TOP10 *E. coli* cells, chemically competent (Invitrogen)
- Luria–Bertani (LB) agar plates + Amp (50 µg/ml)
- LB liquid medium
- LB liquid medium + Amp (50 µg/ml)
- PCR Machine GeneAmp 9600 (PerkinElmer)
- Water bath set at 42°C
- Orbital shaker
- 2 ml 96-Well growth plate (E&K) or 17 × 100 mm (14 ml) sterile culture tubes (VWR)
- Agarose (Research Organics)
- TAE buffer (0.04 M Tris/acetate, 0.01 M EDTA, pH 8)
- Horizon electrophoresis system (Gibco)
- Uncut LIC vector
- Plasmid mini kit (Qiagen)

Method

1. Place 0.5 µl (50 ng) of prepared LIC vector into a PCR plate.
2. Add 1.0 µl (20 ng) of prepared LIC target insert directly to the LIC vector and mix gently.
3. Incubate the plate at room temperature for 5–30 min and then place on ice.
4. Thaw chemically competent TOP10 cells on ice (each tube contains 50 µl, enough for five transformations).
5. Add 10 µl of competent TOP10 cells to the well and leave on ice for 5–30 min.
6. Heat-shock the PCR plate at 42°C for 30–45 s.
7. Add 200 µl of LB medium without antibiotics (pre-warmed to 37°C) to the plate.
8. Incubate the plate without shaking at 37°C for 30 min to 1 h.
9. Plate half of the transformation (100 µl) onto an LB + Amp plate. Invert the plate and incubate overnight at 37°C.
10. Count colonies on the LB + Amp plate the next day. Select four to eight colonies and inoculate each colony into 14 ml sterile culture tubes containing 2.5 ml LB + Amp liquid medium or use a 96-well sterile growth plate containing 1.2 ml LB + Amp liquid medium/well. Grow overnight with shaking at 37°C.
11. Use the Plasmid mini kit (Qiagen) on the overnight cultures.
12. Run 5 µl of miniprep plasmid on a 1% agarose/TAE gel for band-shift analysis to verify the insertion of the PCR target; use 2 µl of uncut vector as a standard (see *Fig. 1*). Select positive clone(s) to send for DNA sequencing or you can wait to sequence the clones after testing for expression (see *Protocol 5*).

18 ■ CHAPTER 2: PROTEIN EXPRESSION IN *ESCHERICHIA COLI*

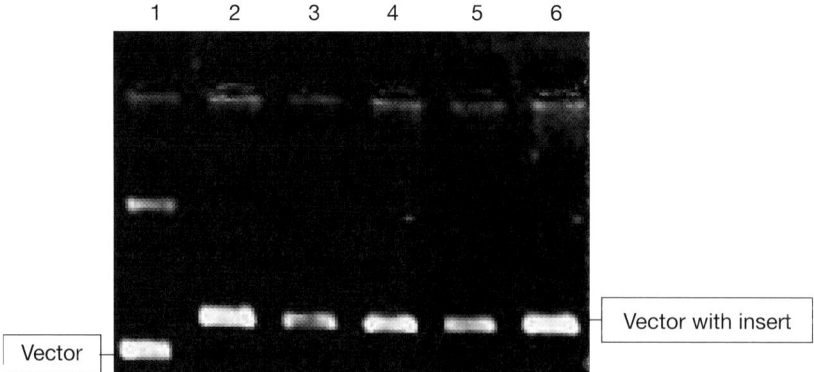

Figure 1. Characterization of clones by agarose gel electrophoresis.
DNA from vector and from transformants is run on a 1.2% agarose/TAE gel to determine which clones contain the insert. *Lane 1*, vector; *lanes 2-6*, clones containing the insert.

2.1.4 Transformation into the expression host

After sequencing the cloned genes, the constructs are transformed into the *E. coli* expression host BL21(DE3)/pSJS1240 (8) and plated onto selective media. The pSJS1240 plasmid with rare *E. coli* tRNA codons has been shown to increase the level of expression of some recombinant proteins.

Protocol 4

Transformation of the LIC construct into the expression host

Equipment and Reagents
- 96-Well PCR plate with silicone cover (E&K)
- 24-Well Cellstar plate (Greiner) with LB + Amp agar medium; store at 4°C and warm to room temperature before use
- *E. coli* competent cells: BL21(DE3)/pSJS1240 (R. Kim)
- Plasmid minipreps of clones
- LB + Amp medium
- LB buffer
- PCR Machine GeneAmp 9600 (Perkin-Elmer)
- Six-channel expandable 1250 µl Pipetman (Matrix Impact)
- 96-Well 2 ml sterile, deep growth plate (E&K)
- Sterile glass beads
- Tweezers

Method

1. Prepare a datasheet that includes the positions of the plasmid samples (from the plasmid miniprep plate) to be transformed on a 24-well agar plate. The 24-well agar plate colonies are used for expression testing. Print out this datasheet and use it as a template for transferring plasmid samples.
2. Thaw the plasmid miniprep plate(s).
3. Pipette 1 µl of plasmid DNA into the assigned 96-well PCR plate, following the datasheet template.
4. Keep the 96-well PCR plate on ice.
5. Thaw the competent cells by placing the tube on ice (10 µl of cells are needed per transformation or approximately 970 µl per 96-well plate).
6. When the cells have thawed completely, gently mix by finger tapping twice.
7. Immediately aliquot 10 µl of cells into each of the wells. Seal with a silicone rubber mat and incubate on ice for 30 min.
8. Set up the PCR machine for a heat-shock program: 30 s at 4°C, 45 s at 42°C, and then 4°C.
9. After the heat-shock is completed, place the plate on ice.
10. Leave the plate on ice for 2 min. Add 100 µl of LB buffer (pre-warmed to 37°C) to each well and then incubate at 37°C for 1 h.
11. Add one sterile glass bead to each well of the 24-well Cellstar plate with flamed tweezers. Each of these wells contains LB + Amp agar.
12. Use the Matrix Impact six-channel expandable 1250 µl Pipetman to pipette 20 µl of the transformation mix into the assigned wells, following the datasheet template.
13. When all additions have been made, swirl the plate so that the beads move around the surface to spread the transformation mix. If 96 plasmids are being transformed, four 24-well Cellstar plates will be needed.
14. Remove the cover and, in a single quick motion, discard the beads into a small plastic tray. Disinfect the beads with 5% bleach and wash. Store and autoclave the beads for reuse. Invert the plate and incubate at 37°C overnight. The next day, record the transformation results.

2.2 Small-scale expression

After transformation into the expression host, two to four colonies are selected and grown in auto-inducing medium (9, 10). Cells are grown in a 96-well deep-well plate overnight, spun down, resuspended, and disrupted by mechanical lysis such as sonication using the Misonix 3000 sonicator (see *Protocol 5a*) or treated with a protein extraction reagent such as Popculture (see *Protocol 5b*). The lysate is analyzed to determine the level of soluble expression of the target by sodium dodecyl sulfate (SDS) polyacrylamide gel electrophoresis (PAGE). Presently, *all steps*, from the PCR for 96 targets to the analysis of the level of expression of the targets, are automated and can be achieved in 3–4 days (7, 11–13).

Protocol 5

(a) 96-Well mini-expression using mechanical lysis

Equipment and Reagents
- 96-Well, sterile deep-well plate (2.0 ml) (E&K)
- 96-Well PCR plate
- 100 mM Tris/HCl (pH 7.5)
- Sterile applicator sticks
- Auto-inducing medium (ZYM) (9, 10)
- Infors shaker (Appropriate Technical Resources)
- Lifeline plate shaker
- Multi-channel pipettor
- Matrix Electrapette pipettor (Matrix Impact)
- Misonix Plate sonicator model S-3000-001 (Misonix)
- Legend centrifuge (Thermo Electron Corp.)
- 2× Sample buffer (80 mM Tris/HCl, pH 6.8, 10% mercaptoethanol, 2% SDS, 10% glycerol, 0.2% bromophenol blue)

Method

1. On day 1, pre-warm the sterile medium (with antibiotics) in a 37°C incubator for 30 min prior to use. Dispense 1.1 ml of medium into as many wells of a deep-well plate as you have clones using the Matrix Electrapette. Inoculate each well with one colony from a fresh transformation plate using a sterile applicator stick. (Use the same stick to spot a LB + antibiotic agar plate and incubate overnight at 37°C. This will be the master plate for future experiments.) Shake the cells at 37°C at 250 r.p.m. for 3–5 h until turbid and then drop the temperature of the incubator to 20°C and let the cells grow overnight.

2. On day 2, take the deep-well plate and spin for 15 min (4°C, 3000 g) in a Legend centrifuge. Shake upside down three times to remove all liquid and then add 100 μl of 100 mM Tris/HCl (pH 7.5) to each well. Resuspend the pellet by mixing for 5 min at speed 1400 in a Lifeline plate shaker.

3. Transfer the samples to a 96-well PCR plate using a multi-channel pipettor. Chill the Misonix sonicator by placing crushed ice in the horn cup for 10 min. Drain the ice and refill the horn cup with ice. Make sure that the bottom of the PCR plate is in direct contact with the water/ice slurry. Sonicate the plate for 1 min, turning it often. Stop the sonication, drain the water, and add more ice to the horn cup. Repeat the process five more times (total of 6 min) until the cells are lysed. Remove 10 μl of each sample to a new PCR plate containing 10 μl of 2× sample buffer/well: this is the **Total protein sample**.

4. Centrifuge the PCR plate for 15 min at 3500 g in the Legend centrifuge. Transfer the supernatant to another 96 well PCR plate using a multi-channel pipette. Take 10 μl of this supernatant and transfer to a plate containing 10 μl of 2× sample buffer: this is the **Cleared lysate sample**.

5. Heat the samples from steps 2 and 3 above for 3 min at 95°C. Load an SDS-polyacrylamide gel. Analyze the gel and note which clones give the highest level of soluble expression (see *Fig. 2*).

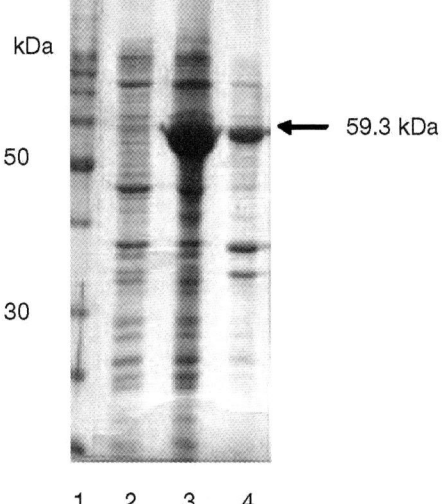

Figure 2. 12% SDS-PAGE of crude *E. coli* cell extract expressing target protein (59.3 kDa). Cells were treated as described in section 2.6. The gel was stained with Coomassie Blue R. *Lane 1*, molecular weight markers as indicated; *lane 2*, uninduced cell extract; *lane 3*, induced cell extract, soluble fraction; *lane 4*, induced cell extract, insoluble fraction.

(b) 96-Well mini-expression using a protein extraction reagent

Equipment and Reagents
- Biomek 2000 (Beckman Coulter)
- PopCulture reagent (EMD Biosciences)
- Benzonase nuclease (25 units/µl; EMD Biosciences)
- rLysozyme (300 000 units/µl; EMD Biosciences)
- 96-Well Unifilter microplate (Whatman)
- Collection plate
- Vacuum unit
- 4% SDS denaturing solution
- SDS-polyacrylamide gel
- Equipment and reagents for running an SDS-polyacrylamide gel

Method
1. Follow day 1 of *Protocol 5(a)*. Spin the 96-well, deep-well plate in the Legend centrifuge for 15 min (4°C, 3000 *g*). Shake the plate upside down three times to remove all liquid.
2. Use the Biomek 2000 to automatically deliver 100 µl aliquots of PopCulture reagent, Benzonase nuclease (2.5 units), and rLysozyme (0.75 units) to each sample. Resuspend the cell pellet.
3. Incubate the 96-well culture plate for 10 min at room temperature.
4. Transfer the lysed cultures to a 96-well filter plate.
5. Apply a vacuum and collect the flow-through containing soluble proteins in a collection plate (**Soluble fraction**).

6. Add 200 µl of 4% SDS denaturing solution to solubilize the inclusion bodies remaining on the 96-well filter plate.
7. Incubate the 96-well filter plate for 10 min at room temperature.
8. Apply a vacuum to collect the solubilized proteins in a second collection plate (**Insoluble fraction**).
9. Run the soluble fraction (10 µl from step 5) and the solubilized insoluble fraction (10 µl from step 6) on an SDS-polyacrylamide gel. Analyze which clones give the highest level of soluble expression.

2.3 Cell paste growth

Upon selection of the clones that show the highest level of soluble expression of target protein of the correct molecular weight, plasmid DNA is prepared and sent for DNA sequencing (if this was not done after *Protocol 3*). The medium used for cell growth has been described by Studier (10). This medium was developed for use with DE3 lysogens (such as BL21(DE3)) that supply T7 RNA polymerase upon induction of the *lac*UV5 promoter, in combination with pET expression vectors (such as pET21a) in which expression of the target gene is controlled by the T7 *lac* promoter. The fully defined noninducing medium (MDAG) is used to eliminate any unintended, spontaneous induction that is known to occur in complex media. This inducing activity is due to small amounts of lactose present in commonly used media such as LB. The auto-inducing medium used for preparing native protein is ZYM-5052 (see *Protocol 6*). Our experience is that the density of cells and levels of expression are often ten times higher when cells are grown in ZYM-5052 than when grown in LB and isopropyl-β-D-thiogalactopyranoside. For preparation of selenomethionine-labeled protein, PASM medium is used (10). Overnight incubation of the cells takes place at 20°C to enhance soluble expression of the targets.

Protocol 6

Cell paste preparation

Equipment and Reagents
All recipes for the media can be found in the paper by Studier (10).
- *E. coli* expression strain BL21(DE3)
- MDAG plates + antibiotics
- MDG medium + antibiotics
- ZYM-5052 medium
- 2.8 l Fernbach flask
- Infors shaker (Appropriate Technical Resources)
- Evolution centrifuge (Thermo Electron)
- 50 mM Tris/HCl (pH 8)

Method
1. On day 1, use the plasmid to transform competent *E. coli* expression strain BL21(DE3). Plate onto MDAG plates + antibiotics and incubate overnight at 37°C.

2. In the middle of the afternoon of day 2, take a colony from the plate and inoculate 10 ml of MDG medium + antibiotics into a 250 ml flask. Shake overnight at 37°C at 200 r.p.m. This starter is stable for 3–4 weeks at 4°C.

3. On day 3, dilute the starter 1 : 1000 into 1 l of ZYM-5052 medium (Fernbach flask, 2.8 l) with the desired antibiotics. Grow at 37°C until OD_{600} = 0.6–0.8. Cool the cells down to 20°C and grow overnight at 20°C at 200 r.p.m.

4. On day 4, spin the cell culture in an SLC 4000 rotor in an Evolution centrifuge at 1722 *g* for 30 min at 4°C. Decant the supernatant.

5. Resuspend the cell pellet in 100 ml of cold 50 mM Tris/HCl (pH 8). Spin the cells at 1722 *g* for 30 min and decant the supernatant. Freeze the cell paste at −80°C.

2.4 Clear lysate preparation

Frozen cell paste is resuspended in lysis buffer selected based on the pI of the protein. Addition of salt or reducing agent to the buffer is based on the particular target (note that metal-chelating resins cannot withstand some reducing agents). Protease inhibitors are added to prevent proteolysis of the target (see *Protocol 7*).

Protocol 7

Cell lysate preparation

Equipment and Reagents
- 50 mM HEPES (pH 8), 0.1 M NaCl, or an applicable lysis buffer (selected based on the pI of the protein)
- Lysozyme, egg white (10 mg/ml; Sigma)
- 0.1 M Phenylmethanesulfonyl fluoride (PMSF) (Sigma)
- DNase I (1 mg/ml; Research Organics)
- 25× Roche protease inhibitor cocktail tablet, EDTA-free (Roche Applied Science)
- Microfluidics M-110L (Microfluidics)
- RC5C-Plus Sorval centrifuge (Thermo Electron Corp.)
- Beckman ultracentrifuge with Ti45 rotor (Beckman Coulter, Inc.)
- Imidazole

Method
1. Resuspend the frozen cell paste at a concentration of 1 g in 5 ml of lysis buffer.
2. After the cell paste is uniformly resuspended, add chicken egg white lysozyme (freshly made, 10 mg/ml) to a final concentration of 0.1 mg/ml. Incubate the cell suspension on a shaking platform or rotating mixer for 20 min at room temperature.
3. Add the following: PMSF to a concentration of 1 mM; DNase I to a concentration of 10 µg/ml; and Roche protease inhibitor cocktail tablet to 1×.
4. Pass the resuspended cells are through a microfluidizer processor (Microfluidics M-110L) and pellet the cell debris by centrifugation at 28 000 *g* for 30 min at 4°C in an RC5C-Plus Sorval centrifuge.
5. Spin the lysate in a Beckman ultracentrifuge Ti45 rotor at 60 000 *g* for 20 min at 4°C to remove the membrane fractions. The sample is then filtered through a 0.2 µm membrane filter and imidazole is added to a concentration of 20 mM.

2.5 Purification of His$_6$-tagged protein

The construction of fusion proteins in which specific affinity tags are added to the protein sequence of interest simplifies the purification by employing affinity chromatography methods (14–16). The following protocol describes the purification of a target that has been tagged with six consecutive histidine residues (His$_6$). There are presently many immobilized metal-affinity columns (IMAC) and bulk resins on the market. This protocol is written for use with the HisTrap HP column, but any other IMAC resin can be substituted. There is a potential for binding of contaminant proteins to the resin, so a low concentration of imidazole is added to the cleared lysate and wash buffers (10–20 mM imidazole). At these low imidazole concentrations, nonspecific, low-affinity background proteins are prevented from binding to the resin, whilst the His$_6$-tagged proteins will still bind strongly. A high-salt wash is included in the protocol to remove background proteins that may be bound to the target. Using the AKTAexplorer, six proteins can be purified sequentially through six individual columns. The automation of protein purification has become widely established due to commercially available products (2, 17, 18).

Protocol 8

Purification of His$_6$-tagged protein

Equipment and Reagents
- AKTAexplorer (GE Healthcare) or comparable chromatography system
- HisTrap HP 5 ml metal-chelating column charged with Ni^{2+} following the manufacturer's protocols (GE Healthcare) or comparable product (column volume is 5 ml)
- Buffer A: 50 mM HEPES, pH 8, 100 mM NaCl, 20 mM imidazole
- Buffer B: 50 mM HEPES, pH 8, 100 mM NaCl, 1000 mM imidazole
- 1 M NaCl/Buffer A
- Cleared lysate from *Protocol 7*
- Broad-range protein standard (Bio-Rad Laboratories)

Method
1. Equilibrate the charged column with 15 column volumes of buffer A at a flow rate of 4 ml/min. All chromatography steps are performed at 4°C.
2. Load the sample at a rate of 2 ml/min. Collect the flow-through for further analysis.
3. After the sample has been loaded, wash the HisTrap HP column with five column volumes of Buffer A.
4. Wash the column with three column volumes of 1 M NaCl/Buffer A to eliminate nonspecific binding proteins and equilibrated with three column volumes of Buffer A.
5. Perform gradient elution in 13 column volumes from 2 to 40% Buffer B (20–400 mM imidazole) at 2 ml/min. Collect 2 ml fractions.
6. Perform a second gradient elution in three column volumes from 40 to 100% Buffer B (400–1000 mM imidazole) at 2 ml/min. Throughout the entire procedure, monitor UV absorbance, flow rate, pH, pressure, and ionic strength, and store the data.
7. Carry out a two column volume wash with Buffer B. The system should then be cleaned and the column stripped and stored according to the manufacturer's specifications.
8. For analysis, the following samples are loaded on an SDS-polyacrylamide gel: 10 µg of load, flow-through, 1 M NaCl wash, eluted fractions, and 10 µl protein standard. Based on the molecular weight of the target, one can then pool those fractions that contain the target protein.

2.6 Tobacco etch virus (TEV) protease digestion of fusion protein

Recombinant proteins are often fused to tags to facilitate purification and to increase yield and solubility. It is usually desirable to obtain the protein free from its fusion partner, as the affinity tag often interferes with functional activity or structural studies. The proteases used most often are factor Xa, enterokinase, and α-thrombin. Recently, a TEV protease mutant (S219V) has been shown to have stringent sequence specificity, as well as being more stable than the wild-type protease (19). For a fusion protein that has not been purified previously, it is a good idea to do a small-scale TEV digestion trial in order to avert large-scale precipitation of the target. This involves cleavage with dialysis in the presence of various buffers representing a range of possible salt concentrations or pHs depending on the characteristics of the protein. The protein is diluted to a concentration of 2 mg/ml or higher. After digestion with TEV protease, the sample is centrifuged and the protein concentration of the supernatant determined. Once the parameters for TEV protease digestion have been determined, large-scale TEV protease cleavage is performed. After TEV protease cleavage, the sample is applied a second time to the IMAC resin. The cleaved target is found in the flow-through, whereas the His_6 tag remains bound to the IMAC resin.

Protocol 9

TEV protease digestion of fusion protein

Equipment and Reagents
- TEV protease (His_6-tagged TEV) prepared from construct pRK793 (19) or AcTEV Protease (Invitrogen)
- 50 ml Microdialysis buttons (Hampton Research)
- Dialysis tubing (Spectrum Labs)
- Dialysis tubing closures
- Appropriate buffers
- 50 ml Conical tubes
- Bradford reagent (Amresco)
- Dissecting microscope

Method
Small-scale digestion with TEV
1. Determine the amount of TEV protease needed to cleave the desired mass of protein. This will depend on the TEV protease preparation used. Load the protein/TEV protease solution into the microdialysis button and place dialysis tubing of the appropriate molecular weight cut-off on top and fix it with an O-ring.

2. Place the button into a 50 ml conical tube containing the dialysis buffer of choice. Place the tube on a rocker for 1–2 h at room temperature. These trials can also be performed at 4°C if protease degradation is suspected. After incubation, observe the button for protein precipitation using a dissecting microscope. The protein solution can be removed from the dialysis button by piercing the membrane with a pipette tip, aspirating the sample, and centrifuging the sample to pellet insoluble material. The supernatant is analyzed by the Bradford method for protein concentration and this is compared with the concentration of the starting solution. This experiment can be done using many different dialysis buffers. The buffer condition with the most soluble protein and complete TEV cleavage is chosen for scaling up.

3. The following are some variables that can be tested:
    - Reduce the protein concentration to ~2 mg/ml.
    - Maintain 50–100 mM imidazole in the dialysis buffer if the protein is eluted from the IMAC resin at >400 mM imidazole. Those proteins that are eluted at high imidazole concentration tend to precipitate when the fusion tag is cut off.
    - Include a condition with a higher salt concentration in the dialysis buffer, such as 0.3 or 0.5 M NaCl.
    - Include a condition with 5–10% glycerol in the dialysis buffer.

Large-scale digestion with TEV

1. It is often helpful to first dilute the protein to ~2 mg/ml prior to cleavage in order to reduce the incidence of post-cleavage precipitation. Add TEV protease to the fusion protein at a ratio of 1:30 (ratio varies depending on the activity of the TEV protease). Load the TEV protease/target solution into the appropriate molecular weight cut-off dialysis bag and place in a large beaker containing the selected buffer and a stir bar. Place on a magnetic stirrer for 24 h at 4°C.

2. After overnight incubation, remove a sample and analyze by SDS-PAGE to verify that cleavage is complete. If cleavage is complete, the cleaved target can then be loaded a second time on a metal chelating column (see *Protocol 8*); the main difference is that the cleaved target will not bind to the IMAC column, whereas the TEV protease (with a His_6 tag) and the His_6 tag will remain bound to the resin.

3. TROUBLESHOOTING

- It is important to include a negative control when checking for expression of the constructs.
- Expression can be tried in different media and at different temperatures in parallel. Often there are striking differences in the levels of expression under different conditions.
- Sequence the expression constructs and, after purification, perform mass spectrometry on the target proteins to make sure that the molecular weight matches the calculated mass.
- Do not use strong reducing agents on the IMAC resin as they will reduce the Ni^{2+} ions and cause them to elute from the resin. Check the handbook that comes with the resin.
- Optimization of the purification process is achieved by matching the capacity of the resin to the amount of tagged protein in the sample. As the His_6-tagged protein has a higher affinity for the IMAC resin than the background proteins, it can fill all of the available binding sites so that very few background proteins will be retained on the resin.

Acknowledgements

We thank the following members of BSGC: Andy DeGiovanni, Barbara Gold, Marlene Henriquez, Candice Huang, Sung-Hou Kim, Raymond Lam, Yun Lou, Natasha Oganesyan, and Hisao Yokota. We appreciate help from David Waugh in providing pRK793 and the protocol for the purification of TEV protease. This work was supported by the National Institutes of Health GM 62412.

4. REFERENCES

1. Yee A, Pardee D, Christendat A, Savchenko A, Edwards AM & Arrowsmith CH (2003) *Acc. Chem. Res.* **36**, 183-189.
2. Kim Y, Dementieva I, Zhou M, *et al.* (2004) *J. Struct. Funct. Genomics*, **5**, 111-118.
3. Stevens RC (2000) *Structure*, **8**, R177-R185.
4. Baneyx F (1999) *Curr. Opin. Biotechnol.* **10**, 411-421.
5. Hannig G & Makrides SC (1998) *Trends Biotechnol.* **16**, 54-60

★★ 6. Aslanidis C & deJong PJ (1990) *Nucleic Acids Res.* **18**, 6069-6074. – *The original publication describing LIC.*

7. Nguyen H, Martinez B, Oganesyan N & Kim R (2004) *J. Struct. Funct. Genomics*, **5**, 23-27.
8. Kim R, Sandler SJ, Goldman S, Yokota H, Clark AJ & Kim S-H (1998) *Biotechnol. Lett.* **20**, 207-210.
9. Grabski A, Drott D & Mehler M (2003) *Innovations*, **16**, 11-13.

★★ 10. Studier W. (2005) *Protein Expr. Purif.* **41**, 207-234. – *A detailed description of auto-inducing media.*

11. Knaust RKC & Norlund P (2001) *Anal. Biochem.* **297**, 79-85.
12. Scheich C, Sievert V & Bussow K (2003) *BMC Biotechnol.* **3**, 12.
13. Braun P & LaBaer J (2003) *Trends Biotechnol.* **21**, 383-388.
14. Winzerling JJ, Berna P & Porath J (1992) *Methods Enzymol.* **4**, 4-13.
15. Crowe J, Dobeli H, Gentz R, Hochuli E, Stuber D & Henco K (1994) In *Methods in Molecular Biology*, vol. 31, pp. 371-387. Edited by AJ Harwood. Humana Press Inc., Totowa, NJ.
16. Gottstein C & Forde R (2002) *Protein Eng.* **15**, 775-777.
17. Page R, Grzechnik SK, Canaves JM, *et al.* (2003) *Acta Crystallogr. D. Biol. Crystallogr.* **59**, 1028-1037.
18. Acton TB, Gunsalus KC, Xiao R, *et al.* (2005) *Methods Enzymol.* **394**, 210-243.

★★ 19. Kapust RB, Tozser J, Fox JD, *et al.* (2001) *Protein Eng.* **14**, 993-1000. – *A protocol for purification of TEV protease.*

CHAPTER 3
Expression engineering of synthetic antibodies using ribosome display

Lydia M. Contreras Martínez and Matthew P. DeLisa

1. INTRODUCTION

Antibodies, and fragments derived from them, are widely used for diagnosis, biological research, and therapy, and their importance is expected to escalate in the coming years (1). Currently, they comprise the majority of recombinant proteins that have been approved by the FDA (18 products currently on the market) and, with many more in various stages of clinical development, the future for these molecules is extremely bright (1–3). The widespread application of antibodies was enabled by the use of hybridoma technology, established in 1975 by Kohler and Milstein (4). Large-scale antibody production via hybridoma technology is based on the fusion of a healthy B cell from the spleen of an animal that has been challenged by a relevant antigen and a cancer cell (myeloma) that can grow indefinitely in culture; in this way, the hybrid cell (hybridoma) multiplies indefinitely whilst producing large amounts of a particular antibody.

Recent advances in genetic engineering have provided the ability to produce antibodies *recombinantly*. This type of expression has been used to improve the low efficacy and the high immunogenicity that has limited the widespread clinical use of antibodies produced by hybridomas. Additionally, recombinant expression has allowed the molecular dissection of naturally produced full-length antibodies into smaller synthetic fragments (2). Antibody fragments (e.g. scAb, scFv, Fab, F(ab')$_2$) can be advantageous for many applications, as they retain the antigen-binding specificity and are more suitable for expression and for protein engineering in bacterial systems. These attributes provide enormous benefits in terms of scaling, engineering, and manufacturing of antibodies. As an example, single-chain Fv antibody fragments (scFvs) consist of covalently linked variable domains of both heavy (V_H) and light (V_L) chains that retain the specificity of the original two-chain Fv domain of the more structurally complex immunoglobulin (IgG) molecule from which they are derived. Despite removal of the IgG constant regions and the introduction of a flexible linker peptide, scFv fragments can still present considerable synthesis challenges. For a detailed discussion of scFv construction from a full antibody sequence, see Holliger *et al.* (2).

Expression Systems: *Methods Express* (M.R. Dyson and Y. Durocher, eds)
© Scion Publishing Limited, 2007

Although eukaryotic systems, such as *Pichia pastoris* and higher mammalian cells, have been used successfully as host systems for recombinant expression of scFvs and other antibody fragment formats (5–7), *Escherichia coli* is currently the host of choice for many antibody expression and engineering efforts. The natural synthetic capabilities of *E. coli* are relatively well understood and easy to manipulate and thus can be exploited more readily. However, a major limitation to the high-yield production of *active* scFv and other antibody formats in *E. coli* is the fact that expression of structurally complex *foreign* proteins in these cells often leads to protein misfolding and aggregation (8, 9). One reason for this phenomenon is that protein disulfide bonds form naturally in the oxidizing environment of the periplasm, but not in the highly reducing environment of the cytoplasm. As disulfide bridges are known to contribute ~4–5 kcal/mol to the overall stability of an scFv (10), lack of disulfide bonds typically results in poor intracellular solubility and loss of activity (11). Thus, expression of disulfide-rich molecules such as scFv or Fab fragments typically cannot be achieved in the *E. coli* cytoplasm.

To promote the formation of disulfide bonds, scFvs are commonly localized to the periplasm via the general secretory or twin-arginine translocation pathways (16–20) where natural or heterologous disulfide bond machinery enhances the likelihood of properly formed disulfide bridges (21, 22). Alternatively, specialized strains of *E. coli* lacking the cytoplasmic reductants *trxB* and *gor* are capable of forming disulfide bonds in cytoplasmic proteins (23) including scFv and Fab fragments (23–25). To improve further the solubility of cytoplasmically expressed scFvs in *E. coli*, a number of general strategies have been explored including, for instance, the co-expression of molecular chaperones (12), lower culture temperatures (13, 14), and genetic fusion to a highly soluble *E. coli* partner protein such as maltose-binding protein (15). A longstanding notion is that significantly higher yields of recombinant protein can be obtained from the cytoplasm relative to the periplasm; however, a growing body of evidence suggests that comparable levels of recombinant protein can be expressed and recovered from the periplasmic compartment. For instance, periplasmic production of an anti-CD18 F(ab')$_2$ antibody fragment joined to a leucine zipper reached 2.5 g/l (26). Thus, great care must be given to the choice of expression system (cytoplasmic versus periplasmic) for a given antibody expression study.

The creation of large combinatorial protein libraries, whereby an antibody's amino acid sequence is evolved for efficient folding, represents an alternative strategy to improve the soluble cytoplasmic expression of scFvs in *E. coli*. The notion of protein or directed evolution also represents a powerful way by which the functional properties (i.e. affinity and specificity) of antibody fragments can be enhanced and exquisitely tailored for desired applications. However, in the presence of millions of mutant proteins, a challenge with any combinatorial library approach is how to *select* variants that fold better than the parental (wild-type) antibody from a vast collection of sequences that show no change or even worsen folding and/or functionality.

Several screening and selection methods using reporter enzymes have been proposed for engineering protein variants that exhibit improved intracellular stability and functionality. Most recently, assays based on β-galactosidase α-complementation (27), green fluorescent protein (28), and the folding quality control of the twin-arginine translocation system (29) have been successful in evaluating cellular protein folding and in the evolution of protein sequences with enhanced cytoplasmic solubility. However, a recent application of the β-galactosidase α-complementation system to the creation of cytoplasmically stable scFv fragments failed to select clones exhibiting higher soluble protein expression levels, but rather selected for higher protease susceptibility of the fusion protein (30). Thus, one must be careful when employing such assays and interpreting the results that arise from their application. *In vitro* display techniques (31) are an alternative strategy that effectively minimize such biases and enable the evolution of scFvs with improved folding, stability, and function. In particular, ribosome display has been demonstrated as a powerful approach for stability engineering of scFv fragments (32–34) and, fortuitously, it has been found that stability-engineered antibody fragments also typically perform better in practical applications (34). This chapter outlines the use of ribosome display as an *in vitro* selection technique for the identification of mutations that improve the stability and/or functionality of recombinant antibody fragments.

2. METHODS AND APPROACHES

2.1 Principles of ribosome display

Ribosomes are especially attractive for protein engineering because of their ability naturally to couple genotype (mRNA) and phenotype (protein) during the process of translation and because they are extremely soluble complexes that are relatively easy to purify. Exploitation of these characteristics has led to the development of *in vitro* ribosome display (35, 36), as well as to the use of the ribosome as a solubility-enhancing partner in protein expression (37, 38).

Originally developed in the form of polysome display (36), ribosome display is based on the fundamental tenet that both the mRNA and the translated protein must stay bound to the ribosome and that the protein must fold *correctly* whilst bound. Some advantages associated with this method include: (i) the built-in ability to introduce mutations (to the recovered DNA) at each cycle, (ii) the fact that elution and purification of the target antibody fragment is not necessary as its corresponding mRNA can be isolated by extraction, and (iii) the possibility that proteins displayed on the ribosome are less prone to aggregation, perhaps because association with ribosomes is likely to have solubility-enhancing effects (38–40). Additionally, this technology is designed to operate completely *in vitro*; thus, it allows the creation of larger and more diverse libraries, as rare codons can be introduced and a transformation step, which often limits cell-based directed evolution experiments, is not required. In fact, whilst some of the largest *in vivo* libraries have been reported to contain 10^{10} library members, the use of *in vitro* ribosome display routinely allows screening of much larger libraries (10^{12}–10^{13}) (41). Likewise, this cell-free technology is better suited for producing toxic or proteolytically sensitive proteins. This represents a considerable advantage when expressing large amounts of recombinant proteins, especially if they are foreign to the host cells.

One limitation of ribosome display is that it only works *in vitro*, where complete translation on all ribosomes is challenging. A further limitation is that in ribosome display (and most other display techniques) the purification and immobilization of the target molecules have to be worked out individually, and binders against every target have to be selected in separate experiments. Additionally, the inability to use this selection method *in vivo* can be particularly problematic when attempting to evolve antibody fragments that require cellular factors such as molecular chaperones and isomerases for efficient folding. Finally, an extra *in vivo* step is needed to ensure that any functional improvements found *in vitro* are in fact reproducible inside host cells, which is important when engineering antibody fragments that eventually need to be manufactured in bacteria in a soluble form or for antibody fragments intended for intracellular functions, so-called intrabodies (2, 42–44).

As indicated by the protocols suggested in this chapter, several experimental modifications have been introduced to ease the application of ribosome display and to improve its efficiency for high-yield synthesis of active antibody fragments. Alternatively, our own efforts have focused on the extension of ribosome display for *in vivo* applications as a way of addressing some of the shortcomings associated with the fact that traditional ribosome display is carried out entirely *in vitro* (L.M. Contreras Martínez and M.P. DeLisa, unpublished observations).

2.2 Experimental overview of ribosome display

As illustrated in *Fig. 1*, ribosome display includes the following steps: (i) amplification and transcription of the DNA sequence of a target antibody fragment, whereby the stop codon is deleted so that the mRNA sequence remains attached to the ribosome; (ii) purification of mRNA; (iii) *in vitro* translation in the presence of different factors that enhance the stability of ternary ribosome–mRNA–protein complexes and aid the disulfide-bond-dependent folding of the translated antibody fragment; (iv) selection of the desired ribosome complexes from the translation mixture by binding of the (functional) antibody fragment to a purified antigen that has been immobilized onto a surface (or to beads in solution); (v) elution of mRNA from the complexes; and (vi) reverse transcription of the mRNA into cDNA, which can then be amplified for further rounds of mutagenesis and selection.

At the end of each selection cycle, the recovered DNA is enriched for those antibody sequences that have the desired property (i.e. ultrahigh affinity for a given antigen). In addition to allowing the selection of evolved antibody fragments, ribosome display also allows the selection of functional antibodies that bind a specific antigen target from naïve (i.e. not stemming from pre-immunized cells) libraries. In each case, the use of ribosome display is possible only when a binding partner (i.e. an antigen) against the antibody fragment is available.

Sections 2.3–2.10 present and evaluate the experimental protocols involved in ribosome display. Although we refer to cloning procedures and library construction methods in this chapter, our focus is on those techniques that are particularly

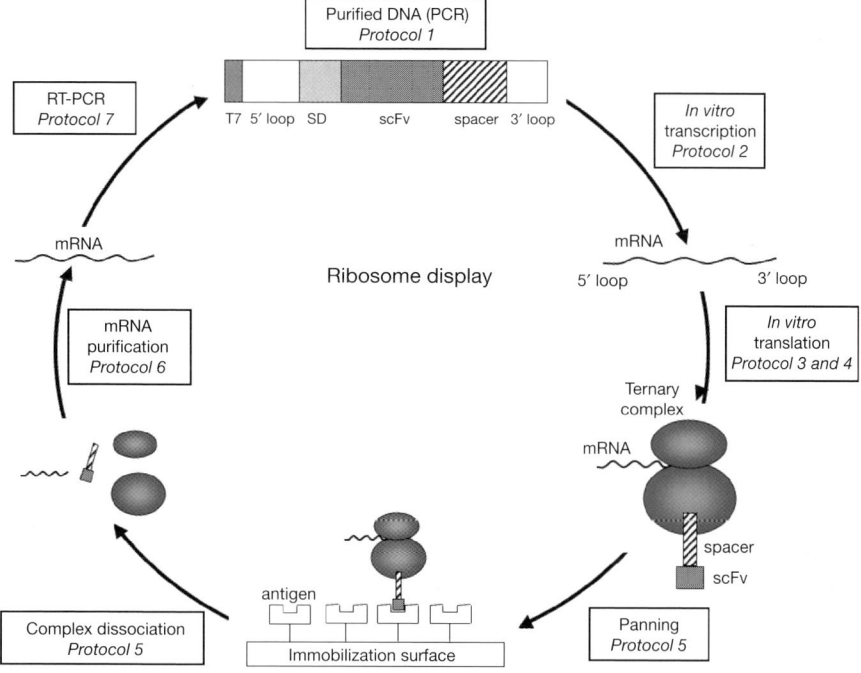

Figure 1. Schematic of ribosome display mutagenesis and the screening cycle.

relevant to the display and selection of scFvs that form part of ribosome ternary complexes.

2.3 ScFv gene sequence and control elements

A generalized schematic of an scFv construct used for ribosome display is illustrated in *Fig. 2* and includes the following important features:

1. A 5′ T7 promoter sequence for DNA transcription by T7 RNA polymerase. This region is not translated into protein.
2. A stable 5′ stem-loop that protects mRNA against degradation by RNases. This region is transcribed into mRNA but not translated into protein.
3. The Shine-Dalgarno (SD) ribosome-binding site sequence (e.g. AGGAGGU) for efficient initiation of *in vitro* translation in prokaryotic systems. A start codon (ATG) follows this sequence, after which the remaining DNA sequence is translated into protein. The Kozak consensus sequence (gccgcc(A/G)cc**AUG**G) is substituted in eukaryotic systems.
4. A 5′ protein epitope (e.g. c-Myc or FLAG epitope tag) allows immunodetection of the scFv.
5. Exclusion of a stop codon (TAG, TGA, TAA) to prevent release of the scFv during translation and thereby allow formation of stable mRNA–ribosome–scFv complexes.

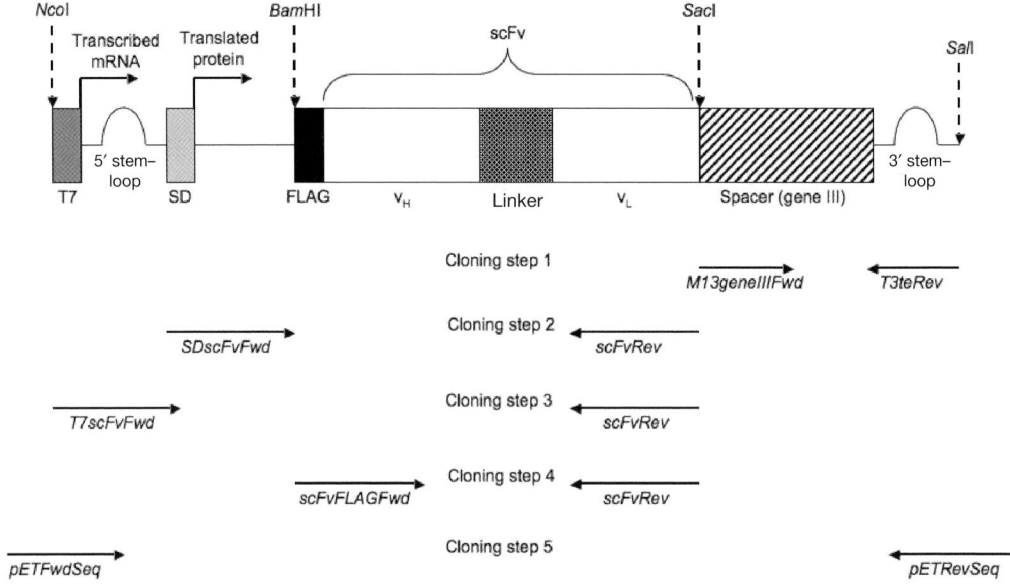

Figure 2. An scFv construct used for ribosome display.

6. A 72–116 amino acid (aa) tether fused to the scFv C terminus to provide sufficient spatial separation between the ribosome and the protein so that the scFv can fold co-translationally, and recognize and bind its target while still attached to the ribosome. The example provided below contains a natural structured spacer sequence derived from aa 211–312 of gene III of bacteriophage M13. Alternatively, synthetic Gly–Ser tethers (with little secondary structure) can be used in place of the naturally occurring spacer sequence. Infrequently used codons are often introduced to the tether sequence to retard translation further and reinforce ribosome stalling. Note that although the choice of tethers has been reported to have no effect on the diversity of the libraries created by ribosome display, the degree of secondary structure provided by the tether might influence the quantity and quality of antibodies generated due to the level of flexibility achieved for folding and prevention of steric hindrance from the ribosome.

7. A second 3′ stem–loop to protect the mRNA against degradation by 3′→5′ exonucleases in the cell extract used for *in vitro* translation. Although in this example we use a stem–loop derived from the early terminator of T3 phage (T3te), it is the presence and not the specific amino acid identity of this secondary structure that is important.

2.4 Construction of an scFv ribosome display vector (pscFvDisplay)

The sequence of human anti-β-galactosidase scFv (scFv13) (11, 30) configured for ribosome display in plasmid pET28a[a] (Novagen) is shown in *Fig. 3*. To make this construct, we suggest several sequential rounds of cloning to avoid the use of very

```
CCATGGCCATACGAAATTAATACGACTCACTATAGGGAGACCACAACGGTTTCCC
 NcoI           T7 Promoter         5'stem loop
TCTAGAAATAATTTTGTTTAACTTTAAGAAGGAGATATATCCGGATCCATG
                           Shine Dalgarno   BamHI Start
gactacaaggacgatgacgacaaggga gccgaggtgcagctggtggagtct
    FLAG epitope tag        scFv13→
ggggqaggcctggtcaagcctggggggtccctgagactctcctgtgcagcct

ctggattcaccttcagtaactatacatgaactgggtccgccaggctccaggg

aaggggctggagtgggtctcatccattagtagtagtagtagttacatatact

acgcagacttcgtgaagggccgattcaccatctccagagacaacgccaagaa

ctcactgtatctgcaaatgaacagcctgagagccgaggacacggctgtttat

tactgtgcgagatccagtattacgattttggtggcggtatggacgtctggg

gcagaggcaccctggtcaccgtctcctcaggtggaggcggttcaggcggagg

tggcagcggcggtggcggatcgcagtctgtgctgactcagcctgcctccgtg

tctgggtctcctggacagtcgatcaccatctcctgcgctggaaccagcagtg

acgttggtggttataactatgtctcctggtaccaacaacacccaggcaaagc

ccccaaactcatgatttatgagggcagtaagcggccctcaggggtttctaat

cgcttctctggctccaagtctggcaacacggcctccctgacaatctctggc

tccaggctgaggacgaggctgattattactgcagctcatatacaaccaggag

cactcgagttttcggcggagggaccaagctggccgtcctaggtgcggccgca
                                         ←scFv13
GAGCTCggctttaatgaggatccattcgtttgtgaatatcaaggccaatcgt
 SacI    spacer-gene III→
ctgacctgcctcaacctcctgtcaatgctggcggcggctctggtggtggttc

tggtggcggctctgagggtggtggctctgagggtggcggttctgagggtggc

ggctctgagggaggcggttccggtggtggctctggttccggtgatttgatt

atgaaaagatggcaaacgctaataagggggctatgaccgaaaatgccgatga

aaacgcgctacagtctgacgctaaaggcaaacttgattctgtcgctactgac

GCTCACCTTCACGGGTGGGCCGTCGAC
  3'stem loop       SalI
```

Figure 3. Human anti-β-galactosidase scFv configured for ribosome display in pET28a (Novagen).

long primers that could lower the cloning efficiency, as outlined below. In summary, step 1 introduces the M13 gene III spacer and the T3te 3' stem–loop to the empty vector, step 2 introduces the target scFv and the SD sequence to the 5' end of the spacer, step 3 introduces the T7 promoter region and the loop to the 5' end of the SD sequence, and step 4 introduces a FLAG epitope tag at the 5' end of the scFv sequence. All of the steps are based on the application of standard molecular cloning techniques, not discussed in this chapter. For specific cloning protocols refer to (45) and (46). Primer sequences are given in *Table 1*.

Table 1. Primers for construction of the pscFvDisplay vector for ribosome display of synthetic antibody fragments

Before using the suggested primers, the specific gene sequence of the scFv of interest must be checked to make sure that it will not be digested by any of the suggested restriction sites (i.e. *Sac*I, *Bam*HI, and *Nco*I). If these particular cut sites affect the sequence of interest, others can be used in their place as long as they are unique to the pET28 vector (or any vector of choice), provide the correct cloning orientation, and do not cut the tether sequence.

Primer	Sequence (5'→3')
M13geneIIIFwd	GCGATG*GAGCTC*GGCTTTAATGAGGATCCATTCGTTTGTGAATATC
T3teRev:	GCGATG*GTCGAC*GGCCCACCCGTGAAGGTGAGCGTCAGTAGCGACAGAATCAAGTTTGC
scFvRev	GCGATGGAGCTC*TGCGGCCGCACCTAGGACGGCC*
scFvFLAGFwd	GCGATGGGATCC**ATG**GACTACAAGGACGATGACGACAAGGGA*GCCGAGGTGCAGC*
SDscFvFwd	GCGATG*CCATGG***AGA**CCACAACGGTTTCCCTCTAGAAATAATTTTGTTTAACTTTAAGAAGGAGATATATCCGGATCCATG*GCCGAGGTGCAGC*
T7scFvFwd	GCGATG*CCATGG*ATACGAAATTAATACGACTCACTATAGGG**AGA**CCACAACGGTTTCCC
pETFwdSe	CGCGAAATTAATACGACTCACTATAGG
pETRevSeq	GCTAGTTATTGCTCAGCGGT

*Sequences between asterisks are specific to this scFv[b].

1. To construct intermediate plasmid pET28-M13geneIIIspacer-T3te, amplify aa 211–312 encoded by the bacteriophage M13 gene III from the *E. coli* phagemid vector M13mp19 (commercially available) with primers *M13geneIIIFwd* and *T3teRev*. Forward primer *M13geneIIIFwd* introduces a *Sac*I restriction site immediately before the 5' end of the spacer sequence. Reverse primer *T3teRev* introduces a stem–loop and a *Sal*I restriction site at the 3' end of the sequence. Digest the PCR product with *Sac*I and *Sal*I and ligate with *Sac*I/*Sal*I-digested pET28a vector.
2. To construct the second intermediate plasmid, pET28-SD-scFv-M13geneIII-spacer-T3te, amplify the scFv of interest using the appropriate template and primers. For this example, we use primers *SDscFvFwd* and *scFvRev*, which anneal to the specific sequence shown in *Fig. 3*. Forward primer *SDscFvFwd* introduces an *Nco*I restriction site, the ribosome-binding site, and a *Bam*HI restriction at the 5' end of the scFv. Reverse primer *scFvRev* introduces a 3'-end *Sac*I restriction site. Digest this PCR product with *Nco*I and *Sac*I and ligate with *Nco*I/*Sac*I-digested pET28-M13geneIIIspacer-T3te (constructed in step 1 above).
3. To construct a third intermediate plasmid, pET28-T7-SD-scFv-M13geneIII-spacer-T3te, reamplify the scFv from pET28-SD-scFv-M13geneIIIspacer-T3te (constructed in step 2 above) with primers *T7scFvFwd* and *scFvRev*. Forward

primer *T7scFvFwd* introduces an *Nco*I restriction site, the T7 initiation site, and a stem–loop at the 5′ end of the scFv sequence. Digest this PCR product with *Nco*I and *Sac*I and ligate with *Nco*I/*Sac*I-digested pET28-SD-scFv-M13geneIIIspacer-T3te (constructed in step 2 above).

4. To construct the final plasmid, pscFvDisplay (illustrated in *Fig. 2*), reamplify the scFv from pET28-T7-SD-scFv-M13geneIIIspacer-T3te (constructed in step 3) with primers *scFvFLAGFwd* and *scFvRev*. Forward primer *scFvFLAGFwd* introduces a *Bam*HI restriction site and a FLAG epitope tag immediately before the 5′ end of the scFv sequence. Digest this PCR product with *Bam*HI and *Sac*I and ligate with *Bam*HI/*Sac*I-digested pET28-T7-SD-scFv-M13geneIIIspacer-T3te (constructed in step 3 above).

5. Sequence the final pscFvDisplay construct with a forward and reverse primer that anneal to pET28 (i.e. *pETFwdSeq* and *pETRevSeq*, respectively) or use the universal T7 primer site to make sure that there are no mistakes in the plasmid, especially with respect to the T7 promoter region and the ribosome-binding sequences that are critical for transcription and translation[c]. Check the scFv sequence carefully for errors that could have been introduced during amplification.

Notes

[a]Although pET28a is used in this example, any other vector of choice may be used. A T7 promoter may be more convenient for the *in vivo* expression and testing of scFvs after all rounds of mutagenesis and selection (note c).

[b]For general use, the annealing part of the primer needs to be adapted to the particular sequence of interest.

[c]We recommend constructing a pscFvTest vector, by ligating *Bam*HI/*Sal*I-digested pscFvDisplay into *Bam*HI/*Sal*I-digested pET28a, and transforming this plasmid into an appropriate cell line that contains the T7 RNA polymerase gene (i.e. BL21(DE3)) to express the scFv. This would allow detection of the antibody fragment by Western blotting (with commercially available anti-FLAG antibody from Stratagene) to confirm that the scFv is expressed appropriately from the constructed plasmid *in vivo*, before continuing on to any *in vitro* work. For Western blotting, we suggest checking for whole-cell expression (and not just the soluble fraction), to guarantee detection of the antibody fragment even if it partitions into the insoluble fraction due to poor solubility in the *E. coli* cytoplasm.

The final pscFvDisplay construct can be used directly for *in vitro* transcription as indicated below, without the need to transform this plasmid into cells. We recommend transformation into any cell line of choice for long-term storage and future generation of additional plasmid.

2.5 Amplification and purification of the DNA target

Following construction of pscFvDisplay (as detailed in section 2.4 above), the scFv DNA target must be amplified from the plasmid for *in vitro* transcription. At this point, mutations can be introduced into the DNA sequence to create a combinatorial library using a number of different methods (e.g. random mutagenesis via error-prone PCR, site-directed mutagenesis, gene shuffling, etc.). Once the DNA library has been amplified, we recommend purification of the PCR product for more efficient transcription. DNA can be purified using a variety of commercially available methods such as a PCR Purification kit (Qiagen), as in *Protocol 1*.

Protocol 1

Purification of DNA from PCR

Equipment and Reagents
- Qiagen PCR Purification kit
- Spectrophotometer
- Microcentrifuge
- 1% Agarose gel containing 10 ng/ml ethidium bromide
- Appropriate DNA molecular mass ladder[a]
- Equipment and reagents for agarose gel electrophoresis including 1× TBE agarose gel running buffer (10× TBE per liter of water: 108.0 g of Tris, 55.0 g of boric acid, 40 ml of 0.5 M EDTA, pH 8.0)
- UV light source and gel documentation system

Method
1. Purify the resulting DNA from the PCR using a Qiagen PCR Purification kit following the manufacturer's instructions with a modification to the elution step[b,c].
2. Quantify the purified DNA by spectrophotometry or another method.
3. Analyze 200–300 ng by agarose gel electrophoresis to assess the DNA quality and degree of degradation, and to ensure that the expected size of the PCR product is obtained[d,e].

Notes
[a]The ladder should span the appropriate size range of the PCR product of interest.
[b]When eluting the DNA, it is beneficial to use DEPC-treated water or RNase-free water to avoid RNase contamination in the subsequent transcription step. Also, ensure that all of the ethanol (from Buffer PE) is removed completely. These aspects are essential for the efficient synthesis and stability of mRNA.
[c]Decreasing the elution volume to 30–50 µl when using the kit is recommended to prevent unnecessary dilution of the DNA when small quantities are expected.
[d]Agarose gel electrophoresis analysis is recommended directly following the PCR amplification step (i.e. prior to PCR purification) to ensure that the correct product has been obtained (at least as measured by size).
[e]The quality of the template DNA generated at this step is extremely important for the efficient generation of full-length mRNA transcripts.

2.6 *In vitro* transcription of a purified DNA library

A number of protocols and commercially available kits exist to carry out coupled transcription/translation reactions. Despite the experimental convenience of such one-step methods, two separate reactions are recommended for scFv sequences that contain disulfide bonds. This is crucial given that the T7 RNA polymerase requires reducing conditions during transcription to maintain its stability; however, reducing conditions disfavor the formation of disulfide bonds needed by scFv antibody fragments to fold properly following translation. Additionally, although *in vitro* transcription is ideally performed for 2-4 h to generate sufficient mRNA, the optimal duration of translation to generate stable ribosome complexes is significantly shorter.

Protocol 2 details the specific steps associated with *in vitro* transcription. Given the importance of preserving the quality of mRNA for the efficiency of protein synthesis (and therefore for the generation of large libraries), we recommend extreme caution when handling mRNA to avoid any degradation due to poor experimental technique.

Protocol 2

In vitro transcription of a DNA library

Equipment and Reagents
- Promega RiboMax Large Scale T7 Production System or Ambion MEGAscript T7 kit
- RNase-free tubes and tips
- Water bath or incubator at 37°C
- Cold (4°C) centrifuge
- Template DNA (5-10 mg from section 2.4)
- Spectrophotometer cell
- Spectrophotometer
- Chloroform
- Phenol
- 10 mM Tris/HCl (pH 7.5)
- 1 mM EDTA, isoamyl alcohol, isopropanol, ethanol (70 and 95%), citrate-saturated phenol:chloroform:isoamyl alcohol (125:24:1; pH 4.7), MicroSpin G-25 columns (GE Healthcare)[a]
- 5 M Ammonium acetate, 100 mM EDTA, isopropanol[b]
- Nuclease-free water

Method
1. Synthesize the mRNA transcript using the RiboMax Large Scale mRNA T7 Production System (Promega) or the MEGAscript T7 kit (Ambion) following the manufacturer's instructions.
2. Remove the DNA template by digestion with RQ1 RNase-free DNase (Promega kit). As indicated by the manufacturer, use 1 unit of DNase per 1 µg of template DNA and incubate the reaction for 15 min at 37°C.

3. Proceed immediately to extract the mRNA by phenol:chloroform precipitation to remove the enzyme and the free (unincorporated) nucleotides.

4. Analyze the purity of the product mRNA by reading the ratio of A_{260}/A_{280}[c]. Quantify the amount of mRNA by determining the A_{260}. An A_{260} reading of 1 corresponds to 44 µg/ml, so the RNA yield can be calculated as A_{260} × dilution factor × 44 = µg/ml RNA. Start by using a 1:30–1:100 dilution ratio (i.e. mix 5 µl of the mRNA product with 145 ml of water for a 1:30 dilution). Higher dilution ratios can be used. Make sure that the spectrophotometer is calibrated with the same solution used for dilution.

5. Store the mRNA transcript at –20 or –70°C in RNase-free water to avoid degradation.

Notes

[a] For the extraction protocol recommended by Promega in the RiboMax T7 Production System user manual.
[b] For the extraction protocol recommended by Ambion in the MEGAscript T7 kit user manual.
[c] For accurate absorbance values, the use of 10 mM Tris/HCl (pH 7.5) is recommended as water is not buffered and A_{260}/A_{280} ratios can vary tremendously based on pH. Pure RNA has an A_{260}/A_{280} ratio of 1.9–2.1 in 10 mM Tris/HCl. For quantification, water (or any other buffer with a neutral pH) should be used, as the relationship between absorbance and concentration is based on an extinction coefficient calculated for RNA at neutral pH.

2.7 *In vitro* translation of a mRNA library

One of the basic components of an *in vitro* translation system is the cell-free extract (S-30 lysate) of *E. coli* that provides the factors required for translation such as ribosomes, tRNA synthetases, tRNAs, and other translational factors. Although the use of commercially available eukaryotic translation systems (i.e. the rabbit reticulocyte lysate) has been described, the S-30 prokaryotic system has been the most studied and optimized, and is usually preferred (39). In fact, although successful scFv libraries have been created and displayed using eukaryotic cell-free translation systems, direct comparison has shown the *E. coli* system to be more efficient for displaying model scFv constructs than the rabbit reticulocyte system (47).

An advantage of a cell-free system is that the addition of molecules to the reaction mix can be controlled to recreate the *E. coli* cytoplasm according to the needs of the target protein. Examples of this include: (i) the addition of varying concentrations of dithiothreitol (DTT) to alter the natural redox state of the cell and (ii) the addition of protein-prolyl *cis-trans* isomerase and eukaryotic protein disulfide isomerase that support the folding of disulfide bond-containing scFvs in the periplasm. Although the use of protein disulfide isomerase is suggested as part of *Protocol 4*, it can be excluded if ribosome display is used to select for disulfide-free functional antibodies, referred to as intrabodies.

Particular attention needs to be paid to RNases when transcription and translation take place separately. The addition of the 5′ and 3′ loops to the gene construct (section 2.3) and the addition of RNase inhibitors (i.e. ribonucleoside-vanadyl complex) both aim to increase the protection of mRNA against the multiple cellular RNases. Additional precaution is taken by using cellular lysate from

E. coli strain MRE600, which lacks one of the predominant ribonucleases against rRNA (RNase I). Nevertheless, the short half-life (30 s to 20 min) (41) in the presence of cell lysate is regarded as the limiting factor for *in vitro* translation; this often results in incomplete protein synthesis.

In addition to the synthesis of full-length proteins, it is important to be able to form stable ribosome complexes during *in vitro* translation. However, natural quality-control mechanisms exist in *E. coli*, whereby a transfer messenger RNA (tmRNA, encoded by the *ssrA* gene) attaches a stop codon to mRNA sequences that lack their own (as in the case of displayed sequences) and tags them for degradation (41). This leads to the forced release of mRNA from ribosomes and, therefore, the destruction of ternary complexes. One way to block the tmRNA activity is by the addition of an antisense *ssrA* oligonucleotide (5'-TTAAGCTGCTAAAGCG TAGTTTTCGTCGTTTGCGACTA-3') that titrates the tmRNA by complementary binding. *Protocol 4*, based on the method of Hanes *et al.* (48), describes the details for the *in vitro* translation reaction using S-30 *E. coli* lysate, generated by use of *Protocol 3*. *Protocol 3* is an adaptation of previously published methods (49).

Although ternary complexes have been shown to be stable under the right conditions (cold temperatures and high (50 mM) Mg^{2+} concentrations) for up to 10 days (41), we suggest proceeding to panning (section 2.9 as soon as possible (within 1–2 days), as we have observed increased loss of protein activity with longer times.

Protocol 3
Preparation of MRE600 *E. coli* S-30 lysate

Equipment and Reagents[a,b]
- Autoclaved growth medium (per liter of water: 5.6 g of KH_2PO_4 (anhydrous), 28.9 g of K_2HPO_4 (anhydrous), 10 g of yeast extract, 10–15 mg of thiamin, and 40 ml of 25% (w/v) filtered glucose solution to be added after autoclaving, when solution has cooled)
- Buffer I (10 mM Tris/HOAc (pH 8.2), 14 mM $Mg(OAc)_2$, 60 mM KCl, 6 mM 2-mercaptoethanol)
- Buffer II (10 mM Tris/HOAc (pH 8.2), 14 mM $Mg(OAc)_2$, 60 mM KCl, 1 mM DTT)
- Buffer III (10 mM Tris/HOAc (pH 8.2), 14 mM $Mg(OAc)_2$, 60 mM KOAc, 1 mM DTT)
- Incubation mix (per 8 ml of water: 6 mmol of Tris/HOAc (pH 8.2), 0.06 mmol of DTT, 0.17 mmol of $Mg(OAc)_2$, 0.6 μmol of each of the 20 amino acids, 0.048 mmol of ATP, 0.54 mmol of Na_3PEP, 16 mg of pyruvate kinase (Sigma))
- DTT
- Analytical balance
- Autoclaved flasks
- Shaker with temperature control
- Centrifuge
- French pressure cell or sonicator (for cell lysis)
- Bio-Rad protein assay or other total protein quantification method
- Spectrophotometer
- Dialysis tubing (with a 6–8 kDa molecular mass cut-off)

Method

1. On day 1, grow two cultures to saturation in 10 ml of growth medium overnight[c].
2. On day 2, inoculate each 10 ml culture into 1 l of growth medium. Grow at 28°C up to mid-exponential phase (OD_{600} ~0.6–0.8).
3. When the cells reach the desired OD_{600}, chill each flask rapidly by swirling the flasks in ice for 1–2 min.
4. Harvest the cells by centrifugation at 4500 r.p.m. (3500 *g*) at 4°C for 15 min.
5. Resuspend the cells in 400 ml of cold Buffer I to wash the cells.
6. Pellet the cells by centrifugation at 4500 r.p.m. (3500 *g*) at 4°C for 15 min and resuspend in 400 ml of cold Buffer I to wash the cells. After this second wash, scrape off any dark material (or debris) on the surface of the pellet.
7. Resuspend the cells *completely* in 50–80 ml of cold Buffer II to begin cell lysis[d].
8. Lyse the cell suspension using a French pressure cell at 4000–8000 p.s.i.
9. Immediately after lysis, add DTT to a final concentration of 1 mM[e].
10. Centrifuge the lysate twice at 30 000 *g* for 30 min at 4°C. Transfer the supernatant to a clean fresh tube.
11. Measure the protein concentration using the Bio-Rad protein assay. Adjust to ~10–15 g/l by diluting with Buffer II, if necessary.
12. Add 1 ml of incubation mix per 7 ml of S-30 lysate. Incubate with shaking in a light-protected vessel at 25°C for 80 min.
13. Dialyze for 18 h at 4°C against 2 l of Buffer III. Use dialysis tubing with a molecular mass cut-off of 6–8 kDa.
14. Aliquot the lysate (~80 ml) in aliquots of 0.5–1 ml. Snap freeze in liquid nitrogen and store at −80°C.

Notes

[a]Make sure that all buffers (except for the incubation mix) are prepared and cooled to 4°C before beginning preparation of the lysate on day 2. We also recommend preparing the dialysis tubing before starting the prep. The incubation mix should be prepared immediately before use.
[b]Where necessary, add 2-mercaptoethanol or DTT just prior to use.
[c]As no antibiotic can be used, we recommend monitoring contamination with a negative (no cell) control.
[d]Use 1 ml of Buffer II per 0.25 g of cells obtained (13–20 g of cells can be expected).
[e]DTT can be excluded from this final buffer for lysate use in translation of proteins containing disulfide bonds.

Protocol 4

In vitro translation

Equipment and Reagents
- 1 M Tris/HOAc (pH 7.5)
- 1 M NH$_4$OAc
- 0.5 M Mg(OAc)$_2$
- 50 mM of each amino acid (Sigma)
- 50 mM ATP
- 50 mM GTP
- 100 mM Cyclic AMP (cAMP; Sigma)
- *E. coli* tRNA (50 mg/ml; Sigma)
- Folinic acid (2 mg/ml)
- 2 M KOAc
- 15% PEG-8000 solution (150 mg/ml)
- Rifampicin (3.3 mg/ml)
- Vanadyl-ribonucleoside complexes (100 mg/ml; Fluka)
- Anti-*ssrA* oligonucleotide
- 30 µM Protein disulfide isomerase (Fluka)
- *E. coli* MRE600 S-30 extract (see *Protocol 3*)
- Template mRNA transcript (from the transcription reaction)
- RNase-free water
- RNase-free tips
- Cold stop buffer (50 mM Tris/HOAc (pH 7.5), 150 mM NaCl, 50 mM Mg(OAc)$_2$, 2.5 mg/ml heparin, 0.1% Tween 20)
- 37°C Water bath or incubator
- Refrigerated (4°C) centrifuge

Method
1. In an RNase-free microcentrifuge tube, mix the following:
   - 11.0 µl of 1 M Tris/HOAc
   - 6.6 µl of 1 M NH$_4$OAc
   - 5.4 µl of 0.5 M Mg(OAc)$_2$
   - 1.54 µl of each 50 mM amino acid solution
   - 8.8 µl of 50 mM ATP
   - 2.2 µl of 50 mM GTP
   - 2.2 µl of 100 mM cAMP
   - 2.2 µl of *E. coli* tRNA (50 mg/ml)
   - 2.2 µl of folinic acid (2 mg/ml)
   - 11.0 µl of 2 M KOAc[a]
   - 22.0 µl of 15% PEG-8000 solution
   - 2.2 µl of rifampicin (3.3 mg/ml)
   - 2.2 µl of vanadyl-ribonucleoside complex solution (100 mg/ml)
   - 2.2 µl of 30 µM protein disulfide isomerase[b]
   - 51.4 µl of *E. coli* S-30 extract

   Add the appropriate volume (i.e. 1–2 µl) of the anti-*ssrA* nucleotide to achieve a final concentration of 3.5 µM. Add ~20 ng of the synthesized mRNA transcript[c]. Add nuclease-free water to give a total final volume of 220 µl.

2. Mix well by gently vortexing or tapping the tube.

3. Incubate at 37°C for 8 min.
4. Immediately stop translation by adding 660 µl of stop buffer.
5. Centrifuge for 5 min at 4°C and 10 000 *g* to remove all insoluble components and collect the supernatant[d] in a clean tube for panning (see *Protocol 5*).

> **Notes**
> [a]Substitution of potassium acetate by potassium glutamate has been reported to improve mRNA yields during display.
> [b]Can be excluded for certain types of experiment (see section 2.7).
> [c]Add mRNA last to the reaction and keep it at 4°C while setting up the reaction.
> [d]The supernatant contains ternary ribosome complexes, encoded protein, and ribosomes. After this point, avoid vortexing the reaction and keep everything cold to preserve the stability of the formed ribosome complexes.

2.8 Affinity selection of ribosome complexes and mRNA isolation

The selection of ribosome complexes displaying scFvs with desired properties requires a purified sample of the cognate antigen. Once an antigen is obtained, selections can be performed either by conjugating the antigen with a tag (e.g. biotin) that can be captured in solution (i.e. by use of streptavidin magnetic beads (45)) or by immobilizing the antigen on a solid (e.g. plastic) surface (50). One of the problems with selection on plastic surfaces is the nonspecific binding of ribosomes and the partial unfolding of the protein that results due to hydrophobic interactions with the plastic. To overcome this problem, improved protocols have introduced blocking steps that use *E. coli* tRNA and 0.2% heparin (which also inhibits nucleases) (41).

In addition to effectively selecting for ribosome complexes that contain scFvs that show phenotypic properties of interest, a second part of the selection process is the ability to obtain their underlying genotype for further analysis and manipulation. Although entire ribosome complexes can be eluted by the addition of a competing ligand (35), this can be challenging when the complexed scFv binds the antigen with high affinity. Instead, EDTA is used as a way of destroying the entire ribosome complex (by chelating Mg^{2+}), thereby releasing the pool of mRNA corresponding to the scFv(s) of interest (i.e. those that remain bound after repeated washing steps). *Protocol 5* details the steps for selection of ribosome complexes by panning on a plastic surface and for mRNA isolation. The suggested protocol has been optimized for use with β-galactosidase, the specific antigen for scFv13 discussed throughout this chapter.

2.9 Reverse transcriptase PCR (RT-PCR)

Although one-step RT-PCR protocols are commercially available, we recommend a two-step reaction, as in *Protocol 7*. For the transcription step, we use a reverse primer (*T3teRev, Table 1*) that anneals to the 3' (loop–spacer) region of the mRNA. This limits transcription to yield full-length DNA fragments that are ready for further rounds of mutagenesis and display, after reintroduction of the 5' untranslated T7 promoter and ribosome-binding site (see *Fig. 2*) via several rounds of amplification.

Although *Protocol 7* suggests the use of multiple PCR rounds to introduce the 5' T7 promoter and ribosome-binding site regions after the transcription step, an alternative method is to clone the collection of amplified DNA fragments (obtained after step 3) back into the original pscFvDisplay vector. This option can minimize the introduction of undesired errors, as fewer PCR rounds are required. With this strategy, it is also possible to replace the use of the long primer *SDscFvFwd* in step 3 by a shorter primer that anneals only to the 5' end of the (tagged) scFv and introduces a *Bam*HI restriction site. In this way, the first PCR product (obtained after step 3) can be digested with *Bam*HI and *Sac*I and ligated into *Bam*HI/*Sac*I-digested pscFvDisplay. This construct can then be used as template for further rounds of mutagenesis and selection. The option of using a shorter primer can be beneficial for troubleshooting RT-PCR, in the event that no product is obtained.

This alternative strategy also allows the use of a different reverse primer that anneals only to the 3' end of the scFv sequence, to replace *T3teRev*. This can alleviate the potential problem of not transcribing truncated/degraded mRNAs that might contain entire scFv sequences of interest but perhaps not the 3' loop or the entire linker sequence to which *T3teRev* anneals. To maximize the efficiency and quality of the RT-PCR, we suggest purification of the isolated mRNA prior to the reaction, as in *Protocol 6*.

Protocol 5

Panning of ribosome complexes and elution of mRNA

Equipment and Reagents
- 96-Well BD Falcon or other ELISA-type plates
- Solution of appropriate antigen in PBS (1 mg/ml)
- 1 M Sterile-filtered $MgCl_2$ solution
- 1× PBS (per liter of water: 8.0 g of NaCl, 0.2 g of KCl, 1.44 g of Na_2HPO_4, 0.24 g of KH_2PO_4)
- Blocking solution (per 50 ml of PBS: 0.5 g of dried milk, 250 µl of 1 M $MgCl_2$, 15 mg of heparin, 2.5 mg of *E. coli* tRNA)
- Washing solution (per 50 ml of PBS: 50 µl of Tween 20, 250 µl of 1 M $MgCl_2$)
- Eluting solution (per 2 ml of PBS: 85 µl of 0.5 M EDTA, 2 µl of RNAsin (Promega))
- Saran paper
- 4°C Refrigerator/cold room
- RNase-free microcentrifuge tubes and tips
- Plate shaker

Method
1. Coat a 96-well BD Falcon plate[a] with 65 µl of a 1–5 ng/ml PBS solution of appropriate antigen (~65 ng per well). Try to avoid bubbles. Wrap in Saran paper and leave at 4°C overnight so that the antigen adheres to the plate.

2. Add 200 μl of PBS and shake at room temperature for 5 min. Discard the PBS by turning the plate upside down abruptly. Remove residual PBS by tapping the plate onto a paper towel a couple of times. Repeat this step three times for a total number of four washes.
3. Add 200 μl of blocking solution to each well. Wrap in Saran paper and incubate at room temperature for 2 h.
4. Add 200 μl of washing solution to each well. Wash four times at room temperature (as indicated in step 2).
5. After the last wash, wrap the plate in Saran paper and incubate at 4°C for 15–20 min[b].
6. Add 50 μl of *cold* blocking solution to all wells. Add an equal volume of the *in vitro* translation supernatant[c] to each well and mix gently by pipetting up and down. Wrap in Saran paper and incubate at 4°C for 1 h[d].
7. Wash for 2–3 min with 200 μl of washing solution at 4°C to remove any unbound complex. Repeat this wash five to six times.
8. To dissociate the mRNA from the complexes, add 100 ml of *cold* eluting solution to each well. Shake gently (to avoid splashing) for 30 min at 4°C. Collect each sample from the wells into *cold* microcentrifuge tubes. Scrape the plate surface with a tip to ensure complete sample removal while collecting the samples.
9. Proceed to mRNA purification (see *Protocol 6*).

Notes

[a] Plan on using 1 well per 50 μl of translation supernatant. It may not be necessary to use the entire plate.

[b] All steps following this one should be done at 4°C and in blocking solution to maximize the stability of the ribosome complexes.

[c] The translation reaction mix can be panned directly without purification.

[d] Incubation time should be optimized for the specific selection of interest (i.e. in a library situation, it might be beneficial to optimize this time to reduce binding of the wild-type sequence).

Protocol 6

Purification of mRNA eluted from selective panning

Equipment and Reagents
- RNeasy Purification kit (Qiagen)
- RNase-free tips
- Spectrophotometer
- Microcentrifuge
- Ethanol

Method
1. If necessary, adjust the volume of all samples collected from panning (see *Protocol 5*) to 100 μl by the addition of RNase-free water. Do a separate purification for each 100 μl sample.
2. Purify the mRNA from the collected sample using the Qiagen RNeasy kit following the manufacturer's instructions, with a modified elution step[a]. Combine the recovered product (from all purifications) into one tube at the end of the elution step. Immediately place the tube on ice.
3. Analyze the quantity and purity of the product mRNA by measuring A_{260} and the A_{260}/A_{280} ratio, as indicated in *Protocol 2* (step 4).
4. Store the purified product at –20°C for long-term storage. Keep on ice while in use.

> **Note**
>
> [a] Elute in 35 μl of RNase-free water. After the first elution, add the eluate back to the column, incubate at room temperature for 5–10 min, and spin at 16 000 *g* for 1 min for increased yield.

Protocol 7

RT-PCR

Equipment and Reagents
- Sensiscript RT kit (Qiagen)
- RNasin (Promega)
- Primer *T3teRev* (0.1 mM)[a]
- 10 mM dNTP mix (60 μl of water, 10 μl of 100 mM dATP, 10 μl of 100 mM dCTP, 10 μl of 100 mM dGTP and 10 μl of 100 ml dTTP)
- 10× ThermoPol Buffer (NEB)
- Vent polymerase (NEB)
- 100 mM $MgSO_4$ (NEB)
- Primer *T7scFvFwd* (0.1 mM solution)[a]
- Primer *SDscFvFwd* (0.1 mM solution)[a]
- Purified template mRNA (obtained from *Protocol 6*)
- Thermocycler
- 37°C Water bath or incubator

- RNase-free tips and tubes
- Equipment for gel electrophoresis (including buffer, ladder, and loading dye)

Method

1. Set up the reverse transcription reaction using the Sensiscript RT kit as recommended by the manufacturer, using ~48 ng of mRNA template and 2 µl of 0.1 mM *T3teRev* primer.

2. Just before the end of the incubation period for the RT reaction (step 1), prepare the following PCR mix (for a 50 µl reaction):
   - 37 µl of water
   - 5 µl of 10× ThermoPol Buffer
   - 2.5 µl of 10 mM dNTP mix
   - 0.75 µl of 100 mM $MgSO_4$
   - 1.0 µl of 0.1 mM *SDscFvFwd*[b]
   - 1.0 µl of 0.1 mM *T3teRev*
   - 1.0 µl of Vent polymerase

3. *Immediately* after the RT reaction is completed, add 2 µl of this reaction mix (as template) to the PCR mix prepared in step 2. Vortex to mix well. Place in the thermocycler and amplify using the following conditions:

Number of cycles	Program
1	95°C for 5 min
30	95°C for 5 min, 54°C for 1 min, and 72°C for 1.5[c] min
1	72°C for 7 min

4. Analyze 5–8 µl of product by gel electrophoresis to make sure that the reaction has been successful.

5. Reamplify the PCR product obtained in step 3 by adding the following:
   - 37 µl of water
   - 5 µl of 10× ThermoPol Buffer
   - 2.5 µl of 10 mM dNTP mix
   - 0.75 µl of 100 mM $MgSO_4$
   - 1 µl of 0.1 mM *T7scFvFwd*[d]
   - 1 µl of 0.1 mM *T3teRev*
   - 1 µl of Vent Polymerase
   - 2 µl of PCR product from step 3 as template

6. Purify the PCR product by following *Protocol 1*.

7. Use the PCR-amplified DNA for the next selection cycle (i.e. by further mutagenesis) or to digest and ligate back into a vector (i.e. pscFvTest), transform into *E. coli*, and isolate the plasmids of individual clones for sequencing analysis and further testing (see section 2.8).

Notes

[a] See section 2.4 and *Table 1*.
[b] This primer reintroduces the nontranslated 5′-end ribosome-binding site.
[c] Estimate 1 min of elongation time per 1000 bp of the DNA sequence.
[d] This primer reintroduces the nontranslated 5′ T7 promoter.

2.10 *In vivo* analysis of selected scFvs

One of the drawbacks of an *in vitro* selection method is that the isolated scFvs may not maintain their phenotype when expressed inside cells. For instance, an antibody fragment that is evolved for improved solubility, or one that is selected from a naïve library for binding to a specific target, might aggregate or be toxic when expressed cytoplasmically during the production or manufacture stage.

For these reasons, we recommend cloning the DNA recovered from the last round of display into a plasmid (i.e. pscFvTest, described in section 2.4), retransforming it into an appropriate cell line such as BL21(DE3) and expressing the corresponding scFv variants inside host cells. Using appropriate controls (i.e. the original wild-type scFv sequence), the intracellular stability and functionality of the isolated mutant scFv(s) can then be evaluated. This should be performed using periplasmic expression via either the secretory or secretory/signal recognition pathways, as correct folding of the antibody fragments will most likely require disulfide bond formation, a process that is normally relegated to the periplasm. If cytoplasmic expression is desired, then a specially engineered strain of *E. coli* (e.g. *trxB gor* mutant) that permits disulfide bond formation in the cytoplasm should be employed. Western blotting and enzyme-linked immunosorbent assay (ELISA) are methods by which expression and activity can be quantified. Given the high diversity of clones obtained in ribosome display, we recommend screening multiple variants (at least 10–15) to ensure that an ample cross-section of the clones obtained in each selection round is evaluated for improvements in intracellular solubility. In our experience, success rates of finding an scFv that binds the target antigen with significantly higher affinity compared with the wild-type sequence range between 15 and 55% (after two rounds of evolution), depending on the specific target antibody sequence, the construction of the display vector (i.e. length and type of tether), and the *in vitro* transcription and translation efficiency.

3. TROUBLESHOOTING

- It is important to handle all mRNA and ribosome samples with extreme care throughout these protocols to avoid degradation and RNase contamination. A few things to keep in mind are: (i) handle mRNA with RNase-free tips and RNase-free equipment; (ii) wear gloves and change them often; (iii) always keep mRNA frozen or on ice when in use; (iv) clean all equipment (including the bench in use) with RNAErase (Bio-101 Systems); and (v) filter sterilize all solutions and bake glassware at 210°C for 3–4 h, rather than autoclaving glassware and solutions.
- If the presence of RNase is suspected, treat the DNA with proteinase K (100 µg/ml) and SDS (0.5%) in 50 mM Tris/HCl (pH 7.5), 5 mM $CaCl_2$ for 30 min at 37°C.
- The best results are obtained when starting with high-quality mRNA and DNA. As the protocols are designed to amplify and transcribe small quantities of

- genomic material, quality is more important than quantity. However, if quality is suspected to be a problem, increasing the amount of starting material can be helpful.
- A positive control (such as an scFv known to be active) can also be included in all of the experiments to ensure that all of the techniques are working properly. This can be particularly helpful if you are having problems with panning, mRNA recovery, and RT-PCR.
- Multiple rounds of mutagenesis and selection might be necessary before an improved scFv variant is found. Do not forget to reintroduce the (untranslated) T7 promoter and ribosome-binding site to the 5′ end of the recovered DNA after each display cycle.

4. REFERENCES

1. Reichert JM, Rosensweig CJ, Faden LB & Dewitz MC (2005) *Nat. Biotechnol.* **23**, 1073–1078.
2. Holliger P & Hudson PJ (2005) *Nat. Biotechnol.* **23**, 1126–1136.
3. Hoogenboom HR (2005) *Nat. Biotechnol.* **23**, 1105–1116.
4. Kohler G & Milstein C (1975) *Nature*, **256**, 495–497.
5. Cupit PM, Whyte JA, Porter AJ, et al. (1999) *Lett. Appl. Microbiol.* **29**, 273–277.
6. Dorai H, McCartney JE, Hudziak RM, et al. (1994) *Biotechnology (N.Y.)*, **12**, 890–897.
7. Verma R, Boleti E & George AJ (1998) *J. Immunol. Methods*, **216**, 165–181.
8. Baneyx F & Mujacic M (2004) *Nat. Biotechnol.* **22**, 1399–1408.
9. Lilie H, Schwarz E & Rudolph R (1998) *Curr. Opin. Biotechnol.* **9**, 497–501.
10. Frisch C, Kolmar H, Schmidt A, et al. (1996) *Fold. Des.* **1**, 431–440.
11. Martineau P, Jones P & Winter G (1998) *J. Mol. Biol.* **280**, 117–127.
12. Wang JD, Herman C, Tipton KA, Gross CA & Weissman JS (2002) *Cell* **111**, 1027–1039.
13. Vasina JA & Baneyx F (1996) *Appl. Environ. Microbiol.* **62**, 1444–1447.
14. Vasina JA & Baneyx F (1997) *Protein Expr. Purif.* **9**, 211–218.
15. Kapust RB & Waugh DS (1999) *Protein Sci.* **8**, 1668–1674.
16. Bothmann H & Pluckthun A (1998) *Nat. Biotechnol.* **16**, 376–380.
17. DeLisa MP, Tullman D & Georgiou G (2003) *Proc. Natl. Acad. Sci. U.S.A.* **100**, 6115–6120.
18. Kazemier B, de Haard H, Boender P, van Gemen B & Hoogenboom H (1996) *J. Immunol. Methods*, **194**, 201–209.
19. Skerra A (1993) *Curr. Opin. Immunol.* **5**, 256–262.
20. Skerra A & Pluckthun A (1991) *Protein Eng.* **4**, 971–979.
21. Kadokura H, Katzen F & Beckwith J (2003) *Annu. Rev. Biochem.* **72**, 111–135.
22. Ritz D & Beckwith J (2001) *Annu. Rev. Microbiol.* **55**, 21–48.
23. Bessette PH, Aslund F, Beckwith J & Georgiou G (1999) *Proc. Natl. Acad. Sci. U.S.A.* **96**, 13703–13708.
24. Jurado P, Ritz D, Beckwith J, de Lorenzo V & Fernandez LA (2002) *J. Mol. Biol.* **320**, 1–10.
25. Venturi M, Seifert C & Hunte C (2002) *J. Mol. Biol.* **315**, 1–8.
26. Georgiou G & Segatori L (2005) *Curr. Opin. Biotechnol.* **16**, 538–545.
27. Wigley WC, Stidham RD, Smith NM, Hunt JF & Thomas PJ (2001) *Nat. Biotechnol.* **19**, 131–136.
28. Waldo GS, Standish BM, Berendzen J & Terwilliger TC (1999) *Nat. Biotechnol.* **17**, 691–695.
29. Fisher AC, Kim W & DeLisa MP (2006) *Protein Sci.* **15**, 449–458.
30. Philibert P & Martineau P (2004) *Microb. Cell Fact.* **3**, 16.
31. Matsuura T & Yomo T (2006) *J. Biosci. Bioeng.* **101**, 449–456.
32. Jermutus L, Honegger A, Schwesinger F, Hanes J & Pluckthun A (2001) *Proc. Natl. Acad. Sci. U.S.A.* **98**, 75–80.

33. Matsuura T & Pluckthun A (2003) *FEBS Lett.* **539**, 24–28.
34. Worn A & Pluckthun A (2001) *J. Mol. Biol.* **305**, 989–1010.
★★ 35. Hanes J & Pluckthun A (1997) *Proc. Natl. Acad. Sci. U.S.A.* **94**, 4937–4942. – *First application of ribosome display for selection of full-length proteins. Includes a detailed description of genetic construction of antibody fragments and selection protocols.*
★★ 36. Mattheakis LC, Bhatt RR & Dower WJ (1994) *Proc. Natl. Acad. Sci. U.S.A.* **91**, 9022–9026. – *The first application of ribosome display to screen peptide libraries.*
37. Kristensen J, Sperling-Petersen HU, Mortensen KK & Sorensen HP (2005) *Int. J. Biol. Macromol.* **37**, 212–217.
38. Sorensen HP, Kristensen JE, Sperling-Petersen HU & Mortensen KK (2004) *Biochem. Biophys. Res. Commun.* **319**, 715–719.
★ 39. Lipovsek D & Pluckthun A (2004) *J. Immunol. Methods*, **290**, 51–67. – *A thorough review of* in vitro *selection methods, with particular focus on the evolution of antibody fragments.*
40. Schimmele B, Grafe N & Pluckthun A (2005) *Protein Eng. Des. Sel.* **18**, 285–294.
★★ 41. Schaffitzel C, Hanes J, Jermutus L & Pluckthun A (1999) *J. Immunol. Methods* **231**, 119–135. – *Includes general background information on ribosome display and conceptual details of all of the experimental protocols described in this chapter.*
42. Lobato MN & Rabbitts TH (2004) *Curr. Mol. Med.* **4**, 519–528.
43. Messer A & McLear J (2006) *BioDrugs*, **20**, 327–333.
44. Miller TW & Messer A (2005) *Mol. Ther.* **12**, 394–401.
★★ 45. O'Brien PM & Aitken R (2001) *Antibody Phage Display – Methods and Protocols.* Humana Press, Totowa, NJ. – *An excellent resource containing protocols for library construction and expression/purification of antibody fragments. It includes selection protocols for display technologies (i.e. using biotinylated agents).*
46. Sambrook J & Russell D (2001) *Molecular Cloning: a Laboratory Manual*, 3rd edn. Cold Spring Harbor Laboratory Press, Cold Spring Harbor, NY.
47. Hanes J, Jermutus L, Schaffitzel C & Pluckthun A (1999) *FEBS Lett.* **450**, 105–110.
★★ 48. Hanes J, Jermutus L, Weber-Bornhauser S, Bosshard HR & Pluckthun A (1998) *Proc. Natl. Acad. Sci. U.S.A.* **95**, 14130–14135. – *A detailed description of the genetic constructs and experimental protocols used in ribosome display.* Protocol 4 *is adapted from this paper.*
★★ 49. Chen HZ & Zubay G (1983) *Methods Enzymol.* **101**, 674–690. – *A thorough discussion of the applications and principles of* in vitro*-coupled transcription/translation.* Protocol 3 *is adapted from this reference.*
★★ 50. Coia G, Pontes-Braz L, Nuttall SD, Hudson PJ & Irving RA (2001) *J. Immunol. Methods*, **254**, 191–197. – *A discussion of all experimental protocols involved in ribosome display including the selection and library recovery through the use of magnetic beads.*

CHAPTER 4
Refolding proteins from inclusion bodies

Renaud Vincentelli

1. INTRODUCTION

Heterologous expression of recombinant protein is the method of choice for high-level production of proteins for functional or structural studies. Unfortunately, despite the choice of *in vivo* and *in vitro* eukaryotic and prokaryotic expression systems available, around 30% of prokaryotic proteins and between 30 and 70% of eukaryotic proteins are still produced in insoluble form. To date, the *Escherichia coli* expression system is the most widely used. Aggregates of misfolded proteins accumulate in the bacteria to form 'inclusion bodies' (IBs) that can often be refolded as active and folded proteins (1).

The lack of a universal refolding buffer and the fact that refolding of IBs is considered to be an art rather than a science has initially slowed down the use of refolding in the protein expression community. When a protein is expressed in insoluble form, most scientists try alternative expression strategies (baculovirus, yeast, mammalian, cell-free, etc.) when quick and cheap refolding trials could be successful. In the last two decades (2), there has been a huge effort towards new methods to improve renaturation of proteins from IBs and the field of refolding is growing rapidly (around 200 scientific papers in 2005). It is now generally accepted that more than 50% of the proteins in IBs could be refolded and there are descriptions of several hundred refolded proteins in the literature. Initial trials were mainly based on long and tedious 'trial and error' strategies, but the effect of several refolding additives and buffers has been studied, sometimes in parallel on the same proteins. Therefore, the protocols are slowly evolving towards the protocols described here, which are cheap, easy, universal, and allow a high throughput, and which should allow any user to determine within a few weeks whether they can refold their protein or whether alternative expression strategies are more likely to help.

2. METHODS AND APPROACHES

In all cases, the first step is to attempt to obtain your protein in soluble form (see *Chapter 2* for the various options to try) as it can immediately be assessed for your needs by activity tests or biophysical characterization. If this fails, you should try to refold your denatured protein.

The refolding of IBs can be divided into four steps: production, solubilization, renaturation, and assessment of the quality of the refolded material (3). *Fig. 1* gives a complete flowchart of the protocols described in this chapter.

The key to the process is the final step: the ability of the user to assess that the protein is correctly refolded (4). Ideally, an activity test should be available. Unfortunately, this is not always the case and we will expose briefly some alternative ways to check the quality of the protein. We will give generic protocols for the production, purification, and solubilization of the IBs.

The protocols used for the renaturation step itself are highly diverse. We will detail briefly the various possibilities and emphasize and detail the ones that are easy to set up and are more likely to work quickly for new users.

The main expression system used for recombinant protein expression is *E. coli*. It is cheap, easy to set up, and gives high protein yields. It has been used for many years and is very flexible. The mechanisms involved that promote protein folding and handle protein misfolding by *E. coli* are well understood (5), but most of the proteins expressed are still insoluble! Almost all proteins that have been successfully refolded have been expressed in this system. There are very few examples of proteins produced in other systems that have been refolded successfully (6). However, there is no reason why you should not try to refold an insoluble protein that has been produced in another system, as, once these proteins have been purified and solubilized, they can go through the same procedure as if they had been produced in *E. coli*.

The more information you have on your protein, the more likely you will be to succeed. In parallel with a literature search, you should also go to the 'Refold' database (7, 8), which concentrates most of the information on refolded proteins, to see whether a similar protein has been refolded and by which means. If there is no information or if you have no idea of the function/family of the protein you are working on, you will have to use a generic approach.

Proteins from most families, including membrane transporters and G-protein-coupled receptors (9, 10), have been refolded for structural and functional studies. Refolding on an industrial scale (11) is common practice, even for therapeutic proteins (12). The quality of the protein is often as good and sometimes more homogeneous than the proteins obtained from alternative expression systems or natural sources.

IBs are dense and insoluble aggregates of misfolded and inactive proteins that accumulate in cells. These unfolded proteins can represent 50% of the total protein content of *E. coli* and are organized aggregates in cells (13) with a biological function (14) that can be observed under an electron microscope (15). Producing proteins in IBs can have some advantages; the proteins are produced in large quantities, are almost pure, and are protected from degradations by proteases. This is also the best way to overexpress proteins that are harmful to cells in their native form.

METHODS AND APPROACHES

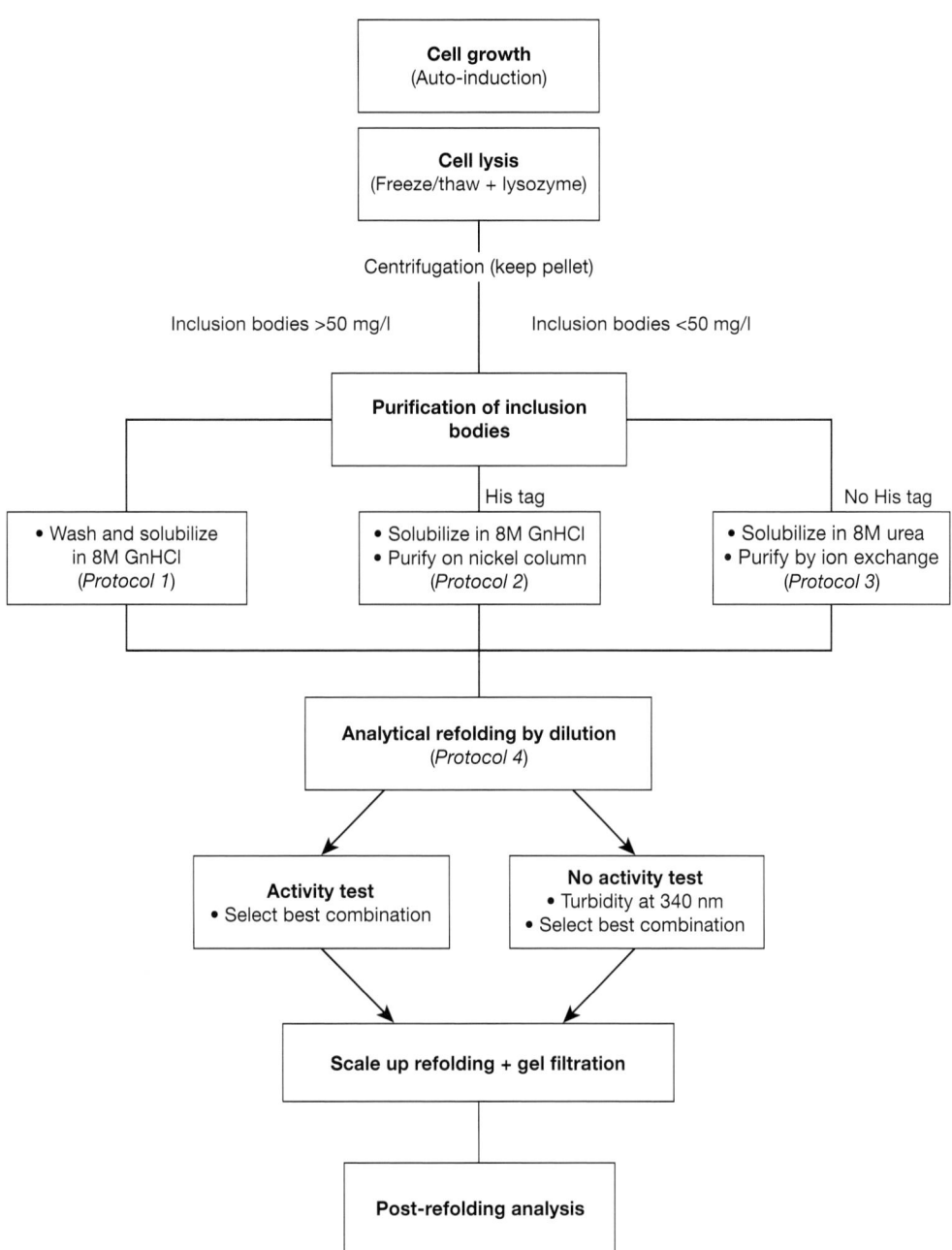

Figure 1. Flowchart of the protocols described in this chapter.

Proteins form IBs for a variety of reasons, not all of which are understood or predictable. These fall into two categories: first, the protein folding problem itself, and secondly, problems relating to the limitation of the *E. coli* expression system.

If the protein requires protein–protein interactions or post-transcriptional modifications to be folded correctly, it will be expressed in an insoluble form. These proteins are unlikely to be refolded unless information on the putative partner or the post-translational modification is known. For protein–protein interactions, the co-expression of both could solve the problem, or eventual refolding of one or several proteins of the complex in parallel could be successful (16). With regard to post-translational modifications, some can be restored using specific strains (17). It should be noted that many eukaryotic proteins that do not require post-transcriptional modification for correct folding or activity have been expressed in a soluble and active form in *E. coli*.

The *E. coli* expression system has some drawbacks, which, if not corrected, will cause the protein to form IBs. These are detailed in Chapter 2, but IBs due to rare *E. coli* codons, incorrectly linked disulfide bridges, or saturation of the *E. coli* expression machinery can be addressed with specific tRNAs, specific chaperones, or expression strains, and therefore these options should be tried before attempting refolding of IBs.

The first steps of any refolding protocol are the production, purification, and solubilization of the IBs.

2.1 Preparation and purification of inclusion bodies

Protocol 1

Preparation and purification of IBs using washing steps[a]

Equipment and Reagents
- Incubator for 2 l shake-flasks, sterile shake-flasks, and culture media (Luria–Bertani (LB) broth and auto-inducing medium (18))
- Centrifuge for 400 ml culture buckets (4000 g) and 50 ml tubes (17 000 g)
- Potter Elvehjem homogenizer (Kimble-Kontes)
- Washing buffer (50 mM Tris/HCl, pH 8, 150 mM NaCl, 5 mM β-mercaptoethanol)
- Lysis buffer (washing buffer + 0.25 mg/ml lysozyme)
- Solubilization buffer (washing buffer + 8 M guanidium hydrochloride)

Method

1. Express your protein in a 1–3 l culture of *E. coli* BL21 (DE3) strain in LB broth or auto-inducing medium (18) at 37°C overnight. For further details, refer to the pET system manual and use Overnight Express auto-inducing medium (Novagen).

2. At the end of the culture growth period, centrifuge the cells (10 min at 4000 g), discard the medium, and resuspend in 50 ml of lysis buffer per litre of initial culture. Freeze for at least 2 h at −80°C (or overnight at −20°C).

3. Thaw the cells; add DNase I (10 µg/ml) and MgSO$_4$ (20 mM) and incubate for 30 min at 37°C (or until the lysate is no longer viscous).

4. Centrifuge for 30 min at 17 000 g and discard the supernatant.

5. Resuspend with the help of a 50 ml Potter homogenizer in the initial volume of washing buffer.

6. Repeat steps 4 and 5 twice.

7. Centrifuge for 30 min at 17 000 g and discard the supernatant.

8. Resuspend the pure IBs with the Potter homogenizer in 20 ml of solubilization buffer containing 8 M guanidium hydrochloride.

9. Centrifuge for 30 min at 17 000 g. *Keep the supernatant* as it contains the solubilized IBs[b].

10. Determine the protein concentration of this fraction (A_{280}) and adjust it to 5 mg/ml.

11. Check the purity by sodium dodecyl sulfate polyacrylamide gel electrophoresis (SDS-PAGE). This fraction is now ready for refolding.

Notes

[a] This protocol and the next ones derive from (19). *Protocol 1* should be used only if you produce 50 mg/l of IBs. If the level is unknown or is below 50 mg/l, go to *Protocol 2*.

[b] After step 9, the pellet should be translucid (mainly composed of membrane parts). If the pellet is still brown and large, repeat steps 8 and 9 to recover more IBs.

Protocol 2

Preparation and purification of IBs for proteins with a His tag

Equipment and Reagents
- Incubator for 2 l shake-flasks, sterile shake-flasks, and culture media (LB broth and auto-inducing medium (18))
- Centrifuge for 400 ml culture buckets (4000 g) and 50 ml tubes (17 000 g)
- Potter Elvehjem homogenizer (Kimble-Kontes)
- HisTrap FF Crude (GE Healthcare)[a]
- AKTA (GE Healthcare) or similar system[b]
- Washing buffer (50 mM Tris/HCl, pH 8, 150 mM NaCl, 5 mM β-mercaptoethanol)
- Lysis buffer (washing buffer + 0.25 mg/ml lysozyme)
- Solubilization buffer (washing buffer + 8 M guanidium hydrochloride)
- Buffer A (50 mM Tris/HCl, pH 8, 150 mM NaCl, 10 mM imidazole, 8 M urea, 5 mM β-mercaptoethanol)

Method

1. Carry out steps 1–3 of *Protocol 1*, then skip steps 4–6 and proceed to steps 7 and 8[c].
2. Centrifuge for 30 min at 17 000 g. *Keep the supernatant*, as it contains the solubilized IBs.
3. Equilibrate a 5 ml HisTrap FF[a] Crude with buffer A.
4. Load the sample onto the column at a rate of 2–5 ml/min according to viscosity.
5. Wash with ten column volumes of buffer A to exchange the guanidium for urea (2–10 ml/min).
6. Wash with five column volumes of buffer A + 50 mM imidazole to reduce non-specific binding (2–10 ml/min).
7. Elute with five column volumes of buffer A + 250 mM imidazole (2–5 ml/min).
8. Determine the protein concentration of this fraction (A_{280}) and adjust it to 5 mg/ml.
9. Check the purity by SDS-PAGE. This fraction is now ready for refolding.

Notes

[a]Any nickel column should work, but if your protein has cysteines, they need to be reduced by adding reducing agent to the buffer. The HisTrap is compatible with low levels of reducing agent. If your nickel column is not compatible with reducing agent, the cysteines can only be reduced at the end using the purified protein in urea. However, most nickel columns have nickel leaks; if you add reducing agent directly to the pool, the traces of nickel present will result in a brown precipitate and the pool may be useless. To remove these traces before reduction, at the end of the nickel step you should use a desalting column such as the HiTrap desalting column or PD10 (GE Healthcare) equilibrated in buffer A and then add the reducing agent to the pool.

[b]If you do not have a chromatographic system, replace the column with a His Gravi trap (GE Healthcare). It should be noted that in many cases the column capacity is reduced when working with samples in urea or guanidium compared with the same column under native conditions, so you may overload the column if you have too many IBs. Try an initial sample and then load accordingly.

[c]If the number of IBs is low, some will be lost during the washing steps before the chromatography.

2.2 Production

If you are planning to refold IBs, this probably means that you have tried to express your protein in a soluble form in *E. coli* and detected your protein only in the pellet of the broken cells. We usually produce a large amount (above 50 mg/l) of IBs in auto-inducing medium (18), purified by simple washing steps (see *Protocol 1*) and then solubilized. If the expression level is low, we advise an initial purification step after the solubilization. The presence of a His tag simplifies the purification of the protein (see *Protocol 2*), but non-His-tagged proteins can also be purified (see *Protocol 3*).

Protocol 3

Preparation and purification of IBs for proteins with no His tag

Equipment and Reagents
- Incubator for 2 l shake-flasks, sterile shake-flasks, and culture media (LB broth and auto-inducing medium (18))
- Centrifuge for 400 ml culture buckets (4000 g) and 50 ml tubes (17 000 g)
- Potter Elvehjem homogenizer (Kimble-Kontes)
- SourceQ or SourceS column (GE healthcare)[a]
- AKTA (GE Healthcare) or similar system
- Washing buffer (50 mM Tris/HCl, 150 mM NaCl, 5 mM β-mercaptoethanol)
- Lysis buffer (washing buffer + 0.25 mg/ml lysozyme)[a]
- Solubilization buffer (washing buffer + 8 M urea)[a]
- Buffer A (50 mM Tris/HCl, 150 mM NaCl, 5 mM β-mercaptoethanol, 8 M urea)[a]
- Buffer B (buffer A + 1 M NaCl)[a]

Method
1. Start with steps 1–3 of *Protocol 1* and then skip steps 4–6 and proceed to steps 7 and 8, but with urea in the solubilization buffer instead of guanidium[b].
2. Centrifuge for 30 min at 17 000 g. *Keep the supernatant*, as it contains the solubilized IBs.
3. Equilibrate a 10 ml SourceQ or SourceS column in buffer A containing 8 M urea.
4. Load the sample on the column at a rate of 2–5 ml/min according to viscosity.
5. Wash with five column volumes of buffer A (5–10 ml/min).
6. Create a gradient in 30 column volumes (5–10 ml/min) from buffer A to buffer B.
7. Check the fractions by SDS-PAGE and pool the fractions of interest.
8. Determine the protein concentration of this pool (A_{280}) and adjust to 5 mg/ml. This fraction is now ready for refolding.

Notes
[a]The pH of the lysis buffer, washing buffer, solubilization buffer, and chromatography buffer as well as the choice of the ion-exchange column are strictly based on the pI of the protein of interest and therefore will not be detailed here. It is possible to perform all of the steps including the solubilization of the proteins in urea in the buffer (pH 8) described here. In this case, for correct binding on the ion-exchange column, the sample needs to be diluted 10–20-fold in buffer A before loading. Be aware that, in many cases, the column capacity is reduced when working with samples in urea compared with the same column under native conditions, so you may overload the column if you have too many IBs. Test a sample the first time and load accordingly.
[b]If the number of IBs is low, some will be lost during the washing steps before the chromatography.

If you cannot detect any protein expression, first check that your construct is correct (by sequencing) and that you are using expression strains compatible with the plasmid you are using. For the most common system (20), the E. coli strains of the BL21 (DE3) series are a good start. If after several trials you cannot detect any expression, you will have to use an alternative expression system or clone your gene into an alternative vector (see Chapter 2).

2.3 Solubilization of inclusion bodies

The interactions between aggregated proteins in IBs are very strong. They are mainly bound together by hydrophobic interactions between residues that would be buried if the proteins were in their native form. Other interactions are involved such as charge–charge interactions, hydrogen bonds, Van der Waals' forces and incorrect intra- and/or intermolecular disulfide bridges. To promote refolding, we first need to break these interactions to suppress aggregation. This is done using a chaotropic agent (8 M urea or 8 M guanidium hydrochloride) or detergents (Triton X-100, Tween 20, etc.) and a reducing agent such as β-mercaptoethanol or dithiothreitol. Note that the most recent generation of nickel affinity resins can resist a low amount of reducing agent. To solubilize IBs, we recommend the use of 8 M guanidium hydrochloride at room temperature; in our hands, it is quicker and more powerful than 8 M urea. Unless you know that you have a large number of IBs (see *Protocol 1*), we suggest carrying out purification under denaturing conditions (see *Protocols 2* and *3*). During the purification, we exchange the 8 M guanidium hydrochloride with 8 M urea (see *Protocol 2*), which can easily be loaded on SDS-polyacrylamide gels for assessment of the purity of the final product. If you do not have a His tag on your construct, you can purify the denatured protein in urea on an ion-exchange column (see *Protocol 3*). In this case, denaturation in guanidium is not recommended (guanidium hydrochloride is charged and might interact with the ion-exchange medium). The amino acid side-chain modification by urea does not make long-term storage advisable (21); therefore, we suggest proceeding with the refolding within several days. The solubilization and purification can be considered to be relatively straightforward steps. Then comes the more adventurous part: refolding and quality control of the final product.

2.4 Refolding of solubilized proteins

In the refolding process, we start with a pure and denatured protein in chaotropic agent. The solubilization agent is removed and replaced by a buffer that will promote the refolding of the protein in its active form. Once correctly refolded, in most cases, the protein can be used in assays or experiments in various buffers.

The method of refolding, protein concentration, buffer composition, length and temperature of the experiment, nature and use of cofactors, etc. all have an impact on the final yield and unfortunately there is no universal method or buffer available.

There are two major procedures for refolding: using solid phase and in solution (22). In the first case, the refolding principle is based on the physical separation of proteins attached to or interacting with a solid phase that protects it from aggregation. This can be a chromatographic medium (23) such as gel filtration (24, 25), an ion-exchange column (26) or, more commonly, a nickel affinity column (27). The quick (for gel filtration) or gradual exchange of buffer allows the protein to refold in the mobile phase.

A chromatographic alternative is to immobilize chaperones on a column, load the partially denatured protein, and elute the refolded protein (28–30). The main advantage of these methods is that the size of the peaks allows determination of the refolding yield. On the other hand, until a suitable buffer is found, each buffer must be tried separately in a 'trial and error' strategy. This requires milligrams of proteins and several days, and there is a risk that the protein will precipitate on the columns. Other alternatives to chromatographic medium such as hydro gels (31) have been successfully used. Because of the time involved in finding the best buffer and the milligram quantities of proteins needed, we do not recommend these methods for a first trial.

One of the first methods used to refold proteins in solution was the dialysis method (32, 33). The denatured protein is dialyzed in one or several gradual steps against the final refolding buffer. Typically, this method requires several days, milligrams of protein, and litres of refolding buffers. If unsuccessful, another buffer must be tried, so this can be a long but sometimes successful strategy.

A recent procedure is using high pressure (34–36) to refold proteins in solution. Unfortunately this procedure requires special equipment to pressurize the IBs and is therefore not easy to set up.

The rest of this chapter and the protocol described (*Protocol 4*) are based on what we think is the easiest way to try and refold your protein if you have little or no experience in refolding: the rapid dilution method (37, 38). The number of proteins that have been refolded using this technique is growing faster than for any of the other techniques because it has several advantages. Once you have your pure and denatured protein (*Protocols 1* to *3*), you need to dilute it at least tenfold in the refolding buffers of your choice. This dilutes out the solubilization agent (urea). Following dilution, if the refolding buffer is optimal, the protein should regain its folded state within seconds. This is the quickest and cheapest option as it does not require any special equipment and the whole procedure usually takes only 1 or 2 days.

Whatever the method of refolding chosen, many parameters influence the refolding success and yield. The temperature of the experiment, protein concentration, and solubilization buffer all have an impact on the refolding. The final buffer can have various pHs, ionic strengths, cofactors and stabilizers, chaperones, redox mixtures, etc. (3, 19).

Initial refolding reports used one or a few refolding buffers such as for the solid-phase refolding strategies (39). However, only with a rapid dilution approach can all of these variables be assessed in parallel and allow a quick refolding buffer screen. Based on sparse matrixes or fractional factorial approaches, several academic groups (40–43) have developed 'folding screens'. Some are in microtiter plates (44) and cover several thousand refolding buffer combinations (19). Recently, several companies have developed their own kits based on the same ideas: for example, Pro-Matrix (Pierce) (45), Fold IT (Hampton Research) (46), and Takara (Sigma). In most cases, after statistical analysis of the results, you can determine the best combination of variables to design the refolding buffer for a scale-up experiment on the same batch of IBs to confirm that the buffer is indeed refolding the protein.

Protocol 4

Renaturation of IBs by rapid dilution[a]

Equipment and reagents
- Refolding kit[b]
- Multi-pipette (optional)
- Microtiter plate reader with 340 nm filter[c]
- Eppendorf tubes and compatible centrifuge
- Filter plate MSDV (Millipore) and vacuum manifold (Millipore)[d]
- Activity test[e]
- Protein concentration device (Millipore)
- Superdex (16/60) 75 or Superdex 200 (GE Healthcare) gel-filtration column[f]
- AKTA (GE Healthcare) or similar system

Method

Analytical refolding

At the end of *Protocols 1–3*:

1. Quickly dilute 5 μl of the solubilized IBs (in urea) in 95 μl of the refolding buffers[b] at 4°C by mixing up and down with a multi-pipette.
2. Check the turbidity at 340 nm in a microtiter plate reader[c] at 4°C. Seal the plate.
3. Check the turbidity again after 18 h at 4°C and proceed to step 4 or 7.

   If an activity test is available:
4. Separate the soluble fraction from the precipitate by centrifugation[d]/filtration[e].
5. Transfer 10–80 μl of each well to the activity test[f].
6. Choose the best hit for scaling up.

   If no activity test is available:
7. Do an analysis of positive hits (A_{340} between 0 and 0.05) and select the best (and simplest) combination of buffer.

Preparative refolding

8. Scale up 2.5–20 mg of the solubilized IB pool (in urea) according to the quantity of solubilized IBs.
9. Dilute this 20-fold drop by drop in a beaker on ice under agitation in the best selected refolding buffer.
10. Check that the solution is not turbid. If the solution is turbid, centrifuge and assay the supernatant for protein concentration. If most of the protein has precipitated, choose an alternative refolding buffer and repeat step 8 again.
11. Concentrate the refolded protein into 2.5 ml.

12. Load on a gel-filtration column[g] equilibrated with refolding buffer at 4°C and collect the elution fractions. Check that the peak elution does not correspond to the void volume of the column (aggregates).

13. Pool the elution fractions, analyze a sample by SDS-PAGE, and concentrate them to the concentration required.

14. Assess the final product using available methods (e.g. activity test, circular dichroism, dynamic light scattering, etc.)

Notes

[a]On an analytical scale, the test gives information on the *solubility* of the refolded material rather than *correct folding*. Soluble aggregates cannot be discriminated from refolded material. The aim of this test is to cut down the possible buffers from 96 to between 1 and 20 (if you are using the kit set up by (19)). If the test gives more than 20 possible hits, concentrate the denatured protein to 10 mg/ml and redo the experiment. The discrimination between soluble aggregate and correctly folded protein will only be detectable during the scale-up at the gel-filtration step.
[b]This protocol can be used with most of the commercial or academic kits available. You can contact the academic team to gain access to their refolding kits.
[c]If you have no access to a microtiter plate reader, transfer all of the samples with no clear precipitate to normal cuvettes and measure A_{340} in 100 µl.
[d]For guidelines and protocols, see (47).
[e]If you have access to an activity test, even partially precipitated wells can sometimes contain correctly refolded material. It is worth checking the nonprecipitated protein of each well in the activity test after removing the aggregated proteins.
[f]Choose the gel-filtration range according to the molecular weight of your protein.

2.5 Analysis of refolded protein

Once you have tried to refold your protein, it is necessary to determine whether the protein is really in its native form on an analytical scale. The easiest way to do this is to use a known activity test. If this assay can be performed on a microscale, you can combine multiple refolding buffer screens with multi-point activity tests and choose the refolding buffer that gives the best activity (48). Alternatives are mainly based on biophysical characterization. They are often more tedious and do not always guarantee correct folding. These include ultracentrifugation and dynamic/static light scattering to assess aggregation state, and circular dichroism and nuclear magnetic resonance (49) to assess correct secondary and tertiary structures. These methods are very powerful but are low throughput, and often require milligrams of protein and expensive equipment.

We suggest following refolding on an analytical scale first by cheap and simple high-throughput alternatives (50, 51) that will eliminate the conditions that do not promote refolding or where the protein precipitates, and emphasize the variables influencing the refolding of your protein. In the first case (50), adapted for all proteins, you can follow protein solubility after refolding by using a spectrophotometer. In the second case (51), which is usable if your protein has a His tag, only the refolded protein is bound to the nickel beads (the same strategy could be applied on an ion-exchange column). The screening of the refolding buffers on an analytical scale can rapidly be followed by a scaled-up experiment. A size-exclusion chromatography step in the optimized refolding buffer will completely remove traces of denaturant and confirm the aggregation state of the refolded protein. After protein concentration, an in-depth characterization of your protein using all of the tools available to you (e.g. SDS-PAGE, native gels, circular dichroism, dynamic light scattering) should confirm that the protein is correctly folded and usable for your project.

2.6 Conclusion

Refolding of IBs is a proven and powerful technique for obtaining milligram quantities of pure and active proteins in *E. coli*. The fact that the refolding conditions are protein specific and that no clear rules apply for finding the optimal protocol have slowed down the use of this technique. The first protocols based on the sometimes long and tedious 'trial and error' strategies are no longer the only way to achieve successful refolding. Following the guidelines and protocols given here, commercial or academic kits covering a large 'refolding space' can be used by any biochemist. There is no need for specific equipment or knowledge, making it worthwhile spending a couple of weeks trying to refold the insoluble protein produced in *E. coli* before going on to longer and more expensive expression systems.

3. REFERENCES

1. Mukhopadhyay A (1997) *Adv. Biochem. Eng. Biotechnol.* **56**, 61–109.
2. Tsuji T, Nakagawa R, Sugimoto N & Fukuhara K (1987) *Biochemistry*, **26**, 3129–3134.
★★ 3. Cabrita LD & Bottomley SP (2004) *Biotechnol. Annu. Rev.* **10**, 31–50. – *A recent and comprehensive review on refolding of IBs.*
4. Kelley RF & Winkler ME (1990) *Genet. Eng. (N.Y.)*, **12**, 1–19.
5. Baneyx F & Mujacic M (2004) *Nat. Biotechnol.* **22**, 1399–1408.
6. Flamand M, Chevalier M, Henchal E, Girard M & Deubel V (1995) *Protein Expr. Purif.* **6**, 519–527.
★★ 7. Buckle AM, Devlin GL, Jodun RA, *et al.* (2005) *Nat. Methods.* **2**, 3. – *REFOLD is the most useful database for refolding protocols*
8. Chow MK, Amin AA, Fulton KF, *et al.* (2006) *Nucleic Acids Res.* **34**, D207–D212.
9. Kiefer H, Maier K & Vogel R (1999) *Biochem. Soc. Trans.* **27**, 908–912.
10. Baneres JL, Martin A, Hullot P, *et al.* (2003) *J. Mol. Biol.* **329**, 801–814.
★★ 11. Clark ED (2001) *Curr. Opin. Biotechnol.* **12**, 202–207. – *One of the few reviews on refolding on an industrial scale.*
★★ 12. Panda AK (2005) *Methods Mol. Biol.* **308**, 155–162. – *A recent review on the refolding of therapeutic proteins.*
13. Carrio M, Gonzalez-Montalban N, Vera A, Villaverde A & Ventura S (2005) *J. Mol. Biol.* **347**, 1025–1037.
14. Villaverde A & Carrio MM (2003) *Biotechnol. Lett.* **25**, 1385–1395.
15. Carrio MM, Cubarsi R & Villaverde A (2000) *FEBS Lett.* **471**, 7–11.
★ 16. Gonzalo P, Lavergne JP & Reboud JP (2001) *J. Biol. Chem.* **276**, 19762–19769. – *A nice demonstration of refolding of complex mixtures.*
17. Wacker M, Linton D, Hitchen PG, *et al.* (2002) *Science*, **298**, 1790–1793.
★★ 18. Studier FW (2005) *Protein Expr. Purif.* **41**, 207–234. – *These auto-induction mixtures are the easiest media to use if you are not familiar with growing bacteria for protein production. These media can be purchased from Novagen ('overnight express').*

19. Vincentelli R, Canaan S, Campanacci V, *et al.* (2004) *Protein Sci.* **13**, 2782–2792.
20. Studier FW, Rosenberg AH, Dunn JJ & Dubendorff JW (1990) *Methods Enzymol.* **185**, 60–89.
21. Gerding JJ, Koppers A, Hagel P & Bloemendal H (1971) *Biochim. Biophys. Acta*, **243**, 375–379.
★★ 22. Swietnicki W (2006) *Curr. Opin. Biotechnol.* **17**, 367–372. – *The most recent and comprehensive review on refolding of IBs. The different refolding strategies are often explained in more detail than in this chapter.*
23. Li M, Su ZG & Janson JC (2004) *Protein Expr. Purif.* **33**, 1–10.
24. Zhang HH, Wang WR, Li QJ & Huang WD (2006) *Immunol. Lett.* **105**, 167–173.
25. Sun QM, Jiang HC, Xu WM, Liu X, Dai CB & Sun MS (2005) *Protein Expr. Purif.* **42**, 278–85.
26. Stempfer G, Holl-Neugebauer B & Rudolph R (1996) *Nat. Biotechnol.* **14**, 329–334.
27. Rehm BHA, Qi QS, Beermann BB, Hinz HJ & Steinbuchel A (2001) *Biochem. J.* **358**, 263–268.
28. Ramon-Luing LA, Cruz-Migoni A, Ruiz-Medrano R, Xoconostle-Cazares B & Ortega-Lopez J (2006) *Biotechnol. Lett.* **28**, 301–307.
29. Oganesyan N, Kim SH & Kim R (2005) *J. Struct. Funct. Genomics*, **6**, 177–182.
30. Altamirano MM, Golbik R, Zahn R, Buckle AM & Fersht AR (1997) *Proc. Natl. Acad. Sci. U.S.A.* **94**, 3576–3578.
31. Cui ZF, Guan YX, Chen JL & Yao SJ (2005) *J. Appl. Polymer Sci.* **96**, 1734–1740.
32. Rudolph R & Lilie H (1996) *FASEB J.* **10**, 49–56.
33. Brown WC, Duncan JA & Campbell JL (1993) *J. Biol. Chem.* **268**, 982–990.
34. Kim YS, Randolph TW, Seefeldt MB & Carpenter JF (2006) *Methods Enzymol.* **413**, 237–253.
35. Schoner BE, Bramlett KS, Guo HH & Burris TP (2005) *Mol. Genet. Metab.* **85**, 318–322.
36. St John RJ, Carpenter JF & Randolph TW (1999) *Proc. Natl. Acad. Sci. U.S.A.* **96**, 13029–13033.
37. Chen Z, Koelsch G, Han HP, *et al.* (1991) *J. Biol. Chem.* **266**, 11718–11725.
38. Cardamone M, Puri NK & Brandon MR (1995) *Biochemistry*, **34**, 5773–5794.
39. Kupper D, Reuter M, Mackeldanz P, *et al.* (1995) *Protein Expr. Purif.* **6**, 1–9.
40. Hofmann A, Tai M, Wong W & Glabe CG (1995) *Anal. Biochem.* **230**, 8–15.
41. Chen GQ & Gouaux E (1997) *Proc. Natl. Acad. Sci. U.S.A.* **94**, 13431–13436.
42. Armstrong N, de Lencastre A & Gouaux E (1999) *Protein Sci.* **8**, 1475–1483.
43. Lindwall G, Chau M, Gardner SR & Kohlstaedt LA (2000) *Protein Eng.* **13**, 67–71.
44. Willis MS, Hogan JK, Prabhakar P, *et al.* (2005) *Protein Sci.* **14**, 1818–1826.
45. Wan L, Zeng L, Chen L, *et al.* (2006) *Protein Expr Purif.* **48**, 307–313.
46. Lamberski JA, Thompson NE & Burgess RR (2006) *Protein Expr. Purif.* **47**, 82–92.
47. Vincentelli R, Canaan S, Offant J, Cambillau C & Bignon C (2005) *Anal Biochem.* **346**, 77–84.
★★ 48. Merli S, Corti A & Cassani G (1995) *Anal. Biochem.* **230**, 85–91. – *One of the first refolding strategies using a multi-parameter approach.*
49. Prestwich GD (1993) *Protein Sci.* **2**, 420–428.
50. Tresaugues L, Collinet B, Minard P, *et al.* (2004) *J. Struct. Funct. Genomics*, **5**, 195–204.
51. Cowieson NP, Wensley B, Listwan P, *et al.* (2006) *Proteomics*, **6**, 1750–1757.

CHAPTER 5

Selection of protein variants with improved expression using green fluorescent protein-derived folding and solubility reporters

Stéphanie Cabantous and Geoffrey S. Waldo

1. INTRODUCTION

This chapter describes methods of screening random mutants or deletion libraries of proteins using a panel of green fluorescent protein (GFP) technologies to monitor protein folding, expression, and solubility. This combinatorial biology strategy has been used successfully to identify new variants of proteins with improved properties that would have been difficult to isolate using conventional methods. Recombinant proteins often misfold and are insoluble when expressed in heterologous hosts. Several strategies for increasing the yield of soluble protein have been described and consist generally of modifying the culture environment (1, 2), co-expressing the target protein with chaperones (DnaK/J, GroEL/S) (3–6), or creating protein fusions with highly soluble proteins that act as a solubility enhancing partner (7–9). Another stratagem involves the recovery of the proteins from inclusion bodies and screening for optimal refolding conditions (10, 11). However, finding the right conditions to maximize soluble protein expression may be time-consuming, costly, and unsuccessful.

An alternative approach is to modify the protein sequence of these recalcitrant protein targets. A large number of variants (mutants or fragments) is created and then screened for improved folding and solubility. The selection method is crucial in directed-evolution strategies. When the protein has a particular function, functional assays can be used to assess improved protein folding. This approach is simple when the enzymatic property is essential for cell viability and growth, such as increasing concentrations of antibiotics (12) or providing an essential nutrient (13). However, when no function is exploitable, function-independent screens of protein folding and stability are essential. Various selection techniques based on display techniques have been established for identification of proteins with improved solubility or stability, such as phage display (14), ribosome display, and mRNA display (15). These methods are combined with one or several secondary screens for protein stability and aggregation. These include resistance to proteolysis (16), accessibility of an affinity tag (17), or screening for reduced hydrophobicity (18). Although these methods allow screening of large libraries, they have been limited so far to the selection of small proteins or proteins with low multimerization (i.e. monomers and dimers). An alternative for selecting folded proteins consists of fusions to folding reporters. Chloramphenicol acetyl transferase has been described for monitoring protein folding by selection based on chloramphenicol resistance (19), but the method is hindered by the trimeric nature of the reporter, which may lead to the formation of higher-order aggregates in fusion with multimeric proteins. We have demonstrated that fluorescence of GFP fusions is positively correlated with a lack of misfolding of the target protein (20). This property has been used in conjunction with a DNA shuffling technique (21) to evolve and crystallize multimeric proteins that were insoluble when overexpressed in *Escherichia coli* cells (22–25).

This chapter outlines the application of an improved version of the GFP folding reporter technology, termed 'GFP insertion'. The target protein is inserted into the scaffolding of the GFP to avoid the selection of false-positive artifacts from internal cryptic ribosome-binding sites in the target RNA. In addition, we have created a set of four different reporters that sense different degrees of protein misfolding. Variants with improved folding are selected using the reporter with appropriate stringency based on the brightness of the GFP insertion fusions. The optimal clones are recombined for several rounds of evolution and then screened for solubility using the split-GFP solubility reporter (26).

2. METHODS AND APPROACHES

2.1 GFP insertion technology

The topological orientation of the gene in the GFP folding reporter (C-terminally fused to target proteins) allows efficient discrimination of variants bearing stop codons and possible frame shifts. However, regions of the coding sequence that are similar to the canonical Shine–Dalgarno motif may occur during evolution, thus creating new starting points for the internal translation of short products. These truncated polypeptides are often better folded and more soluble than the wild-type protein. Such truncated fragments lead to the selection of false-positive clones with bright GFP fusion fluorescence (see *Fig. 1a*). One strategy to avoid this problem is to sandwich the protein of interest between the two halves of the GFP so that the presence of a cryptic ribosome-binding site will produce two separate translation products (each fused to half of the GFP) that are less likely to complement and fluoresce, especially if one of the two GFP fragments is attached to an insoluble or poorly folded target protein domain (see *Fig. 1b*). In the GFP insertion topology, the two GFP-derived domains GFP1 (aa 173–238) and GFP2 (aa 1–173) flank a DNA cassette containing the test gene and lead to the

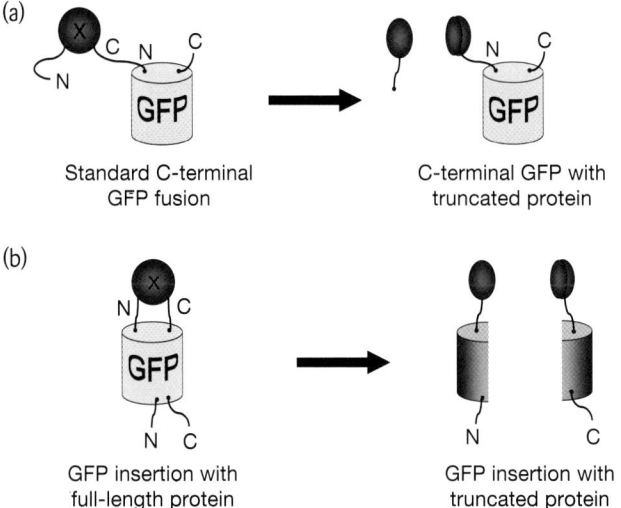

Figure 1. Strategy for discrimination of an internal ribosome-binding site using GFP. (*a*) A soluble truncated protein originating from an internal ribosome-binding site can give a bright fluorescent fusion with the C-terminal GFP folding reporter. (*b*) Insertion of a target protein in the GFP scaffolding: the internal ribosome-binding site produces two halves of GFP that are less likely to produce a full-length functional GFP, especially if one of the insert fragments interferes with the folding fused GFP domain.

expressed fusion (GFP1)-L_1-(X)-L_2-(GFP2), where L_1 and L_2 are flexible (GGGS)$_2$ linkers (see *Fig. 2a*). We engineered four reporter vectors with differing sensitivity to protein misfolding. Hybrids of the traditional folding reporter GFP (20) and the more robust superfolder GFP (27) combine fragments from both superfolder and folding reporter variants (see *Fig. 2b*). These GFP insertion vectors can used to discriminate between target proteins exhibiting even subtle differences in folding interference (see *Fig. 2c*, also available in the color section).

2.2 Principles of GFP insertion

First, the insertion GFP vector with the appropriate stringency is identified. The starting gene, coding for an insoluble protein, is cloned in all four vectors in order to determine which one gives the minimum detectable fluorescence signal. This resulting vector is used as receiving plasmid to clone a library of variants from the insoluble gene. After expression of the library in the GFP insertion vector, optima with improved folding appear brighter than the wild-type insoluble gene. These mutants constitute the pool (typically 24–48 individual clones) of variants that will be chosen and recombined for another round of evolution. Prior to beginning the second round of evolution, the target gene mutants from the pooled optima (24–48 clones) are subcloned into the four GFP insertion vectors to test which one gives the best stringency for continuing the next round of evolution. Typically, for each pool in a given reporter vector, we observe in the library four populations of clones of various fluorescence levels from very bright to very faint (see *Fig. 3*, top right image, also available

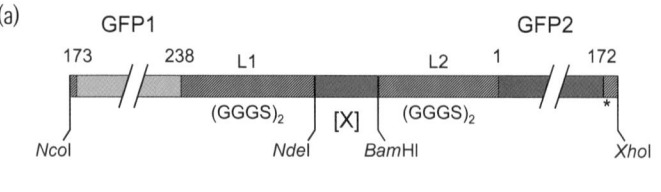

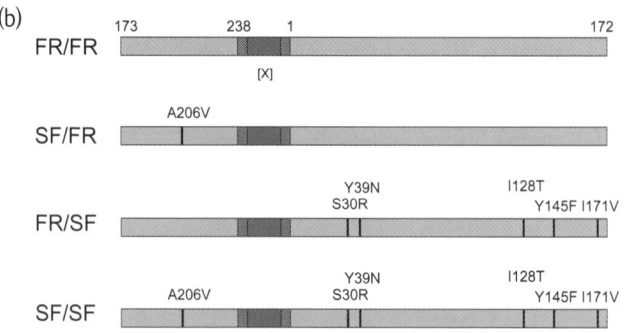

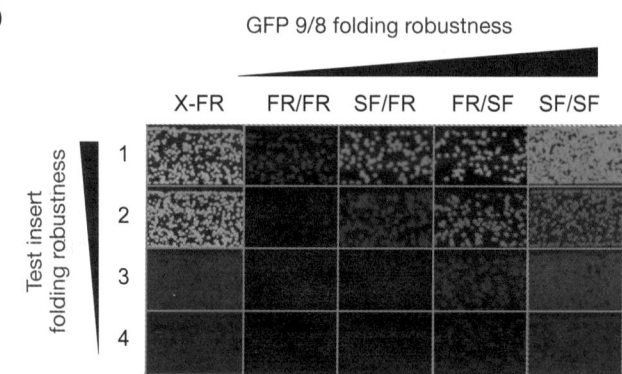

Figure 2. GFP insertion vector topology and reporter stringency (see page xx for color version).
(a) The cloning cassette of the GFP insertion vector. Two GFP fragments (GFP1 and GFP2) flank a DNA cloning cassette bearing the NdeI and BamHI restriction sites for insert test proteins, separated by two (GGGS)$_2$ linkers. (b) Schematic representations of GFP insertion constructs with various stringencies. The designations to the left of each cassette refer to the GFP scaffolding from which the fragments that bracket the test protein were obtained. FR, folding reporter GFP; SF, superfolder GFP. Thus, FR/FR indicates that both GFP fragments came from folding reporter GFP, for example. The corresponding SF mutations are indicated above the constructs. (c) Reporter stringency for X-FR (conventional C-terminal folding reporter GFP) and the four GFP insertion vectors. The reporter abbreviation above each column designates the identity of the GFP variant from which the flanking fragment is derived. Each row corresponds to one of four *Pyrobaculum aerophilum* test proteins with different solubility levels (#1 is fully soluble, #2 and #3 are partially soluble, and #4 is totally insoluble). The least-stringent vectors (SF/SF) are able to detect most proteins. As the number of SF mutations decreases, they detect proteins with increased solubility, as for the most stringent FR/FR.

in the color section, and *Protocol 5*). Generally, the first evolution cycles use vectors with lower stringency and vectors with increased stringency are used gradually during the evolution cycles as the protein becomes more soluble. This stepwise process results in an improved selection power compared with the original C-terminal GFP folding reporter, enabling selection of variants that produce more soluble protein, whilst reducing the number of artifacts linked to new translation starts.

2.3 Methodology

A flow diagram of the methods involved is shown in *Fig. 3*. The gene encoding an insoluble protein is amplified by PCR using primers specific for the vector. The most efficient mutagenesis method for improving protein folding is the DNA shuffling

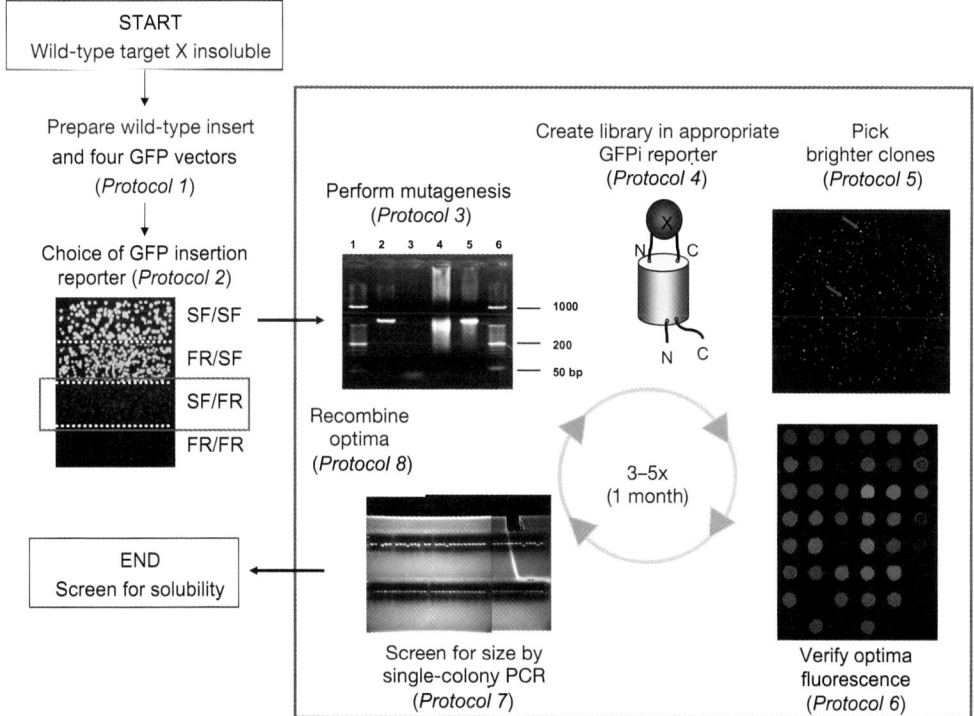

Figure 3. Directed evolution of protein folding (see page xxi for color version).
The most suitable GFP insertion vector is identified and then libraries of target genes encoding insoluble targets are subjected to random mutagenesis by DNA shuffling: a pool of homologous genes (*lane 2*) is fragmented into small ~20 bp fragments with DNase I (*lane 3*), which are reassembled by primerless PCR into full-length genes (*lane 4*) and reamplified with vector-specific primers to yield the full-length gene (*lane 5*). The purified mutant library is digested by restriction endonuclease, sized by preparative gel electrophoresis, and cloned into the appropriate GFP folding reporter (*Protocol 4*). Variants with improved folding robustness are associated with brighter fluorescent colonies (see image of *in vivo* induction plate, *Protocol 5*), whose fluorescence is verified by spotting individual clones onto nitrocellulose membrane (*Protocol 6*). Optima clones are screened for insert length by single-colony PCR and analytical agarose gel electrophoresis (*Protocol 7*). Full-length clones with improved folding are recombined for another cycle of evolution (*Protocol 8*).

recombination method. This method has been developed by Stemmer *et al.* (12) and has been applied to improve protein folding of GFP (28). *Protocol 1* describes how to prepare the wild-type gene for cloning into the GFP reporters, and the preparation of each of the four GFP reporter vectors to receive inserts. The wild-type insert is cloned into each of the four GFP reporter vectors, and the vector best suited for directed evolution is chosen, as described in *Protocol 2*. *Protocol 3* describes how to generate a pool of variants of a gene encoding an insoluble protein using DNA shuffling. The method can be applied to any gene, although some limitations have been noticed for genes >5 kb or containing large tracts of closely homologous nucleotide sequences. The amplified insert and the four GFP insertion vectors are digested using the same endonucleases (*Protocol 3*). DNA from the reassembled pool of genes is cloned into the appropriate GFP insertion vector and transformed into an *E. coli* strain with high transformation efficiency to generate a library of $\sim 5 \times 10^6$ variants or may be transformed directly into BL21(DE3) chemically competent cells to obtain a diversity of about 20 000 clones, which is usually sufficient for selecting genes of about 1 kb (*Protocol 4*). The selection of single clones that develop the brightest fluorescence is performed after induction of colonies on membranes resting on plates (*Protocol 5*). The fluorescence of each optimum is analyzed in more detail using imaging methods and optima are ranked by the highest fluorescence (*Protocol 6*). We have noticed in an earlier study (29) that the presence of homologous sequences in the starting gene may lead to recombination artifacts during DNA shuffling such as deletions or translocations. *Protocol 7* describes the size analysis of each clone selected with GFP screening by single-colony PCR to identify and eliminate these DNA deletion artifacts (see *Fig. 4*). Finally, the brightest variants (*Protocol 6*) identified as full-length genes by single-colony PCR (*Protocol 7*) are then pooled together and plasmids are isolated for another round of evolution (*Protocol 8*). After the DNA shuffling step (*Protocol 3*), a subset of the new DNA pool of genes may be cloned in the set of GFP insertion vectors (*Protocol 2*) to choose the appropriate reporter for screening the library at round 2. The process is repeated until there is no further increase in fluorescence (typically three to four rounds), or the target gene exhibits the desired improvement in folding.

2.4 Selecting optima with improved solubility

Once the folding of the protein has been improved, optima are evaluated for solubility. The fluorescence of the GFP insertion fusion protein is a good indicator of lack of misfolding by the test protein, and proteins with a lower tendency to misfold are often more soluble when expressed in a non-fusion form without the GFP domains. However, protein insolubility due to slow aggregation processes potentially may evade detection by the GFP folding reporter. For accurate assessment of the soluble fraction of these protein variants in an efficient, high-throughput manner, we developed a two-step solubility screen using a split-GFP solubility reporter (26). The advantage of this screen is that it can be used first to screen solubility *in vivo*, and precise solubility can be assessed *in vitro*, without requiring sodium dodecyl sulfate polyacrylamide gel electrophoresis (SDS-PAGE) quantification (for detailed protocol, see 30). We have applied the split-GFP system in

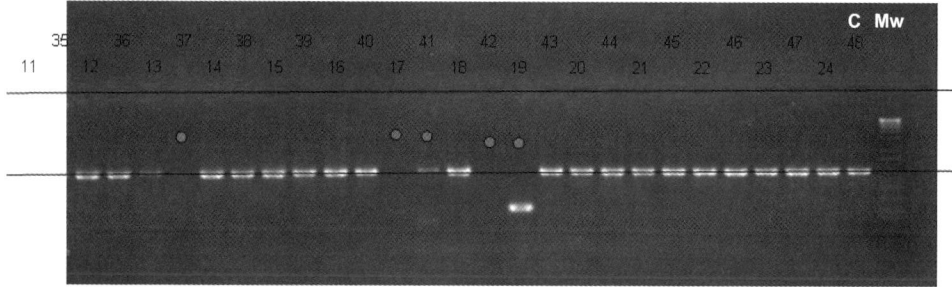

Figure 4. Analytical gel of single-colony PCR screen (see page xxii for color version). PCR products for optima clones and the positive control (C, lane 48). Red dots indicate clones to discard (truncated or missing sites).

directed-evolution experiments in combination with the GFP insertion reporters on four proteins of *Mycobacterium tuberculosis*. N-terminal His$_6$-tagged wild-type proteins Rv2895c, Rv1202, Rv2388c, and Rv0341 were expressed as inclusion bodies in *E. coli* BL21(DE3) at either 27 or 37°C, as determined by SDS-PAGE of fractionated lysates (data not shown). For each gene, a round of directed evolution was performed using the GFP insertion vector SF/FR, and a second round was done using the GFP insertion vector FR/FR. For the final round, libraries of variants expressed from the most stringent vector were bright for most genes (see *Fig. 5*, top left, also available in the color section). One protein, Rv2833c, was slow to evolve, and variants from the fourth round still showed faint colonies when expressed in the FR/FR insertion vector (see *Fig. 5*, protein #3). Forty-eight clones were picked from each final expression library and subcloned by restriction endonuclease digestion as a pool directly from the GFP insertion vector into the split-GFP vector system and then screened for solubility. A solubility screening was performed *in vivo* on a set of 5000 colonies (to ensure redundant representation of the 48 optima in each library) by sequential induction of GFP 11-tagged protein and the GFP 1–10 detection fragment (for detailed protocol, see 30). Sixteen clones from each protein library were picked and grown in liquid culture. Soluble and pellet fractions were assessed for the 16 optima of each library by a second screening *in vitro*, leading to precise solubility determination for each optima (see *Fig. 5*, lower images). Proteins 1, 2, and 4 displayed most fluorescence in the soluble *in vitro* assay. The split-GFP *in vivo* and *in vitro* screens are an efficient route to identify appropriate clones for large-scale expression based on the highest soluble fluorescence value (for additional applications, see 29).

Protocol 1

Preparing wild-type insert and GFP insertion vectors

Equipment and Reagents
- Vent polymerase, buffer and reagents (NEB)
- 10 mM dNTPS (2.5 mM each of dATP, dCTP, dGTP, and dTTP)
- 5.0 µM forward gene-specific NdeI oligonucleotide 1 (5'-AGATATACATATGXXXXXXXXXX XXXXXXXXXXXXXX-3') where X represents gene-specific sequence and the ATG of CATATG is in frame with the gene
- 5.0 µM reverse GFP-specific oligonucleotide 2 (5'-AATTCGGATCCXXXXXXXXXXXXXXXX XXXXXXXX-3') where X represents gene-specific sequence (reverse complement) and GGATCC is in frame with the gene. The oligonucleotide should not contain a stop codon
- Sterile, distilled water (ddH$_2$O)
- QIAquick PCR purification and extraction purification kits (Qiagen)
- NdeI and BamHI restriction endonucleases (20 units/µl; NEB)
- Calf intestinal alkaline phosphatase (NEB)
- 0.5 mM MgCl$_2$ + 100% ethanol solution
- 1.5% Agarose gel
- 6× Type III loading buffer (0.25%, w/v, bromophenol blue, 0.25%, w/v, xylene cyanol FF, 30% glycerol in water)
- Equipment and reagents for agarose gel electrophoresis including 1× TAE agarose gel running buffer (40 mM Tris base, 20 mM acetic acid, 1 mM EDTA)
- Staining solution: 10 ng/ml ethidium bromide
- Thermocycler

Method
1. Prepare a PCR by combining:
   - 66.78 µl of ddH$_2$O
   - 10 µl of 10× Vent polymerase buffer
   - 12 µl of dNTPs
   - 0.88 µl of MgSO$_4$ stock (NEB)
   - 1.34 µl of Vent polymerase (NEB)
   - 4 µl of forward gene-specific NdeI oligonucleotide 1
   - 4 µl of reverse gene-specific BamHI oligonucleotide 2
   - 1 µl of 10–100 ng/µl template (i.e. 20-fold-diluted Qiagen 3 ml plasmid prep)

2. Place the PCR tube in the thermocycler after the initial melting temp has been reached (below) and run the following program:

Number of cycles	Program
1	3 min at 97°C
25	25 s at 96°C, 25 s at 56°C, and 1 min/kb at 72°C
1	3 min at 97°C
1	Hold at 2°C

3. Run the PCR sample on a 1.5% agarose gel in TAE buffer. Load 4 µl of the reaction to which 2 µl of 6× loading buffer has been added. Resolve at 120 V for 30 min. Stain and photograph. Clean the PCR product using a QIAquick PCR purification kit following the manufacturer's instructions. Elute with 75 µl of 10 mM Tris/HCl (pH 8.0).

4. Assemble the following in a microcentrifuge tube, mix well by vortexing gently, centrifuge briefly, and incubate at 37°C for 2 h:

- 50.0 µl of cleaned PCR product[a]
- 7.2 µl of 10× *Bam*HI buffer
- 3.6 µl of 20× BSA (dilute the 100× NEB stock)
- 3.6 µl of *Bam*HI
- 3.6 µl of *Nde*I

5. In four separate microcentrifuge tubes, one for each type of GFP insertion vector, combine 50.0 µl of GFP insertion vector, the enzymes and buffer reagents as indicated in the insert digest (step 4), and add 0.2 µl of calf intestinal alkaline phosphatase. Mix well by vortexing gently, pellet by brief centrifugation, and incubate at 37°C for 2 h.

6. Clean up the digested insert (step 4) and vectors (step 5) using a QIAquick PCR purification kit following the manufacturer's instructions[b] and elute with 50 µl of 10 mM Tris/HCl (pH 8.3). Store the vector digests for later use (see *Protocol 2*, step 1, and *Protocol 4*, step 1).

7. Precipitate the DNA insert by adding 750 µl of a solution of 0.5 mM $MgCl_2$ + 100% ethanol. Vortex and incubate at −20°C for 10 min.

8. Centrifuge at 14 000 *g* for 10 min, with the hinge out. Decant the supernatant and repeat the centrifugation procedure for 3 min. Remove traces of supernatant by pipetting; do not scrape the hinge side.

9. Dry in an air flow for 5 min, leaving a trace of liquid. Add 20 µl of 10 mM Tris/HCl (pH 8.5). Pipette the walls of the tube to wash down bound DNA and mix by vortexing.

10. Mix with 8 µl of 6× Type III loading buffer and load on a 1.5% agarose gel (large combs). Resolve at 150 V for 20–30 min.

11. Stain, visualize, and excise the slice. Perform the QIAquick gel extraction protocol following the manufacturer's instructions, with minor modifications[c]. Elute with 45 µl 10 mM Tris/HCl (pH 8.5).

Notes

[a]The the gene should not contain internal *Nde*I or *Bam*HI sites. Primer-directed mutagenesis can be used to silence internal *Nde*I or *Bam*HI sites by silent codon mutagenesis, if desired.

[b]For inserts, it is essential to perform a PCR clean-up with a PB/QIAquick column prior to ethanol precipitation and gel purification.

[c]Perform two additional washes with 500 µl of QG (gel dissolution buffer) to remove traces of agarose.

Protocol 2

Choice of GFP insertion reporter

Equipment and Reagents
- 5× T4 DNA ligase buffer (Invitrogen)
- T4 DNA ligase (400 units/µl; NEB)
- *E. coli* BL21(DE3) strain made chemically competent by a standard protocol
- SOC recovery medium (SOB + 20 mM glucose, 10 mM MgCl) (31)
- 130 mm diameter supported nitrocellulose membranes (GE Osmonics)
- LB/agar Bauer plates (150 × 15 mm) containing 35 µg/ml kanamycin, and induction plates containing 35 µg/ml kanamycin, 1 mM IPTG
- Heat block or thermocycler
- Illumatool Lighting system with excitation filter 488 nm (blue) and a glass colored 520 nm long-pass filter (Lightools Research)
- Digital camera, such as the Olympus C-5060 on a photographic boom (Edmund Scientific)

Method

1. Combine the following in a 0.2 ml PCR tube for each vector and incubate for 1 h at 30°C in a PCR block with a heated lid:
   - 3.8 µl of doubly digested purified insert DNA (see *Protocol 1*, step 11)
   - 1.0 µl of doubly digested dephosphorylated GFP insertion vector DNA (see *Protocol 1*)
   - 1.0 µl of 5× T4 DNA ligase buffer
   - 0.3 µl of T4 DNA ligase (400 units/µl)

2. Transform into BL21(DE3) by mixing 2 µl of the ligation (step 1) with 40 µl of chemically competent BL21(DE3) cells held on ice. Heat shock the cells for 1 min at 37°C[a].

3. Recover each reaction in 1 ml of SOC medium and incubate for 1 h at 37°C with shaking at ~250 r.p.m. Plate two dilutions of the recovery reaction onto nitrocellulose membranes on selective LB/agar to obtain single colonies[b].

4. Invert and incubate overnight at 32°C until the colonies are ~1 mm in diameter.

5. Transfer the membrane bearing colonies face up onto a pre-warmed induction plate. Incubate at the desired temperature (typically 37°C[c]) for 2–3 h.

6. Visualize cell fluorescence using 488 nm excitation and 520 nm emission filters[d,e]. Choose the reporter vector that gives fluorescence that is as low as possible, but clearly detectable, with the starting gene (see *Figs 2c* and *3*, and *Protocol 2*). For the first cycle only, save two clones of the selected area and make a freezer stock. Inoculate a 3 ml overnight culture and perform a plasmid prep. This will serve as the template for DNA shuffling (see *Protocol 3*).

Notes

[a] A PCR machine may be used to perform the chemical transformation using the program: cycle 1: −1°C for 30 min, cycle 2: 37°C for 1 min, cycle 3: −1°C soak.

[b] Plate 300 µl of the recovery reaction and dilute 200 µl of the recovery reaction with 800 µl of SOC. Plate 300 µl of the recovery reaction and diluted the suspension separately onto nitrocellulose membranes.

[c] If the protein is extremely poorly folded, a lower temperature of induction may be required to observe fluorescence.

^dIlluminate the plate with the Illumatool to visualize the fluorescent clones. An alternative home-made solution to visualize green fluorescence is to use two filters: a blue excitation ROSCO filter (fit into a slide projector or cover a small halogen lamp) and a yellow-green emission ROSCO filter (place in a three-finger clamp on a ring stand). A better home-made version consists of a 488 nm 50.8 mm square filter from Edmund Scientific that fits in a standard 35 mm slide projector, and a 520 nm 50.8 mm square filter for viewing the fluorescence.

[e]Photograph the colonies using a suitable digital camera such as the Olympus C-5060. Take a series of exposures ranging from 1/30 s to 2 s to ensure that all colony phenotypes (bright to faint) are within the linear dynamic range of the camera under at least one of the exposure conditions.

Protocol 3

DNA shuffling mutagenesis

Equipment and Reagents
- Vent exo+ polymerase and reagents (NEB)
- 10 mM dNTPS (2.5 mM each of dATP, dCTP, dGTP, and dTTP)
- 5.0 µM forward GFP-specific oligonucleotide 1 (5'-GATATAACTAGTAAGCGTGACCACATG GTCCTTCTTGAG-3')
- 5.0 µM reverse GFP-specific oligonucleotide 2 (5'-TACTTCGGTACCTCCAGTGAAAAGTTCTT CTCCTTTGCT-3')
- Sterile, distilled water (ddH$_2$O)
- DNase I (1 unit/µl; Invitrogen)
- Solution of 100 mM CoCl$_2$ (from CoCl$_2$.6H$_2$O; Sigma)
- Solution of 500 mM Tris/HCl (pH 7.4)
- 6× Type III loading buffer (0.25%, w/v, bromophenol blue, 0.25%, w/v, xylene cyanol FF, 30% glycerol in water)
- QIAquick PCR purification kit (Qiagen)
- Centrisep columns (Princeton Separations)
- *Pfu* exo– polymerase (Stratagene)
- Microcentrifuge
- Thermocycler
- 1.5% Agarose gel (large format)
- Equipment and reagents for agarose gel electrophoresis including 1× TAE agarose gel running buffer
- Staining solution: 10 ng/ml ethidium bromide

Method

1. Prepare a PCR by combining:
   - 1 µl of double-stranded template plasmid[a]
   - 66.78 µl of ddH$_2$O
   - 10 µl of 10× Vent polymerase buffer
   - 12 µl of dNTPs
   - 0.88 µl of MgSO$_4$ stock (NEB)
   - 1.34 µl of Vent exo+ polymerase (NEB)
   - 4 µl of forward GFP-specific *Nde*I oligonucleotide 1
   - 4 µl of reverse GFP-specific *Bam*HI oligonucleotide 2

2. Place the PCR tube in the thermocycler after the initial melting temp has been reached (see below) and run the following program:

Number of cycles	Program
1	3 min at 97°C
25	25 s at 96°C, 25 s at 56°C, and 1 min/kb at 72°C
1	5 min at 72°C
1	Hold at 2°C

3. Run a PCR sample on 1.5% agarose gel in TAE buffer. Load 4 µl of reaction to which 2 µl of 6× loading buffer has been added. Resolve at 120 V for 30 min. Stain and photograph. Clean the PCR product using a QIAquick PCR purification kit following the manufacturer's instructions. Elute with 100 µl of 10 mM Tris/HCl (pH 8.0).

4. Assemble the following in a PCR tube[b] and mix by pipetting:
   - 90 µl of cleaned PCR product
   - 10 µl of 500 mM Tris/HCl (pH 7.4)
   - 2 µl of 100 mM $CoCl_2$

5. Dilute the DNase I stock tenfold in 50 mM Tris/HCl (pH 7.4).

6. Pre-incubate the samples (step 4) in a thermocycler at 15°C for 5 min and add 2 µl of diluted DNase I. Pipette by mixing and initiate the digest for 10 min at 15°C. Stop the digest by increasing the temperature to 90°C for 3 min.

7. Clean the DNase reaction using a Centrisep gel filtration column according to the manufacturer's instructions[c].

8. Perform a PCR without primers according to Stemmer's protocol (21). Assemble two 25 µl reactions using two concentrations of cleaned DNA fragments in PCR tubes on ice (the low concentration is indicated in parentheses):
   - 17.5 µl (12.5 µl) of 'cleaned' DNA fragments
   - – (5.0) of ddH_2O
   - 2.5 µl of 10× *Pfu* recombinant polymerase buffer
   - 4 µl of 10 mM dNTPs
   - 0.5 µl of 10 mM $MgCl_2$
   - 1.4 µl of *Pfu* exo– polymerase

9. Incubate the samples using the following PCR program:

Number of cycles	Program
1	3 min at 97°C
35	25 s at 96°C, 25 s at 56°C, and 1 min + 5 s/cycle at 72°C
1	5 min at 72°C
1	Hold at 2°C

10. Analyze the products by 1.5% agarose gel electrophoresis. The ideal reaction will yield a product whose modulus is at or slightly below[d] the starting template molecular mass (see *Fig. 3* inset – *Protocol 3*).

11. Fill in the ends. Perform a standard PCR using the same specific primers used in the starting PCR[b]. Run a diagnostic gel using 3 µl of PCR product. The total quantity of DNA from the PCR should be 5–10 µg[e].

12. Digest and gel purify the PCR product (step 11) using *Protocol 1*, step 4, and then *Protocol 1*, steps 6–11[f].

Notes

[a] The initial library PCR is performed using the wild-type gene in the chosen GFP insertion vector (see *Protocol 2*, step 6), using vector-specific primers. This gives a flanking sequence so that mutagenesis can occur throughout the insert sequence. In the subsequent directed evolution cycles, the PCR template corresponds to a plasmid pool of genes selected in the GFP insertion (see *Protocol 8*, step 6).

[b] May be performed at room temperature.

[c] Use 3.3 mM Tris/HCl (pH 8.5) buffer for column hydration.

[d] Occasionally the product will be heavier than the expected mass (greater than the molecular mass of the starting template). This can be due to some single-stranded product during reassembly, or ramification. This product will still work for the fillout reaction (step 8).

[e] For inserts, it is essential to perform a PCR clean-up with a PB/QIAquick column prior to ethanol precipitation and gel purification.

[f] Perform two additional washes with 500 µl of QG (gel dissolution buffer) to remove traces of agarose.

Protocol 4

Construction of a cDNA insert library in a GFP folding reporter vector plasmid

Equipment and Reagents
- 5× T4 DNA ligase buffer (Invitrogen)[a], T4 DNA ligase (400 unit/µl; NEB)
- *E. coli* BL21(DE3) strain made chemically competent
- SOC recovery media (SOB + 20 mM glucose, 10 mM MgCl) (31)
- Three LB/agar Bauer plates (150 × 15 mm) containing 35 µg/ml kanamycin
- 15 ml Falcon tubes or similar sterile screw-top plastic tubes (Fisher Scientific)

Method

1. Perform a large-scale ligation of the insert into the digested, dephosphorylated GFP reporter vector in a PCR cycling tube by incubating the following mixture for 12 h at 16°C:
   - 38 µl of gel-purified digested insert (see *Protocol 3*, step 12)
   - 9 µl of dephosphorylated vector digest (see *Protocol 1*, step 5)
   - 9 µl of 5× T4 DNA ligase buffer (Invitrogen)
   - 5 µl of T4 DNA ligase (NEB)

2. Transfer the ligation into a microcentrifuge tube[b]. Precipitate the ligation as indicated in steps 7–9 of *Protocol 1*.

3. Transform the ligation in BL21(DE3) cells[c]. In four 0.2 ml PCR tubes, mix 2.5 µl of cleaned ligation (step 2) with 40 µl of chemically competent BL21(DE3) cells held on ice. Heat shock the cells for 1 min at 37°C.

4. Recover each reaction in 1 ml of SOC in a 12 ml culture tube at 37°C for 1.5 h.

5. Combine the four recovery cultures in one microcentrifuge tube and resuspend in 1 ml of SOC.

6. Plate the library onto a selective Bauer plate. Plate a 1 : 400 dilution of the library on a counting plate (two serial dilutions of 2 µl of the 800 µl pooled transformations).

7. Incubate for 12–16 h at 37°C.
8. Resuspend the cells from the overnight plate with LB medium using a plating tool[d].
9. Determine the optical density of the plate wash by diluting some concentrated cell resuspension 20-fold in LB medium and measuring the OD at 600 nm.
10. Dilute a 1 ml aliquot of the original concentrated resuspension to an OD of 1.0 in 20% glycerol/LB.
11. Aliquot into small PCR tubes or 0.5 ml Eppendorfs[e] and freeze into a dry ice/ethanol slurry.

Notes

[a]The ligase buffer from Invitrogen contains PEG 4000, which increases the ligation efficiency.

[b]It is essential to wash the walls of the PCR ligation tube with 10 μl of Tris/HCl (pH 8.5) until all of the DNA has been resuspended. Add this additional aliquot to the microcentrifuge tube prior to ethanol precipitation.

[c]A DH10B library may be made if a diversity of >20 000 clones is needed. Combine 100 μl of thawed DH10B with 50 μl of pre-chilled ddH$_2$O and 10 μl of resuspended, ligated DNA. Transform by electroporation according to the manufacturer's protocol. Recover each transformation in 1 ml of pre-warmed SOC in 12 ml culture tubes and shake at 350 r.p.m. at 37°C for 1.5 h. Combine the three recovery cultures in one microcentrifuge tube and resuspend in 1 ml of SOC. Plate the library onto an LB/agar Bauer selective media plate with 35 μg/ml kanamycin. Plate a 1:400 dilution of the library on a counting plate. Between 200 and 1000 colonies should be expected on the counting plate; the primary library should be a lawn. Resuspend the entire cell lawn with a spreader in 12 ml of LB onto a Kirby Bauer library plate, vortex in a 15 ml Falcon tube for approximately ten 10 s pulses (high), and perform a Qiagen plasmid prep on 750 μl of cell suspension (the cell mass should be equivalent to a 3 ml overnight liquid plasmid prep, i.e. ~75–100 mg cell pellet). Perform three transformations of 4 μl of plasmid prep in BL21(DE3) cells as indicated in step 3.

[d]The cells are viable if the resuspension/dilution is carried out in less than 30 min.

[e]An alternative method is to lay the tube(s) on their side(s) on a wet paper towel stack sitting on foil on the floor of a –80°C freezer.

Protocol 5
Selection of clones for protein folding

Equipment and Reagents
- 4 LB/agar Bauer plates (100 × 15 mm) containing 35 µg/ml kanamycin
- 4 LB/agar Bauer plates (100 × 15 mm) containing 35 µg/ml kanamycin and 1 mM isopropyl β-D-thiogalactopyranoside (IPTG)
- 132 mm diameter supported nitrocellulose membranes (GE Osmonics)
- ColiRollers plating beads (Novagen)
- Illumatool lighting system with excitation filter 488 nm (blue) and a glass colored 520 nm long-pass filter (Lightools Research)
- Boekel replicator
- 96-Well tissue culture plates

Method
1. Rapidly thaw the freezer aliquots of viable cells (see *Protocol 4*, step 11). Perform two sequential 350-fold serial dilutions in 800 µl of LB (2.3 µl into 800 µl of LB).
2. Plate 800 µl of the last dilution onto the nitrocellulose membrane using ColiRollers plating beads on an LB/agar Bauer selective plate, dry in an air flow for 5 min, and grow overnight at 32°C.
3. Pre-warm an LB/agar selective plate containing 1 mM IPTG.
4. Move the membrane (colony side up) from the overnight plate (step 2) to the pre-warmed induction plate (step 3).
5. Incubate at the desired temperature (typically 37°C, or 32°C for poor folders) for 4 h[a].
6. Illuminate the plate with the Illumatool to visualize the fluorescent clones (see *Fig. 3* inset – *Protocol 5*).
7. Pick the 44 brightest[b] colonies into LB/kanamycin medium (175 µl/well) in a 96-well tissue culture plate. Include the wild-type gene cloned in the GFP insertion reporter as a control in an available well (i.e. the 45th well[c]).
8. Grow overnight at 32°C. The following morning, prepare a 20% glycerol/LB copy in a 96-well tissue culture plate for –80°C library storage using a sterile Boekel replicator (five transfers from the overnight plate to the storage plate).

Notes
[a]Longer incubation times may be necessary to observe fluorescent clones.
[b]Desirable clones will have these characteristics: (i) be brighter than the wild-type GFP fusion; and (ii) be as bright or brighter at 27°C relative to 37°C. Avoid colonies that become 'unrealistically' bright (>3 standard deviations from the 'consensus'). These are likely to be deletion or truncation mutants. Artifacts to avoid include: internal starts of translation, short fragments, and stable RNA stem–loops (brighter at 37°C than at 27°C).
[c]The objective is to use blocks of 48 wells of a 96-well tissue culture plate, one block of 48 for each round of evolution. Forty-four wells are used for optima and four are for controls (wild type as well as top optimum from up to three round(s) of evolution).

Protocol 6

Verification of optima fluorescence

Equipment and Reagents
- LB/agar Bauer plates (150 × 15 mm) containing 35 µg/ml kanamycin
- LB/agar Bauer plates (150 × 15 mm) containing 35 µg/ml kanamycin and 1 mM IPTG
- 132 mm diameter supported nitrocellulose membranes (GE Osmonics)
- Boekel replicator
- Illumatool lighting system with excitation filter 488 nm (blue) and a glass colored 520 nm long-pass filter
- Digital camera, such as the Olympus C-5060 on a photographic boom (Edmund Scientific)

Method
1. Label two fresh 132 mm nitrocellulose membranes. One plate will be induced at 37°C and the other at 27°C.
2. Place the freezer storage plate (see *Protocol 5*, step 8) on the bench for 2 min at room temperature to thaw slightly. Transfer some of the cell suspension to the surface of the membrane using the Boekel replicator. Grow overnight at 30°C.
3. Transfer the membranes from the overnight plates to the induction plates. Incubate one plate at 37°C and the other at 27°C until cell fluorescence develops (~4 h).
4. Photograph the plates using a digital camera[a] (see *Fig. 3* inset – *Protocol 6*).
5. Quantify the green fluorescence using appropriate image analysis software[a]. Export the output fluorescence to a spreadsheet (i.e. Microsoft EXCEL) to sort the clones based on total integrated counts[b].

Notes
[a]Photograph the colonies using a suitable digital camera such as the Olympus C-5060 on a secure mount such as a camera boom (Edmund Scientific). Take a series of exposures ranging from 1/30 s to 2 s to ensure that all colony phenotypes (bright to faint) are within the linear dynamic range of the camera under at least one of the exposure conditions.

[b]Digital photographs can be analyzed using the blob analysis tool of the public-domain image analysis program NIH IMAGE (http://rsb.info.nih.gov/nih-image[5.1]) or the Windows-compatible version of NIH IMAGE developed by Scion (http://www.scioncorp.com/pages/scion_image_windows.htm[5.2]). Use a suitable image processing program to separate the RGB channels and select the monochrome green channel for subsequent analysis.

Protocol 7

Single-colony PCR

Equipment and Reagents
- LB/agar Bauer plates (150 × 15 mm) containing 35 µg/ml kanamycin
- Vent polymerase, buffer, and reagents (NEB)
- 10 mM dNTPS (2.5 mM each of dATP, dCTP, dGTP, and dTTP)
- 5.0 µM forward GFP-specific oligonucleotide 1 (5′-GATATAACTAGTAAGCGTGACCACATG GTCCTTCTTGAG-3′)
- 5.0 µM reverse GFP-specific oligonucleotide 2 (5′-TACTTCGGTACCTCCAGTGAAAAGTTCTT CTCCTTTGCT-3′)
- Sterile, distilled water (ddH$_2$O)
- 96-Well PCR plate (Robbins Scientific)
- 1.5% Agarose gel (large format)
- 6× Type III loading buffer (0.25%, w/v, bromophenol blue, 0.25%, w/v, xylene cyanol FF, 30% glycerol in water)
- Equipment and reagents for agarose gel electrophoresis including 1× TAE agarose gel running buffer
- Staining solution: 10 ng/ml ethidium bromide
- Microcentrifuge
- Thermocycler

Method
1. Prepare a PCR master mix as follows:
   - 449 µl of ddH$_2$O
   - 65 µl of 10× Vent polymerase buffer
   - 78 µl of dNTPs
   - 5.7 µl of MgSO$_4$ stock (NEB)
   - 8.7 µl of Vent polymerase (NEB)
   - 26 µl of forward GFP-specific oligonucleotide 1
   - 26 µl of reverse GFP-specific oligonucleotide 2

2. Transfer 12 µl of PCR mix to each of 48 wells of a Robbins PCR plate on ice.

3. Pipette 1.5 µl of the diluted glycerol storage plate (see *Protocol 5*, step 8) into each well with a multi-channel pipette[a].

4. Place the tubes in the thermocycler after the initial melting temperature has been reached (see below) and run the following program:

Number of cycles	Program
1	8 min at 97°C
25	25 s at 96°C, 25 s at 56°C, and 1 min/kb at 72°C
1	5 min at 72°C
1	Hold at 2°C

5. Run the PCR samples[b] on a 1.5% agarose gel in TAE buffer. Load 8 µl of each reaction to which 2 µl of 6× loading buffer has been added. Resolve at 120 V for 30 min.

6. Stain the gel with ethidium bromide solution and photograph.

7. Export the image for size analysis (see *Fig. 4*, also available in the color section). Discard truncated artifacts[c].

CHAPTER 5: SELECTION OF PROTEIN VARIANTS WITH IMPROVED EXPRESSION

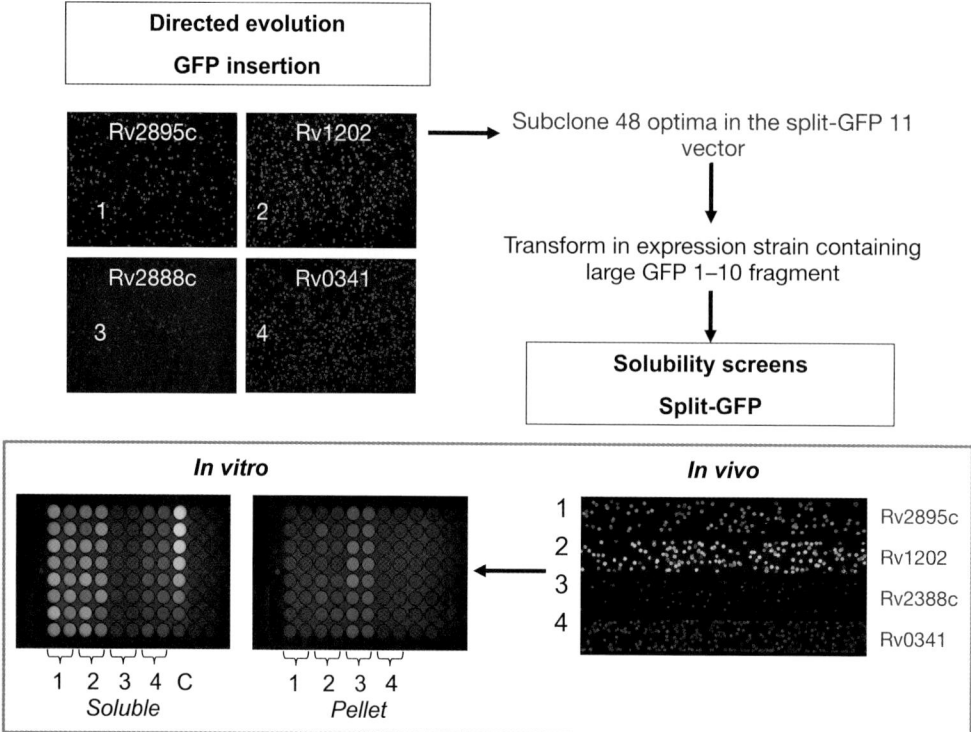

Figure 5. Assessing solubility on variants with improved folding (see page xxii for color version).
At the fourth round of directed evolution in the GFP folding reporter vector, the library of *M. tuberculosis* protein variants display stable levels of fluorescence (top left). Forty-eight optima are selected from the brightest colonies and the variants are subcloned into the split-GFP 11 solubility reporter vector and transformed into an *E. coli* strain containing the large, complementing GFP 1–10 fragment. The four *M. tuberculosis* targets are screened for solubility *in vivo* (bottom right) and the 16 brightest clones are grown and induced in 96-well plates. Soluble and pellet fractions are assayed *in vitro* by adding the large, complementing GFP fragment assay reagent for precise quantification. Protein #3 (Rv2388c) is an example of slow evolution that requires a change in strategy (faint in both GFP insertion and split-GFP reporters). Proteins 1, 2, and 4 display most fluorescence in the soluble fraction (bottom left).

Notes

[a] If the experiment is performed consecutively after *Protocol 5*, a new fresh 20% glycerol/LB replica can be made from an overnight growth of the glycerol storage plate (see *Protocol 5*, step 8). Do not totally thaw the master glycerol storage plate when starting cultures. Thaw frozen glycerol stock culture plates for a maximum of 2 min at room temperature.
[b] If the wild-type control is not included in the main glycerol storage plate, set up a control PCR using the starting plasmid with the insert in the GFP insertion reporter as a control. Load this amplicon in an outside lane as a size standard.
[c] The red dots in *Fig. 5* indicate clones to discard (truncated or missing sites) relative to the positive control.

Protocol 8

Recombining optima for the next round of evolution

Equipment and Reagents
- LB/agar Bauer plates (150 × 15 mm) containing 35 μg/ml kanamycin
- 132 mm diameter supported nitrocellulose membranes (GE Osmonics)
- Boekel replicator
- Microcentrifuge

Method
1. For each optimum, combine data from the fluorescence measurements (see *Protocol 6*, step 5) and size analysis (see *Protocol 7*, step 7).
2. Pick the most fluorescent 20% of the clones that correspond to full-length cDNA inserts.
3. Copy the entire 96-well plate from the master glycerol 96-well storage plate (see *Protocol 5*, step 8) onto a nitrocellulose membrane on a Bauer selective plate. Grow overnight at 30°C.
4. Aspirate the cell mass from the optima (step 2) on the membrane and combine in a microcentrifuge tube filled with 750 μl LB. Resuspend and pellet a sufficient cell mass to equal a typical overnight 3 ml culture (~75–100 mg of cell paste).
5. Discard the supernatant and proceed with a standard plasmid prep according to the manufacturer's instructions.
6. Perform a PCR for the next round of evolution using this plasmid as the DNA template[a] (see *Protocol 3*, step 1).

Notes
[a] Perform two 50 μl reactions to ensure having sufficient DNA material for the DNase I digest (see *Protocol 3*, step 1). Use forward and reverse oligonucleotides specific to the GFP scaffolding (see *Protocol 3*, Equipment and Reagents).

3. TROUBLESHOOTING

- It is recommended that you verify expression of the protein that will be evolved. GFP insertion constructs can be used in any BL21(DE3) strain bearing the T7 RNA polymerase lysogen.
- If, after three rounds of mutagenesis and selection, the folding and solubility of the protein are not improved, other approaches towards making the diversity library may be tried, such as expressing a fragment library of the target protein or searching for folding partners.
- The appearance of numerous deletion artifacts may be deleterious for efficient evolution. It is recommended that you identify the source of artifacts by sequencing and that you change the DNA sequence accordingly to reduce recombination events (for example, by PCR-directed 'silent' codon mutation).

4. REFERENCES

1. Georgiou G & Valax P (1996) *Curr. Opin. Biotechnol.* **7**, 190–197.
2. Hockney RC (1994) *Trends Biotechnol.* **12**, 456–463.
3. Yokoyama K, Kikuchi Y & Yasueda H (1998) *Biosci. Biotechnol. Biochem.* **62**, 1205–1210.
4. Lee KH, Kim HS, Jeong HS & Lee YS (2002) *Biochem. Biophys. Res. Commun.* **298**, 216–224.
5. Chen Y, Song J, Sui SF & Wang DN (2003) *Protein Expr. Purif.* **32**, 221–231.
6. Baneyx F & Palumbo JL (2003) *Methods Mol. Biol.* **205**, 171–197.
7. Fox JD, Kapust RB & Waugh DS (2001) *Protein Sci.* **10**, 622–630.
8. Fox JD & Waugh DS (2003) *Methods Mol. Biol.* **205**, 99–117.
9. Malakhov MP, Mattern MR, Drinker M, Weeks SD & Butt TR (2004) *J. Struct. Funct. Genomics*, **5**, 75–86.
10. Armstrong N, de Lencastre A & Gouaux E (1999) *Protein Sci.* **8**, 1475–1483.
11. Tresaugues L, Collinet B, Minard P, *et al.* (2004) *J. Struct. Funct. Genomics*, **5**, 195–204.
12. Stemmer WP (1994) *Nature*, **70**, 389–391.
13. Naki D, Paech C, Ganshaw G & Schellenberger V (1998) *Appl. Microbiol. Biotechnol.* **49**, 290–294.
14. Jung S, Honegger A & Pluckthun A (1999) *J. Mol. Biol.* **294**, 163–180.
15. Roberts RW (1999) *Curr. Opin. Chem. Biol.* **3**, 268–273.
16. Kristensen P & Winter G (1998) *Fold. Des.* **3**, 321–328.
17. Finucane MD, Tuna M, Lees JH & Woolfson DN (1999) *Biochemistry*, **38**, 11604–11612.
18. Matsuura T & Pluckthun A (2003) *FEBS Lett.* **539**, 24–28.
19. Maxwell KL, Mittermaier AK, Forman-Kay JD & Davidson AR (1999) *Protein Sci.* **8**, 1908–1911.
20. Waldo GS, Standish BM, Berendzen J & Terwilliger TC (1999) *Nat. Biotechnol.* **17**, 691–695.
21. Stemmer WP (1994) *Proc. Natl. Acad. Sci. U.S.A.* **91**, 10747–10751.
★★ 22. Pedelacq JD, Piltch E, Liong EC, *et al.* (2002) *Nat. Biotechnol.* **20**, 927–932. – *A demonstration that the GFP folding reporter can be used to evolve multimeric proteins.*
23. Yang JK, Park MS, Waldo GS & Suh SW (2003) *Proc. Natl. Acad. Sci. U.S.A.* **100**, 455–460.
★★ 24. Wurth C, Guimard NK & Hecht MH (2002) *J. Mol. Biol.* **319**, 1279–1290. – *A demonstration that the GFP folding reporter can be used to select more soluble variants of a protein associated with slow aggregation.*
25. Pedelacq JD, Waldo GS, Cabantous S, Liong EC & Terwilliger TC (2005) *Protein Sci.* **14**, 2562–2573.
26. Cabantous S, Terwilliger TC, Waldo GS (2005) *Nat. Biotechnol.* **23**, 102–107.
27. Pedelacq JD, Cabantous S, Tran T, Terwilliger TC & Waldo GS (2006) *Nat. Biotechnol.* **24**, 79–88.
28. Crameri A, Whitehorn EA, Tate E & Stemmer WP (1996) *Nat. Biotechnol.* **14**, 315–319.
★ 29. Cabantous S, Pedelacq JD, Mark BL, Naranjo C, Terwilliger TC & Waldo GS (2005) *J. Struct. Funct. Genomics*, **6**, 113–119. – *One example of an application using the combination of GFP insertion and split-GFP technologies to evolve soluble proteins.*
★★★ 30. Cabantous S & Waldo GS (2006) *Nat. Methods*, **3**, 845–854. – *A detailed protocol for performing solubility assays using split-GFP.*
31. Sambrook J, Fritsch EF & Maniatis T. (1989) *Molecular Cloning: a Laboratory Manual*, 2nd edn. Cold Spring Harbor Laboratory Press, Cold Spring Harbor, NY.

CHAPTER 6
Protein expression in the wheat-germ cell-free system

Tatsuya Sawasaki and Yaeta Endo

1. INTRODUCTION

Modern proteomics presupposes the availability of protein expression technology that can produce a wide variety of proteins (a) with natural folding and (b) in sufficient amounts. Currently, three strategies are used for protein production: chemical synthesis, *in vivo* expression, and cell-free protein synthesis. The first two methods have severe limitations in that chemical synthesis is not practical for the synthesis of long peptides (1) and that *in vivo* expression can only produce those proteins that do not affect the physiology of the host cell (2–4). On the other hand, cell-free translation systems can synthesize proteins with high accuracy and speed approaching *in vivo* rates (5, 6), and they can express proteins that seriously interfere with cell physiology. One of the main obstacles in the cell-free systems is the short life of translational activity and thus low yield in protein production. To overcome this limitation, Spirin *et al.* (7) proposed a continuous-flow cell-free (CFCF) translation method, in which a solution containing amino acids and energy sources is supplied to the translation chamber through a semi-permeable membrane. In fact, when operated in the CFCF mode, the popular *Escherichia coli* cell-free system gives much higher productivity than in a conventional batch mode. The *E. coli* cell-free system, however, still has inherent limitations because of the prokaryotic nature of its translation and folding mechanisms; multi-domain proteins, found more often in eukaryotes than in prokaryotes, tend to misfold in prokaryotic systems, whether *in vivo* or *in vitro* (8). It has been suggested that the eukaryotic translation and folding machinery has been optimized through evolution to facilitate co-translational domain folding. In fact, there is a difference in the rate of peptide growth on ribosomes between eukaryotes and bacteria, being five to ten times slower in the former than in the latter (8, 9). Another popular cell-free system is the rabbit reticulocyte cell-free system. Derived from eukaryotic animal cells, it is expected to have the ability to synthesize eukaryotic proteins in a folded state. However, because of its low efficiency, it cannot be used to produce preparative amounts of proteins. This limitation is due to the codon usage in reticulocytes, which is optimized for globin synthesis, and a significant amount of ribonuclease M in the lysate, which is released from the plasma membrane during the lysis. Another limitation is the high endogenous hemoglobin content, which obstructs product purification. Other cell-free systems derived from eukaryotes such as yeast cells, tumor cells, and insects or insect cell lines have been used widely to produce eukaryotic multi-domain proteins in active forms, but all suffer from low productivity. One of the most convenient and promising eukaryotic cell-free translation systems is conceivably the one based on wheat embryos in which all of the components of translation except for mRNA are stored in a dried state, ready for protein synthesis as soon as germination starts. Originating from a eukaryotic plant source, it is expected to have the ability to synthesize eukaryotic multi-domain proteins in a folded state. The conventional wheat-germ system, however, is plagued by its short life and, as a result, it is as inefficient as other cell-free systems. This shortcoming was overcome by the advent of a new wheat-germ cell-free system. In this chapter, we describe how the practical cell-free system is built and review its application to today's functional and structural biology (10).

2. METHODS AND APPROACHES

2.1 Principles of the new wheat-germ cell-free system

We concentrated on wheat-germ cell-free systems because they have numerous advantages such as low cost, easy availability in large amounts, and the capacity to synthesize high-molecular-mass proteins. Moreover, they are eukaryotic systems and hence more suitable for the expression of eukaryotic proteins. After we discovered that the mechanism of action of the ricin toxin is ribosome inactivation (11–13), many other ribosome-inactivating proteins (RIPs) with an identical mechanism of action have been found in higher plants (14). Most commonly, these toxins are single-chain proteins, and they inhibit protein synthesis by removing a single adenine residue in a universally conserved stem–loop structure of 28S rRNA (11–14). Although the biological function of RIPs is not known, it is generally believed that they are important for cell defense (14). The most widely studied example is an antiviral effect during infection by several plant viruses (15). As originally proposed by Ready et al. (16), the explanation for the antiviral activity of RIPs is that when a cell wall is damaged, the RIP is released into the cytosol where it inactivates ribosomes, thereby preventing virus replication. Tritin, found in wheat seeds and thought to be localized mainly in the endosperm, is such a single-chain RIP (17). To improve protein synthesis in the wheat-germ cell-free system, we started with the hypothesis that the embryonic ribosomes are in fact susceptible to tritin. In this case, contamination of wheat-germ preparations with tritin-containing endosperm fragments would be fatal. Accordingly, we prepared our cell-free system from highly washed embryos, and indeed found that the system became far more active. In order to maximize the throughput of the system for practical use, other elemental techniques were developed. These were (i) eliminating both the 5′ ^{7}mGpppG (cap) and poly(A) tail by optimizing the 5′- and 3′-untranslated regions

METHODS AND APPROACHES

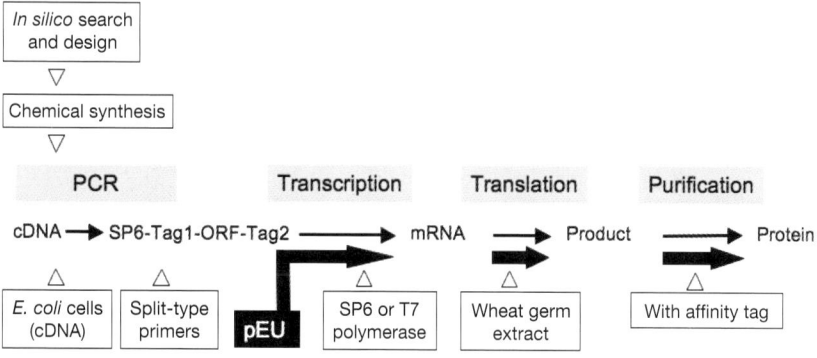

Figure 1. Procedures for protein synthesis using the wheat-germ cell-free system.

(UTRs) of mRNA, thereby increasing translation initiation and the stability of the template (18); (ii) designing the split primers for the polymerase chain reaction (PCR) to generate transcription templates that can minimize artificial generation of the products (18), a technique that permits high-throughput construction of DNA templates directly from *E. coli* cells carrying cDNAs, with the time-consuming cloning steps bypassed; (iii) constructing an expression vector pEU (18) specialized for the mass production of proteins, which reduces the introduction of inevitable mutations during the PCR and abates the cost; (iv) developing the transcription and translation reaction in one tube in which no mRNA purification step is needed (19); and (v) inventing the bilayer reaction, which enables us to perform CFCF mode translation without using a membrane for the high-throughput production (20). By combining all these elemental techniques, we established two cell-free protocols for practical use (see *Fig. 1*). One is the protocol for materializing genetic information in parallel, which consists of (i) *in silico* selection of suitable genes from the database, (ii) construction of templates for transcription by the split-primer PCR, (iii) transcription, and (iv) bilayer translation in which the solution resulting from transcription is used directly as the mRNA source. For the production of proteins with tags such as for purification or for the reporter or for both, DNA constructs can also be generated using split-primer PCR. The other protocol is for massive preparation (see *Fig. 1*, bold arrows), which consists of (i) selection of suitable gene products from the parallel production described above and subsequent functional screening, (ii) cloning of the genes from entry vectors into pEU and transcription of mRNA, (iii) fine tuning of translational conditions such as the concentrations of pertinent ions, and (iv) protein production that incorporates either the bilayer or discontinuous batch translation method. Proteins with a desired purification tag can easily be affinity purified, and the tag portion can be removed by proteolytic digestion at a designed linker sequence if desired.

2.2 Preparation of the extract from wheat embryos

In our pursuit of ways to improve the wheat-germ cell-free translation system, we found that conventional wheat-germ extracts contained the RNA *N*-glycosidase

Unwashed  Washed

Figure 2. Wheat embryos (see page xxiii for color version).
Extensive washing of wheat embryos eliminates endosperm contaminants, visible as a white color on the unwashed embryos, producing clean, washed embryos.

tritin and other inhibitors of translation such as thionin, ribonucleases, deoxyribonucleases, and proteases, which are suspected of being involved in a suicide system targeting the translational machinery. We also ascertained that these inhibitors originated from the endosperm (21, 22). Extensive washing of wheat embryos was found to eliminate these endosperm contaminants to produce clean embryos (see *Fig. 2*, also available in the color section) for extracts with a high degree of stability and activity (22). In fact, this new translation system proved to be stable not only in its activity and longevity but also in storage; it can be stored for years without loss of activity, even in a lyophilized state. The extensive washing of wheat embryos to eliminate endosperm contaminants gave birth to an extract with high stability and activity. The high efficiency of our system can be attributed to at least two factors (23): high initiation, elongation, and termination rates (efficient usage and recycling of all of the translation factors, ribosomes, mRNA, tRNA, etc.), and low endogenous ribonuclease activity (retention of heavy polysomes for a prolonged time). These characteristics, observed in test tubes, demonstrated that the new system utilized the properties of the embryos.

Protocol 1

Isolation of wheat embryos and preparation of wheat embryo extract (22)

Equipment and Reagents
- Wheat seeds[a]
- Roter Speed Mill (model pulverisette 14; Fritsh)
- Sieve (710–850 μm mesh)
- Cyclohexane
- Carbon tetrachloride
- NP-40
- Milli-Q water, freshly prepared
- Bronson model 2210 sonicator (Yamato)
- Extraction buffer: 40 mM HEPES/KOH (pH 7.6), 100 mM potassium acetate, 5 mM magnesium acetate, 2 mM calcium chloride, 4 mM dithiothreitol (DTT), and 0.3 mM of each of the 20 amino acids
- Sephadex G-25 (fine) column (Amersham Biosciences)
- Translational substrate buffer (TSB): 30 mM HEPES/KOH (pH 7.8), 100 mM potassium acetate, 2.7 mM magnesium acetate, 0.4 mM spermidine, 2.5 mM DTT, 0.3 mM amino acid mix, 1.2 mM ATP, 0.25 mM GTP, and 16 mM creatine phosphate (all reagents from Wako Pure Chemical Industries)
- Amicon Ultra Centrifugal Filter Devices (molecular mass cut-off 10 000; regenerated cellulose; Millipore Corp.)

Method
1. Grind the wheat seeds in a mill.
2. Collect embryos by sieving through a 710–850 μm mesh.
3. Select the intact embryos by solvent flotation using the solvents cyclohexane and carbon tetrachloride at a ratio of 1 : 2.5 (v/v).
4. Dry overnight in a fume hood.
5. Wash three times with 10 vols of Milli-Q water with vigorous stirring.
6. Sonicate for 3 min in a 0.5% NP-40 solution in a sonicator.
7. Grind 5 g of isolated wheat embryo to a fine powder in liquid nitrogen.
8. Add 5 ml of extraction buffer and vortex the mixture briefly.
9. Centrifuge the embryo homogenate (30 000 g, 30 min) and retain the supernatant (about 8 ml).
10. Gel filtrate the supernatant using a Sephadex G-25 column (22 ml, 1 cm (R) × 7 cm (H)), equilibrated with 2 vol of extraction buffer.
11. Repeat the gel filtration as above but using TSB as the equilibration buffer.
12. Collect the void fraction.
13. Centrifuge the fraction (30 000 g, 10 min) and retain the final supernatant.
14. Concentrate the fraction to approx. one-half to one-third of the volume using an Amicon Ultra-15 (10K) filter unit according to the manufacture's instructions at 10°C.
15. Adjust to 240 A_{260} units/ml with TSB.
16. Divide into small aliquots and store at –80°C until use.

Note

[a] Any strain can be used.

2.3 Development of the 5'UTR of mRNA to enhance translation

The 5' cap and 3' poly(A) tail found on almost all eukaryotic mRNAs play crucial roles in the regulation of gene expression: they control mRNA translational efficiency, stability, and localization. However, for *in vitro* translation, the use of the cap and poly(A) tail is problematic. First, the cap molecule itself binds to initiation factor eIF4E and thus is a strong competitor of the capped mRNA, resulting in inhibition of translation initiation. Thus, the cap analog has to be eliminated from transcripts prior to the translation and this is the primary reason for the difficulties encountered in the construction of efficient coupled transcription-translation cell-free systems from eukaryotic cells. In addition, cap-containing RNA fragments accumulated by ribonuclease(s) attack during the reaction compete against native-form mRNA for the initiation factor. Furthermore, the optimum concentration of capped mRNA is generally in a narrow range, and has to be determined accurately prior to the start of the reaction for efficient translation. This is also problematic, especially when protein synthesis is to be carried out for high-throughput materialization from a large number of genes, and for production of large amounts of protein in the continuous system (18), as supplement of mRNA is needed during the reaction. The cap analog for mRNA synthesis is costly and this is another drawback. To compound the above problems, long plasmid-encoded poly(dT/dA) is unstable during replication in host cells. To solve these problems, we endeavored to find 5'- and 3'UTRs that enhance mRNA translation in the absence of the cap and poly(A) tail, based on a vector of pSP65. We found that the Ω sequence (from tobacco mosaic virus) at the 5' end, with the additional three-base 5'-GAA-3' sequence (thus, GAA-Ω) had the highest template activity among the sequences examined. We went further to create a shorter 5' enhancer than the Ω sequence. Taking advantage of the high translation activity of this new wheat-germ system, we succeeded in selecting a 52 nt sequence from a random nucleotide sequence pool that showed a similar level of enhancing activity as Ω. One of the nucleotide sequences of the enhancers, named E02, is: 5'-GAACUC ACCUAUCUCAUACAACUUUCAACUUCCUAUUUCUACACAAAACAUUUCCCUAC-3' (24). We next examined the effect of the 3'UTR by varying the 3'UTR with GAA-Ω as with the 5'UTR, and found that translation did not depend on a specific sequence or poly(A) tail, but depended only on the length of the 3'UTR. Based on these results, we designed efficient 5' (Ω, E01 or E02) (18, 24) and 3' (longer than 300 nt) UTRs that enhanced the template activity of the mRNA. These mRNAs worked well with a wider concentration range, whilst the capped mRNA showed a typical narrow optimum concentration, where one has to optimize mRNA concentration on a case-by-case basis.

2.4 Expression vector pEU

Based on these findings, we constructed an efficient expression vector derived from pSP65, pEU (plasmid of Ehime University) (18), that included GAA-Ω or GAA-E02 at the 5' end in addition to promoter and multi-cloning sites (see *Fig. 3a*). By adding mRNA transcribed from circular pEU without prior linearization

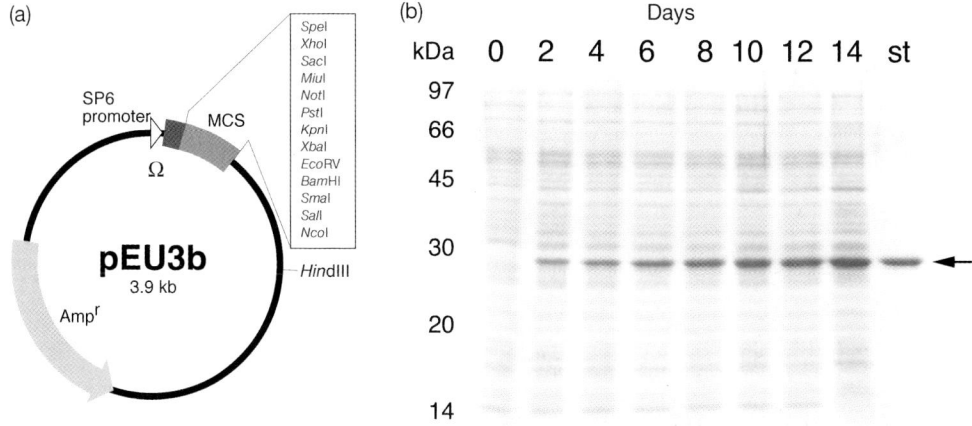

Figure 3. A cell-free expression vector and demonstration of its performance.
(*a*) Schematic illustration of pEU. (*b*) SDS-PAGE analysis of GFP produced during 14 days of reaction. mRNA produced by transcription of circular pEU was used for the translation reaction in the dialysis membrane system and was added every 48 h. A 0.1 μl aliquot of the mixture was run on the gel and protein bands were stained with Coomassie Brilliant Blue. The arrow shows GFP and lane st designates an authentic GFP band (0.5 μg).

into the dialysis cell-free system, a large amount of protein (9.7 mg of GFP in 1 ml) could be produced. The translation reaction continued for surprisingly long periods, up to 14 days (see *Fig. 3b*). Although supplementation of mRNA was required every 2 days, the amount of protein produced was more than that of the endogenous proteins. This result supported our previous notion that the translation machinery itself is inherently robust and stable (22).

2.5 PCR-directed generation of DNA template (18)

Although pEU is useful for large-scale production of a specific protein, the procedure includes laborious molecular cloning steps. This limits the throughput of the cell-free protein production system when it is applied to the screening of many different protein sequences for attractive characteristics. Thus, we searched for a PCR primer set with which the above UTR elements could be introduced into any given cDNA sequence so that the PCR product could be used directly for the transcription reaction without purification. The upstream primer(s) should have an SP6 promoter and GAA, whilst the downstream primer can be any sequence from the vector carrying the cDNA. We started by examining a conventional PCR method using four primers, one of which had a complete sequence covering the promoter (25, 26) to introduce the SP6 promoter into several different genes. However, the resulting products were not good templates for transcription when used without purification. In fact, agarose gel analyses revealed that this method gave mostly shorter transcripts and resulted in the production of small peptides (data not shown). This could have been due to the generation of primer–dimer

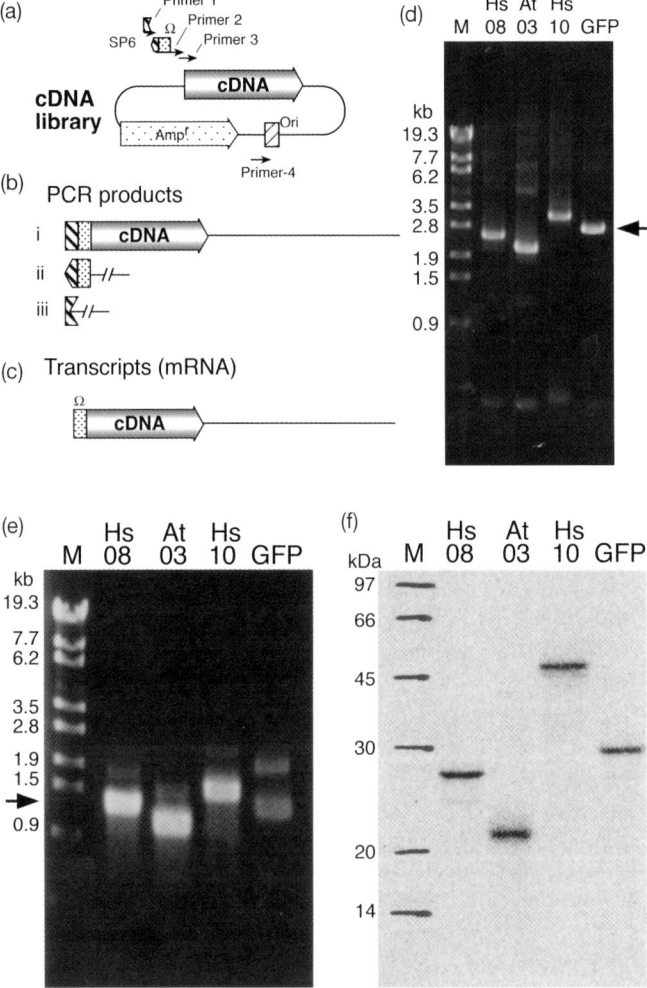

Figure 4. PCR-based expression of cDNA information using a new set of primers. (a) Design of the split primers for the introduction of the required UTRs into cDNA sequences. (b, c) Expected PCR-generated DNAs and mRNA, respectively. (d) PCR-generated DNA. (e) A 10 μl aliquot from a PCR sample was used for the 100 μl transcription reaction and transcripts were analyzed. (f) All of the transcript was used for batch mode translation (50 μl, 4 h), and products were analyzed by autoradiography.

(25) artifacts. Thus, we designed a set of primers so that the complete promoter sequence was not contained in any one of the primers, but was generated only when the sequences of two primers were correctly joined (see *Fig. 4a* and *b*), which led to the production of the complete mRNA (see *Fig. 4c*). Each PCR product (see *Fig. 4d*) by this split-primer method produced the mRNA with an apparent correct size (see *Fig. 4e*) and gave a single radioactive protein band on the autoradiogram indicating the expected size when translated (see *Fig. 4f*).

Protocol 2

Split-primer PCR (18)

Equipment and Reagents
- ExTaq DNA polymerase (Takara Bio)
- dNTPs and buffer (supplied with the polymerase)
- PCR thermocycler (model MP; Takara Bio)
- Primer 1: 5′-GCGTAGCATTTAGGTGACACT-3′ (the underlined sequence indicates the 5′ half of the promoter)[a] (Invitrogen)
- Primer 2: 5′-*GGTGACACT*ATAGAAGTATTTTTACAACAATTACCAACAACAACAACAAACAACAA CAACATTACATTTTACATTCTACAACTA*CCACCCACCACCACCAATG*-3′ (the underlined sequence indicates the 3′ half of the promoter and the sequence in italic denotes the annealing region of primer 1 or primer 3)[b] (Sigma)
- Primer 3: 5′-CCACCCACCACCACCAatgnnnnnnnnnnnnnnnnnn-3′ (the 5′-coding region of the target gene is indicated in lowercase) (Invitrogen)
- Primer 4: 5′-AGCGTCAGACCCCGTAGAAA-3′ (based on reported PCR procedures) (Invitrogen)

Method
1. For the first PCR, prepare a 60 µl PCR by combining 200 µM of each dNTP, 1.5 units of ExTaq DNA polymerase, primer 3 and primer 4 to 10 nM, 3 ng of plasmid or 3 µl of *E. coli* (overnight culture) as template, and the buffer supplied by the manufacturer to a concentration of 1×.
2. Incubate in the thermocycler using the following cycle: denaturation for 4 min at 94°C, followed by 30 cycles of 98°C for 10 s, 55°C for 30 s, and 72°C for 5 min, depending on the length of the gene (1 kb per min).
3. In order to equip the first PCR-amplified target gene with the Ω and SP6 promoter sequences, carry out a second PCR in a volume of 30 µl using 3 µl of the first PCR product (without any purification), primer 1 and primer 4 at final concentrations of 100 nM each, and primer 2 at 1 nM, using the same cycling conditions as in the first PCR.

Notes
[a] The design of primers as shown in *Fig. 4* for the split-primer PCR technique involves four main steps: (i) designing the target-specific primer (primer 3) in such a way that its 3′ end has target-specific sequence (5′ terminus of the open reading frame) and its 5′ end contains a region of the 5′-terminal site of the Ω sequence; (ii) ensuring that primer 2 has the full-length Ω sequence and a region of the SP6 promoter sequence at its 5′ end; (c) ensuring that primer 1 contains part of the SP6 promoter and part of the overlapping sequence at the 3′ end; (iv) ensuring that primer 4 is specific for the sequence of the 3′UTR of the vector. Therefore, for each clone of cDNA, primer 3 is the only specific primer covering the 5′-end region of the open reading frame, and the remaining primers are common to all cDNAs.
[b] We described the use of the Ω sequence here, but enhancers E01 and E02 that we selected can be utilized equally well (24).

2.6 Preparation of mRNA

Protocol 3 describes a small-scale conventional batch transcription reaction. For the production of large amounts of mRNA, the transcription reaction can be carried out using the continuous-flow principle. The same transcription mixture is placed into the dialysis bag or dialysis cup and then immersed in a tenfold volume of a solution containing all of the ingredients except for the DNA, polymerase, and human nuclease inhibitor. The transcription reaction continues approximately seven times longer than the batch transcription. During an overnight incubation, we usually obtain 7–10 mg of mRNA per ml of reaction mixture, which corresponds to a yield that is an order of magnitude higher than the batch reaction and results in a significant reduction in costs. We rationalize that this prolonged transcription reaction could be caused by the law of mass action by pushing and pulling the reaction equilibrium point with substrate and byproduct, respectively.

Protocol 3

Preparation of mRNA

Equipment and Reagents
- 5× Transcription buffer (TB): 400 mM HEPES/KOH (pH 7.8), 80 mM magnesium acetate, 10 mM spermidine, and 50 mM DTT (Wako Pure Chemical Industries)
- NTP mix: a solution containing 25 mM each of ATP, GTP, CTP, and UTP (Wako Pure Chemical Industries)
- SP6 RNA polymerase (80 units/μl; Promega)
- RNasin (80 units/μl; Promega)
- Milli-Q water, freshly prepared
- 7.5 M Ammonium acetate (Sigma)
- ~99 and 70% Ethanol (Sigma)
- TSB (see *Protocol 1*)

Method
1. Prepare the following reaction mixture for transcription[a]:
   - 10 μl of 5× TB
   - 6 μl of 25 mM NTP mix (final conc. 3 mM)
   - 0.625 μl of SP6 RNA polymerase (final conc. 1 unit/μl)
   - 0.625 μl of RNasin (final conc. 1 unit/μl)
   - template DNA to give a final concentration of 1/10 vol. PCR product (>10 μg/ml) or 5 μg of high-quality circular plasmid
   - Milli-Q water to a final volume of 50 μl

2. Incubate the reaction mixture at 37°C for 3 h.

3. Centrifuge the mixture (6000 *g*, 1 min) and retain the supernatant.

4. Add 3 vols of Milli-Q water, 1/7.5 vol. of 7.5 M ammonium acetate and 2.5 vols of ~99% ethanol, mix well, and incubate on ice for 10 min.

5. Centrifuge the mixture (20 000 *g*, 5 min) and discard the supernatant.

6. Wash the pellet with 500 μl of 70% ethanol and then spin for 5 min at 20 000 *g*.

7. Dissolve the mRNA pellet in 30 μl of TSB (about 1.7 mg mRNA/ml) for batch, bilayer, or dialysis mode of translation.

Note
[a]Approximately 50 μg/50 μl of RNA is synthesized using this reaction.

2.7 Translation

Three protein synthesis reaction protocols are available, depending on the experimental aim.

2.7.1 CFCF mode and bilayer reaction for genome-wide biochemical annotation

The CFCF method described by Spirin et al. (7) is well suited for efficient protein production using the efficient wheat-germ cell-free system. However, it may not be suitable for the massive screening of gene products from comprehensive cDNA libraries; equipped with a semi-permeable membrane in the reaction chamber, the CFCF apparatus is mechanically complicated to manage. Biochemical analyses

(a)

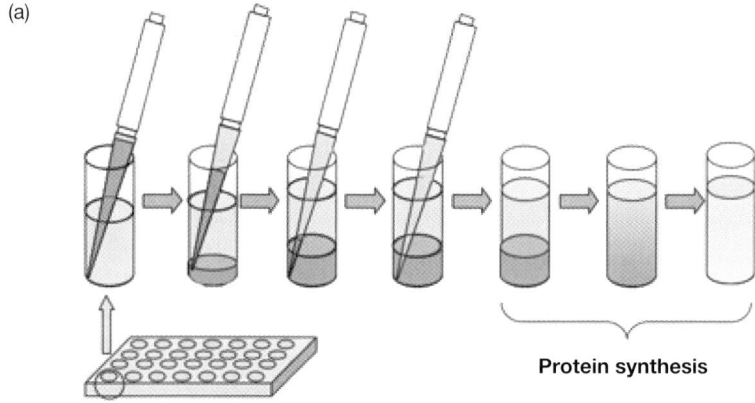

Protein synthesis

(b)

Incubate at 17°C for 60 min → Transfer the reaction mixture to Amicon Ultra filter → Centrifuge for 30 min ay 10°C 5000 **g** → Discard the flow-through → Add TBS to bring to the original volume

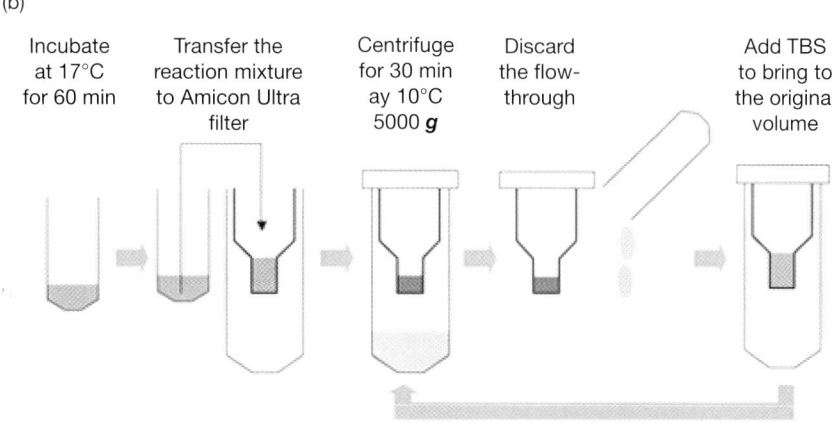

Repeat this process six times

Figure 5. Novel cell-free synthesis reactions.
(a) The bilayer reaction for functional protein analysis. (b) The discontinuous batch reaction.

such as the functional analysis of enzymes require only submicrogram amounts per assay. However, regular batch-mode reactions cannot synthesize a sufficient amount of protein for such analyses, as they can work only for an hour or so. We therefore felt the need to invent a reaction system for analyzing gene products from cDNA libraries that was simpler and more cost-effective than the CFCF method and yet capable of synthesizing larger amounts of protein than batch-mode reactions. Subsequently, we developed the bilayer translation method as illustrated in *Fig. 5(a)*. The method does not require a membrane (20), as the translation mixture is simply overlaid with the substrate mixture. This new system worked for more than ten times longer than the batch mode reaction, yielding submilligram amounts of protein. The longevity of the bilayer method can be explained by the fact that it enables the continuous supply of substrates, and the removal (dilution) of small byproducts through a phase between the translation mixture and substrate mixture by diffusion. It performed well to synthesize sufficient amounts of proteins for functional analysis and solubility determination of gene products from cDNA libraries.

2.7.2 Discontinuous batch reaction for large-scale production

The bilayer reaction method is useful for producing up to milligram amounts of protein, but may not be suitable for larger-scale production. The CFCF-based reaction method, however, had a weak point: it required a long incubation time and thus was not very practical, particularly from the standpoint of general stability of proteins. We therefore tried to develop another reaction method to reduce the production time (27). We were guided by the experimental evidence that, in batched reactions, halted translation in a translation mixture could be restarted with the same initial reaction velocity as a new batch when the mixture was subjected to ultrafiltration and the ingredients were replaced with fresh substrates (results not shown). We took advantage of this characteristic of the cleansed wheat-extract system and devised a discontinuous batch translation method, whose principle is depicted in *Fig. 5(b)*. This method worked efficiently and 35 mg of green fluorescent protein (GFP) was produced in 5 h in a reaction volume of 50 ml.

Protocol 4

Wheat cell-free translation

Equipment and Reagents
- Creatine kinase (1 mg/ml; Roche)
- TSB (see *Protocol 1*)
- Saran wrap
- Dialysis cup (molecular mass cut-off 12 000; Biotech International)
- 24-Well microplate (Whatman)
- U-shaped or flat-bottomed microplate
- Amicon Ultra Centrifugal Filter Devices (see *Protocol 1*)
- Synthesized RNA dissolved in TSB (about 1.7 mg/ml, from *Protocol 3*)[a]

Method

CFCF mode of translation using a dialysis cup[b]

1. Prepare 50 μl of translational reaction mixture by combining 8.5 μl of wheat-embryo extract (final concentration of 40 A_{260} units/ml), 2 μl of 1 mg/ml creatine kinase, 10.5 μl of TSB and 30 μl of the mRNA from *Protocol 3*.
2. Wash the dialysis cup three times with 200 μl of Milli-Q water.
3. Add 2 ml of TSB to a Whatman 24-well plate.
4. Add the reaction mixture to the dialysis cup.
5. Place the cup into the Whatman plate.
6. Wrap with Saran wrap to avoid evaporation.
7. Keep the 24-well plate in an incubator at 17°C for 24 h[c]. To keep the reaction for another 24 h, add mRNA as in step 1.

Bilayer translation (20)

1. Prepare 25 μl of translational reaction mixture by combining 6.5 μl of wheat-embryo extract (final concentration of 60 A_{260} units/ml), 1 μl of 1 mg/ml creatine kinase, 5.5 μl of TSB and 12 μl of the mRNA from *Protocol 3*.
2. Add 125 μl of TSB to a microplate well.
4. Carefully place 25 μl of the reaction mixture into the bottom of the microplate well.
5. Place a coverlet on the plate and then wrap with the Saran wrap to avoid evaporation.
6. Keep the plate in an incubator at 17°C for 18 h without shaking.

Discontinuous batch reaction

1. Construct an 80 A_{260} system[d].
2. Keep the plate in an incubator at 17°C for 60 min as for the regular batch translation reaction.
3. Transfer the reaction mixture to an Amicon Ultra Centrifugal Filter.
4. Centrifuge to reduce the volume of the reaction mixture to half of the original volume (this usually takes 30 min at 5000 *g*).
5. Add the translation substrate solution to restore the original volume[e].
6. Repeat this process (from steps 2 to 5) six times[f].

Notes

[a]The final mRNA concentration in the system is proportional to the extract concentration; thus, for example, 1 mg/ml of mRNA for the 40 A_{260} CFCF system, and 800 μg/ml for the 60 A_{260} bilayer system.
[b]This protocol is for translating the proteins on a preparative scale starting from 50 μl to 2 ml of the reaction volume. The level of protein synthesis over time was analyzed by sodium dodecyl sulfate polyacrylamide gel electrophoresis (SDS-PAGE) and is shown in *Fig. 3(b)*.
[c]Incubation temperature is another important factor that affects both productivity and quality of products as mentioned in section 3 below. Generally, incubation at 17°C gives the highest productivity, although this is dependent on the given gene.
[d]The reaction volume is dependent on your purpose.
[e]Through this process, the system gains fresh substrate and at the same time decreases any by-products that inhibit translation.
[f]One can restart the reaction by supplying fresh mRNA, and it will then work for a further six rounds.

2.8 Adapting the CFCF reaction for transcription and translation in one tube

For our study, we used purified mRNA as the translation template, which is common practice in eukaryotic cell-free protein synthesis. The purification step, however, was cumbersome and time-consuming. To examine the possibility of skipping this step, we tested the use of the whole transcription mixture containing mRNA along with NTPs, a high concentration of magnesium ion, spermidine, DTT, and pyrophosphate/phosphate in the bilayer reaction system. Surprisingly, we found that direct addition of the whole mixture produced template activity almost identical to that of the purified mRNA (see *Protocol 3*). Under the conditions employed, the concentration of magnesium ion was 12.06 mM, which was far more than the well-known optimum concentration of around 3 mM (28, 29). We followed up this unusual observation by titrating the magnesium ion to find that the system in fact worked well with a wide range of magnesium ion concentrations, from 5 to 18 mM, in the presence of NTPs and 1.3 mM spermidine. This apparently wide range of magnesium concentrations may have been due to high concentrations of NTPs, as NTPs are known to reduce free magnesium ion concentration. In fact, experiments similar to the above except for the use of purified mRNA have confirmed that typical values of the optimum magnesium concentration lie within a narrow range (22, 28). The new system that performs transcription and translation in one tube without mRNA purification also worked well in the batch reaction (19). Note: We called this 'transcription and translation in one tube' rather than 'coupled translation', as there is little evidence showing the formation of DNA/RNA polymerase/mRNA/ribosome complexes, even in the *E. coli* cell-free system.

Protocol 5

Transcription and translation in one tube (19)

Equipment and Reagents
- 5× TB, NTPs, SP6 RNA polymerase and RNasin (see *Protocol 3*)
- Milli-Q water, freshly prepared
- TSB (see *Protocol 1*)
- Creatine kinase (2 or 20 mg/ml; Roche)
- U-shaped or flat-bottomed microplate for small-scale preparation
- Six-well titer plate for large-scale preparation (Whatman)

Method
1. Prepare a transcription reaction by adding the reagents as given in *Table 1* for either a large-scale or small-scale preparation, except for wheat embryo and creatine kinase.

Table 1. Transcription reactions for large- and small-scale preparations

Reagent	Small-scale prep	Large-scale prep	Final conc.
Template DNA	x μl	x μl	1/5 vol. (HC PCR)[a] 100 ng/μl (CP)[b]
5× TB	5 μl	50 μl	1×
NTP mix (25 mM)	2.5 μl	25 μl	2.5 mM for each NTP
SP6 RNA polymerase (80 000 units/ml)	0.3 μl	3 μl	1 unit/μl
RNasin (80 000 units/ml)	0.3 μl	3 μl	1 unit/μl
Milli-Q water	To final vol. of 24 μl	To final vol. of 249 μl	
Wheat embryo extract	25 μl	250 μl	120 A_{260}/ml
Creatine kinase (2 mg/ml)	1 μl	–	40 μg/ml
Creatine kinase (20 mg/ml)	–	1 μl	40 μg/ml
Total volume	50 μl	500 μl	

2. Incubate the reaction mixture at 37°C for 6 h.
3. Add the wheat-embryo extract, and 1 μl of creatine kinase, and then mix well.
4. Prepare 550 μl of TSB for the small-scale prep in the U-shaped 96-well titer plate well or 5.5 ml in the six-well plate for the large-scale prep.
5. Due to the higher density of the step 3 mix compared with the step 4 mix, two layers should be clearly visible. Carefully pipette 50 μl (or 500 μl for the large-scale prep) of the reaction mixture at the bottom of the titer plate well by inserting a pipette tip to the bottom of the well.
6. Place a sealing film and then a coverlet on the plate to avoid evaporation.
7. Keep the plate in an incubator at 4–26°C for 8 to >20 h without shaking[c].

Notes
[a]Highly concentrated PCR product (>50 μg/ml).
[b]High-quality circular plasmid.
[c]Incubation time is dependent on the reaction temperature; for example, >20 h at 4°C, 18 h at 17°C, or 8 h at 26°C.

2.9 Applications based on the cell-free system

2.9.1 Functional biology

In order to evaluate the performance of the cell-free method in high-throughput screening, we carried out parallel protein syntheses from 27 genes originating from *E. coli* cells carrying the cDNAs (20). In 50 out of a total of 54 cases (authentic and fusion proteins with glutathione *S*-transferase (GST)), a clearly visible Coomassie Brilliant Blue-stained protein band was obtained. The yield after a 36 h incubation was estimated by densitometric scanning of the bands using bovine serum albumin as the standard and was found to be 0.1–2.3 mg/ml of the reaction volume in the dialysis cup mode reaction (CFCF) (see *Table 2*). Some gene products were recovered in a soluble form in the supernatant and others in the precipitate phase after centrifugation at 30 000 *g* for 15 min. It is worth mentioning here that we could not detect any dependency of the productivity and solubility of the proteins on the gene source. Furthermore, the system had little preference in codon usage, which is a prerequisite for genome-wide protein expression. In fact, we could express the malaria proteins from 76% AT-rich cDNAs of *Plasmodium falciparum* (30–32). The most important requirement for an expression system, however, is that it produces a high-quality product. To verify the quality of the proteins and the general applicability of the wheat-germ cell-free system, we tested a subset of product proteins for their functional activity (see *Fig. 6*). The *PHOT1* gene from *Arabidopsis thaliana* encodes a protein of 120 kDa, which binds noncovalently to flavin mononucleotide (FMN) and presumably acts as a chromophore in light-dependent autophosphorylation (33). As shown in *Fig. 6(a)*, the protein that was responsible for the blue-light-responsive activity could be produced only when the translation was carried out in the presence of FMN (34). This result demonstrated that a prosthetic group is required during translation and folding into the holoenzyme, which supports the notion that co-translational folding takes place on eukaryotic ribosomes (8). *Fig. 6(b)* shows the results of the preparation and measurement of activity of human kinases. Each of the 11 kinases was synthesized in a GST-fused form, affinity purified with a GST column, separated by SDS-PAGE, and stained with Coomassie Brilliant Blue. When the enzymes were incubated with [γ-^{32}P]ATP in the presence of the well-known substrates histone H1 and myelin basic protein, subsequent autoradiography clearly demonstrated that each enzyme had its own unique substrate specificity (34). This suggested that each product might be enzymatically active. *Fig. 6(c)* shows another example in which the polyhedrins of baculoviruses synthesized in a cell-free system were active enough to be assembled *in vitro* into polyhedra (35). We also attempted to express certain genes whose proteins, if synthesized in *E. coli* cells, exhibited little activity. Our cell-free system proved able to produce some of these difficult proteins in an active form without the refolding process. Examples of such successful protein synthesis have already been reported elsewhere and are summarized in *Table 3*. Although the system has not yet been optimized for the production of full-sized membrane proteins in a soluble, folded state, there is the possibility of combining them with detergents or liposomes (41, 42).

Table 2. Example of proteins synthesized in the PCR-directed wheat-germ cell-free system

Proteins encoded by cDNAs	GenBank/ annotation no.	Molecular mass (Da)	Authentic Total (mg)	Authentic Sup. (%)	Fusion Total (mg)	Fusion Sup. (%)	Clone name
Arabidopsis (from GenBank and MIPS[a])							
Chlorophyll a/b-binding protein	X56062	25 995	0.2	30	0.5	90	At01
Agamous-like gene 9 (AGL9)	AF015552	29 065	0.7	30	0.8	90	At02
Flowering locus T (FT)	AF152096	19 808	0.3	100	0.8	100	At03
HY5	AB005456	18 462	0.4	90	1.5	90	At04
Flowering locus F (FLF)	AF116527	21 864	0.2	100	0.4	100	At05
Hypothetical protein	At1g69630[a]	11 311	0.1	40	0.1	100	At06
Arabidopsis (from a commercial cDNA library[b])							
Putative heat-shock protein 40	AL021749	38 189	1.8	30	1.0	80	At07
Heat-shock protein 70-3	Y17053	71 144	0.9	100	ND	ND	At08
Putative S-adenosylmethionine synthetase	AY037214	42 793	1.5	100	0.6	100	At09
NADPH thioredoxin reductase	Z23108	40 635	0.1	10	0.5	20	At10
Putative ACC oxidase	AF370155	36 677	1.0	10	1.2	100	At11
Putative fructokinase	AF387001	35 276	1.0	10	0.6	100	At12
Rubisco activase	X14212	51 981	0.4	20	0.6	80	At13
Glutaredoxin	At4g15660[a]	11 311	ND	ND	0.4	80	At14
Chlorophyllase 2	AF134302	34 902	0.2	10	0.4	70	At15
Human (from GenBank)							
Neuron-specific γ-2 enolase	M22349	47 266	1.0	100	0.5	100	Hs01
Zeta-crystallin/quinone reductase	L13278	35 205	2.3	80	1.3	100	Hs02
X11-like protein	AB014719	82 480	0.5	100	0.2	80	Hs03
Importin α1	NM_002266	57 859	ND	ND	0.2	30	Hs04
Glyceraldehyde-3-phosphate dehydrogenase	M17851	36 051	0.4	70	0.9	100	Hs05
Enolase 3[c]	NM_001976	46 956	1.7	80	0.9	100	Hs06
APBA3[c,d]	NM_004886	61 451	0.9	100	0.2	100	Hs07
JAK-binding protein[c]	NM_003745	23 550	0.4	30	0.2	100	Hs08
Human (from a commercial cDNA library[e])							
Phosphoglycerate kinase 1	XM_010102	43 965	1.0	100	0.7	100	Hs09
b-Actin	X00351	41 735	1.3	100	0.3	100	Hs10
Hypothetical protein FLJ10652	XM_006938	41 539	–[d]	–[d]	0.2	10	Hs11
Hypothetical protein FLJ10559	XM_001479	35 237	0.7	50	1.0	70	Hs12

[a]MIPS *Arabidopsis thaliana* database MAtDB (http://mips.gsf.de/proj/plant/jsf/athal/index.jsp[6.3])
[b,e]Lambda ZAP II Library (products of Stratagene cat. nos 937010 and 936204, respectively).
[c]These genes were cloned from tissue (heart, brain, kidney, liver, and placenta) cDNAs (BioChain Institute, cat. no. 0516001).
[d]Amyloid beta (A4) precursor protein-binding, family A, member 3.
ND, Not detectable; Sup., supernatant.

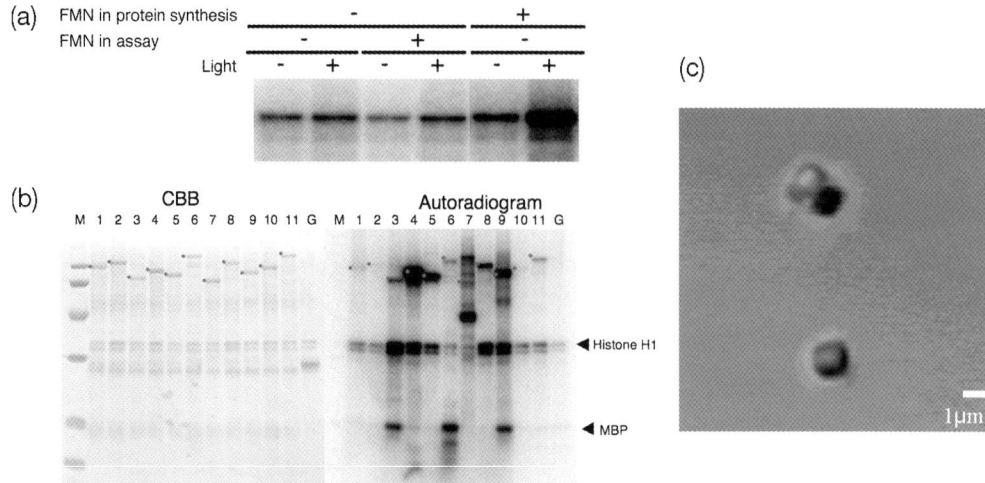

Figure 6. Quality of eukaryotic gene products expressed in the wheat-germ cell-free system.
(a) Blue-light-induced autophosphorylation activity of the *PHOT1* gene product. (b) Activity of 11 human protein kinases. Each affinity-purified kinase with GST fused to its N terminus was treated with λ protein phosphatase and incubated with [γ-^{32}P]ATP in the presence of added histone H1 (from bovine calf thymus) and myelin basic protein (from bovine brain). Samples were separated by SDS-PAGE, stained with Coomassie Brilliant Blue for analysis (left panel), and subjected to autoradiography (right panel). (c) After the translation of polyhedrin gene from baculoviruses in the cell-free system, the translation mixture was kept for 48 h at 4°C. Microscopic observation revealed the formation of small polyhedra.

Table 3. Difficult proteins expressed in the wheat-germ cell-free system

Species	Activity	Gene name	Reference
Cypovirus	Crystalline particle formation	VP3	35
Bacillus	Restriction enzyme	Endonuclease (*Bam*HI)	20
Thermococcus	DNA synthesis	KOD DNA polymerase	20
Aquifex	Methyltransferase	tRNA (Gm18) methyltransferase	36
Arthrobacter	Sarcosine oxidation	Sarcosine oxidase	20
Arabidopsis	Protein phosphorylation by light	*PHOT1*	34
Rice	Anthranilate synthase	Anthranilate synthase α-subunit	37
Rice	Shkimate kinase	Shkimate kinase	38
Rice	EPSP synthase	EPSP synthase	38
Mouse	Immunoglobulin induction	Osteopontin	39
Human	DNA binding	c-*fos*	40
Human	DNA binding	c-*jun*	40

2.9.2 Structural biology

One of the bottlenecks in the high-throughput structural determination of proteins is the step to produce proteins in a folded state. The other limiting step is amino acid-specific, selective labeling for assigning the signals collected in the ^1H-^{15}N heteronuclear single quantum coherence spectra in nuclear magnetic resonance (NMR) spectroscopy, or selenomethionine labeling for multi-wavelength anomalous diffraction (MAD) phasing in X-ray crystallography. Cell-free systems have one general advantage over cell expression systems: cell-free products for NMR measurements do not need extensive purification, as none of the endogenous proteins are labeled during the translation. Cell-free systems derived from *E. coli*, however, contain high levels of amino acid-metabolizing enzymes, as the extracts are prepared from cells in the exponential growth phase. As a result, many types of inhibitor of amino acid metabolism or even suitable mutant strains have to be added to avoid scrambling of the label among amino acids. In the wheat-based system, in contrast, most of these enzyme activities were expected to be very low, as the embryos are in hibernation. Nevertheless, we found that two transaminases and one synthetase were active during the translation reaction. Our subsequent search for their inhibitors was successful in that the interconversions of amino-^{15}N between alanine and glutamine (alanine transaminase) and between glutamine and aspartic acid (asparatate transaminase) were completely inhibited by adding β-chloro-L-alanine and L-methionine sulfoximine, and that the other leaking pathway between glutamine and aspartic acid (glutamine synthetase) was suppressed by amino-oxyacetate (43). Incorporation of amino acids into protein is not very high, even in our system. This issue of economy could be solved by recycling the amino acids left in the mixture using the conventional method for purifying amino acids after translation. This is another attractive feature that benefits the NMR field. The improved methodology may pave the way for a high-throughput protein synthesis system suited to estimating the degree of folding (44) and intermolecular interactions between ligands and the protein of interest, as well as assigning signals in the NMR field. A fruitful result of the application of the wheat-germ cell-free protein synthesis system to NMR structure determination can be found at http://www.uwstructuralgenomics.org/[6.1] (45, 46). The results of X-ray analyses including selenomethionine labeling for MAD phasing have been published (47).

2.9.3 Robotic automation and protein array

The improved wheat-germ cell-free translation system is especially powerful in the high-throughput production of eukaryotic multi-domain proteins in a folded state. We have recently succeeded in the robotic automation of two protocols, one for genome-wide synthesis adapting the bilayer translation unit (26), and the other for large-scale production using the discontinuous batch mode reaction (27). The machines (http://www.cfsciences.com[6.2]) can perform all of the steps preceding protein purification, holding the promise of increasing the throughput and decreasing the cost of protein production. Protein microarrays, which have enormous potential in the biosciences and medical fields, are one of the promising applications of the cell-free system. Recent progress in elemental technologies in this field is remarkable (48, 49), but the goal is still distant. The difficulty is due mainly to the diversity of the biochemical properties of proteins. One of the most difficult tasks is to establish a technique for storage, i.e. a technique to keep each immobilized protein intact in such a tiny space, which would be totally different from that for DNA chips. In this light, we propose the possible use of the cell-free system to overcome such difficulties. A prospective scheme is to take advantage of the stability and capability of the system and store all of the ingredients involved in transcription and translation in a lyophilized form together with the DNA template. Adding water prior to the preparation of chips would then start the production of proteins fused to immobilization tags. Another potential application of the wheat-germ cell-free system in protein chip technology is for the production of viral polyhedral proteins. Foreign proteins with 79 N-terminal residues (a sequence from *Bombyx mori* cytoplasmic polyhedrosis virus VP3) can be immobilized in virus-like particles that resist dehydration and higher temperatures (35). The accumulated information on the structures and functions of gene products should revolutionize our understanding of biology and fundamentally alter the practice of medicine, as well as influencing other industries.

3. TROUBLESHOOTING

- It is important to include a positive control in the transcription and translation steps. A gene encoding GFP may provide a good control.
- When little protein production can be seen, the nucleotide sequence of the open reading frame of the gene should be confirmed by direct sequencing. In our experience, over 90% of problems derive from the wrong nucleotide sequence in the open reading frame, the lack of an initiation codon, or the generation of termination codons in the original gene.
- For the production of protein from genes containing a protease motif, the translation reaction should be carried out at very low temperature to avoid digestion of the product. To this end, it is recommended that the translation and subsequent purification should be done at 4°C rather than 17°C or higher. The wheat-germ system retains sufficient activity at lower temperatures (which may reflect the embryo's nature) to allow the bilayer reaction to be performed at 4°C, although productivity may be decreased by 15%.
- For expression of apo-form proteins, carry out the reaction in the presence of partners such as a prosthetic group.
- For the synthesis of protein kinase, transcription factors, nuclear receptors, and other proteins, which have (or might have) binding site(s) for other molecules (this can be checked by a prior search *in silico*), perform a short translation reaction incubation to decrease the level of expressed protein. This trick may help minimize aggregation of the product, and often helps to recover the proteins in an active form.
- To synthesize proteins with disulfide bonds, run a similar reaction but without DTT in the reaction mixture.

4. REFERENCES

1. Blaschke UK, Silberstein J & Muir TW (2000) *Methods. Enzymol.* **328**, 478-496.
2. Henrich B, Lubitz W & Plapp R (1982) *Mol. Gen. Genet.* **185**, 493-497.
3. Goff SA & Goldberg AL (1987) *J. Biol. Chem.* **262**, 4508-4515.
4. Chrunyk BA, Evans J, Lillquist J, Young P & Wetzel R (1993) *J. Biol. Chem.* **268**, 18053-18061.
5. Kurland CG (1982) *Cell*, **28**, 201-202.
6. Pavlov MY & Ehrenberg M (1996) *Arch. Biochem. Biophys.* **328**, 9-16.
7. Spirin AS, Baranov VI, Ryabova LA, Ovodov SY & Alakhov YB (1988) *Science*, **242**, 1162-1164.
8. Netzer WJ & Hartl FU (1997) *Nature*, **388**, 343-349.
9. Hartl FU & Hayer-Hartl M (2002) *Science*, **295**, 1852-1858.
10. Endo Y & Sawasaki T (2006) *Curr. Opin. Biotechnol.* **17**, 373-380.
11. Endo Y, Mitsui K, Motizuki M & Tsurugi K (1987) *J. Biol. Chem.* **262**, 5908-5912.
12. Endo Y & Tsurugi K (1987) *J. Biol. Chem.* **262**, 8128-8130.
13. Wool IG, Gluck A & Endo Y (1992) *Trends Biochem. Sci.* **17**, 266-269.
14. Barbieri L, Battelli MG & Stirpe F (1993) *Biochim. Biophys. Acta*, **1154**, 237-282.
15. Taylor S, Massiah A, Lomonossoff G, Roberts LM, Lord JM & Hartley M (1994) *Plant J.* **5**, 827-835.
16. Ready MP, Brown DT & Robertus JD (1986) *Proc. Natl. Acad. Sci. U.S.A.* **83**, 5053-5056.
17. Massiah AJ & Hartley MR (1995) *Planta*, **197**, 633-640.
★ 18. Sawasaki T, Ogasawara T, Morishita R & Endo Y (2002) *Proc. Natl. Acad. Sci. U.S.A.* **99**, 14652-14657. – *To establish a high-throughput cell-free expression system, the authors developed a series of essential elementary technologies such as the initiation enhancer sequence, split-primer PCR, and expression vectors specialized for the wheat cell-free system.*
★ 19. Sawasaki T, Morishita R, Gouda MD & Endo Y (2005) *Methods Mol. Biol.* **310**, 131-144. – *A wheat-germ cell-free protein production method based on transcription and translation in one tube.*
★ 20. Sawasaki T, Hasegawa Y, Tsuchimochi M, et al. (2002) *FEBS Lett.* **514**, 102-105. – *Development of the bilayer translation reaction method for high-throughput materialization from cDNAs.*
21. Ogasawara T, Sawasaki T, Morishita R, Ozawa A, Madin K & Endo Y (1999) *EMBO J.* **18**, 6522-6531.
★ 22. Madin K, Sawasaki T, Ogasawara T & Endo Y (2000) *Proc. Natl. Acad. Sci. U.S.A.* **97**, 559-564. – *Clarification of the cause of instability of the cell-free translation reaction, and development of a method for preparing a highly efficient and yet robust extract from wheat embryos.*
23. Madin K, Sawasaki T, Kamura N, et al. (2004) *FEBS Lett.* **562**, 155-159
24. Kamura N, Sawasaki T, Kasahara Y, Takai K & Endo Y (2005) *Bioorg. Med. Chem. Lett.* **15**, 5402-5406.
25. Sambrook J & Russell DW (2001) *Molecular Cloning: a Laboratory Manual*, 3rd edn. Cold Spring Harbor Laboratory Press, Cold Spring Harbor, NY.
★ 26. Endo Y & Sawasaki T (2004) *J. Struct. Funct. Genomics*, **5**, 45-57. – *Development of a robot that enables high-throughput screening of genetic information, based on the bilayer reaction method.*
★ 27. Vinarov DA, Loushin Newman CL & Markley JL (2006) *FEBS J.* **273**, 4160-4169. – *Description of robotic automation, based on the discontinuous batch translation method developed by Sawasaki et al. (19), enabling a high yield of protein in a short reaction time.*
28. Roberts BE & Paterson BM (1973) *Proc. Natl. Acad. Sci. U.S.A.* **70**, 2330-2334.
29. Craig D, Howell MT, Gibbs CL, Hunt T & Jackson RJ (1992) *Nucleic Acids Res.* **20**, 4987-4995.
30. Aguiar JC, LaBaer J, Blair PL, et al. (2004) *Genome Res.* **14**, 2076-2082.
31. Zou L, Miles AP, Wang J & Stowers AW (2003) *Vaccine*, **21**, 1650-1657.
32. Tsuboi T (2005) *Seikagaku*, **77**, 646-649.
33. Sakai T, Kagawa T, Kasahara M, et al. (2001) *Proc. Natl. Acad. Sci. U.S.A.* **98**, 6969-6974.

★ 34. Sawasaki T, Hasegawa Y, Morishita R, Seki M, Shinozaki K & Endo Y (2004) *Phytochemistry,* **65**, 1549–1555. – *Demonstration that addition of the prosthetic group FMN to the translation system was essential to synthesize the active* PHOT1 *gene product.*
★ 35. Ikeda K, Nakazawa H, Shimo-Oka A, *et al.* (2006) *Proteomics,* **6**, 54–66. – *This work showed that polyhedrin synthesized in the wheat-germ cell-free system from the baculovirus gene could assemble into polyhedra. Using the cell-free system, it was shown that a sequence of 79 N-terminal residues of the* Bombyx mori *cytoplasmic polyhedrosis virus VP3 protein was useful for immobilizing foreign proteins on the surface of polyhedra.*
36. Klammt C, Lohr F, Schafer B, *et al.* (2004) *Eur. J. Biochem.* **271**, 568–580.
37. Pornillos O, Chen YJ, Chen AP & Chang G (2005) *Science,* **310**, 1950–1953.
38. Morita EH, Shimizu M, Ogasawara T, Endo Y, Tanaka R & Kohno T (2004) *J. Biomol. NMR,* **30**, 37–45.
39. Morita EH, Sawasaki T, Tanaka R, Endo Y & Kohno T (2003) *Protein Sci.* **12**, 1216–1221.
★ 40. Vinarov DA, Lytle BL, Peterson FC, Tyler EM, Volkman BF & Markley JL (2004) *Nat. Methods* **1**, 149–153. – *Description of a wheat-germ cell-free platform for protein production that supports efficient NMR structural studies of eukaryotic proteins and offers advantages including a cost-effective performance over* E. coli *cell-based methods.*
★ 41. Vinarov DA & Markley JL (2005) *Expert Rev. Proteomics,* **2**, 49–55. – *A review of the development of an automated platform for NMR-based structural proteomics that employs wheat-germ extract for cell-free production of labeled protein.*
42. Zhu H & Snyder M (2003) *Curr. Opin. Chem. Biol.* **7**, 55–63.
43. Merkel JS, Michaud GA, Salcius M, Schweitzer B & Predki PF (2005) *Curr. Opin. Biotechnol.* **16**, 447–452.
44. Hori H, Kubota S, Watanabe K, *et al.* (2003) *J. Biol. Chem.* **278**, 25081–25090.
45. Kanno T, Kasai K, Ikejiri-Kanno Y, Wakasa K & Tozawa Y (2004) *Plant Mol. Biol.* **54**, 11–22.
46. Kasai K, Kanno T, Akita M, Ikejiri-Kanno Y, Wakasa K & Tozawa Y (2005) *Planta,* **222**, 438–447.
47. Miyazono K, Watanabe M, Kosinski J, *et al.* (2007) *Nucleic Acids Res.* **35**, 1908–1918.
48. Miyazaki T, Ono M, Qu WM, *et al.* (2005) *Eur. J. Immunol.* **35**, 1510–1520.
49. Miyamoto-Sato E, Takashima H, Fuse S, *et al.* (2003) *Nucleic Acids Res.* **31**, e78.

CHAPTER 7

Saccharomyces cerevisiae: a microbial eukaryotic expression system

Christine Lang

1. INTRODUCTION

The yeast *Saccharomyces cerevisiae* is widely recognized and used as a robust host for recombinant protein expression. This is due to the following features:

- It allows easy scaling up of production.
- It has powerful genetic tools available that allow fast and convenient construction of customized host strains.
- It allows both integrative and plasmid-based expression.
- It offers a whole range of constitutive and regulated promoters.

Recombinant proteins are expressed both as intracellular proteins – soluble or membrane bound – and as extracellular proteins. *S. cerevisiae* possesses a eukaryotic secretory pathway, which allows post-translational modifications via secretion. *S. cerevisiae* has also been shown to produce active and functional recombinant proteins for both enzymatic and functional assays, as well as for structural biology. In many cases, this is due to the fact that *S. cerevisiae* is able to modify proteins faithfully post-translationally, not only with respect to glycosylation, but also with respect to phosphorylation and methylation. In addition, *S. cerevisiae* has proved to be a reliable model for mammalian diseases and metabolic pathways. This has led to its frequent use as a model eukaryote to understand the function and properties of many mammalian proteins in recent years (1). *S. cerevisiae*-based models have also been developed for G-protein-coupled pathways (2, 3).

The host–vector systems for this organism are exceedingly well developed and a wealth of knowledge on genetics and physiology has been accumulated.

Once a strain producing the desired protein has been established, methods and processes for scaling up production are readily available that are based on traditional baker's yeast production processes or modern technical processes for recombinant commercial proteins and enzymes. These can be applied to increase the yield of a recombinant protein further (4, 5). Also in this respect, an enormous amount of data and experience has been gathered during the last two decades to allow scaling up and yield improvement in laboratory-scale fermenters. Thus, batches of difficult-to-express proteins can be obtained that allow functional and structural characterization of protein targets (6–8).

This chapter introduces the methods and techniques currently used for exploiting the convenient and efficient yeast expression system. It will focus on vector design for replicative and integrative transformation, choosing the promoter, and strain design, and will provide protocols for transformation, strain characterization using the polymerase chain reaction (PCR), and expression analysis.

2. METHODS AND APPROACHES

2.1 Vectors and promoters

The most commonly used vectors are *Escherichia coli*/yeast multi-copy shuttle plasmids, based on the yeast endogenous plasmid 2 μm DNA (see *Fig. 1*). In addition, there are vector systems that rely on autonomously replicating sequences and on integrative transformation. For a basic review of transformation in yeast, see (9).

The yeast 2 μm DNA is a 6318 bp plasmid. Most *S. cerevisiae* strains carry this plasmid at a copy number of about 60 per cell. The 2 μm DNA-based plasmids exhibit a high transformation rate and show a high meiotic and mitotic stability. These plasmids lacking any foreign DNA have been described as being extremely stable at high copy number, and are employed in some of the highly productive

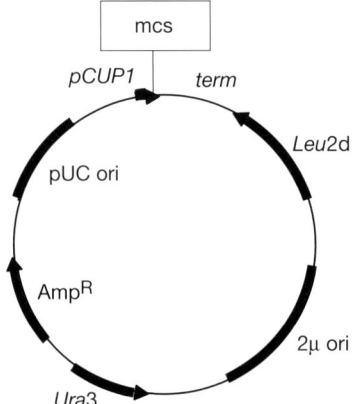

Figure 1. pYEXBX, a typical yeast expression vector.
pYEXBX contains *E. coli* replication and selection domains (pUC ori, Amp^R), the *S. cerevisiae* replication origin (2 μ ori), *S. cerevisiae* selection markers (*URA*3 and *LEU*2d), and an *S. cerevisiae* expression cassette consisting of the yeast promoter *CUP1* and a transcriptional terminator (*term*). Genes to be expressed are inserted via a multiple cloning site (mcs) at the promoter sequence.

industrial processes that use *S. cerevisiae* to produce human albumin at yields of up to 5 g/l of secreted protein and human α1-antitrypsin at 40% (w/v) total soluble protein (10, 11).

Integration of foreign DNA into the genome can in some cases be advantageous when high stability of a strain is aimed for and when more than one heterologous gene is to be introduced in one strain. Integration is based on homologous recombination processes and is achieved by a simple recombinational integration process that requires not more than 30–50 bp of homology on the 5′ and the 3′ site of integration (12, 13). Thus, integrating a single copy of a target sequence can in many cases be achieved by using PCR-derived fragments harboring the recombination sequences flanking the desired expression cassette and a suitable marker gene. Even multiple integration, which is recommended for high-level expression of recombinant genes, can be achieved by choosing the rDNA (14) or delta sequences (15, 16) as the recombination target sites.

Vector systems have been described that allow easy cloning and shuffling of genes between plasmids with various tag sequences (17) and with restriction-independent cloning sites (e.g. the GATEWAY system; 18).

The selection markers most widely used are complementing genes for amino acid or nucleotide auxotrophies. Studies have been made to evaluate the selective fitness of the most commonly applied markers (19). Alternative selection markers are based on dominant resistance traits; the most robust systems are the G418/geneticin resistance and the hygromycin B resistance systems.

Autonomously replicating vectors can be forced to be maintained at a higher copy number by modifying the promoter of the selection marker and making it less active. This requires a high copy number of the corresponding plasmid to produce sufficient amounts of the required amino acid or nucleotide (20, 21).

An elegant procedure for reusing selection markers for multiple subsequent integration events is based on the bacteriophage Cre-Lox recombination system, which has a high level of integration frequency and leaves only a 34 bp loxP site as a recombination footprint after recovery of the selection marker by enzyme-mediated recombination (22, 23).

Expression of recombinant proteins can be either designed as constitutive or regulated by induction. High-level expression has been reported in either case. The best-known constitutive promoters are the *Ashbya*-derived TEF promoter (24) and the ADH1 promoter (25, 26). Among the regulated promoters, the MFα and GAL promoters are the most widely used (27). In addition to the standard constructions, new developments show that even these can be optimized for higher yield (28, 29) or tighter regulation (30). The CUP1 promoter is an example of a promoter system that is regulated but does not exhibit a tight repression (31–33).

A set of expression vectors harboring different promoters and selection markers is available at the American Type Culture Collection (ATCC; for example, nos 87669 and 87728). Basic yeast/*E. coli* shuttle vectors, such as Yep13, YEp14, and YIp5, are also deposited in strain collections (Yeast Genetic Stock Center at ATCC and the German Collection of Microorganisms and Cell Cultures (DSMZ)).

Finally, the expression vector may contain a secretion leader sequence to direct the protein to be expressed by the secretion machinery and enable its release into the growth medium. Whilst different sequences have been described and used successfully, the secretion leader of the alpha mating type factor has proved to be the most robust and most versatile sequence for heterologous proteins (34, 35). In some cases, synthetic leaders have been employed successfully (36).

2.2 Strains

Standard host strains for most expression solutions are readily available in strain collections (Yeast Genetic Stock Center at ATCC and DSMZ). These are haploid strains bearing auxotrophies that can be complemented with the most commonly used selection markers. Examples are strains AH22 (ATCC 38626, DSM 3820), GRF18 (ATCC 64667, DSM 3797), and DBY747 (ATCC 44774). Choosing the right host strain for the target protein can greatly influence expression yield and quality (37).

In some cases, protease-deficient host strains are employed to enhance the stability of the expression product (see *Fig. 2*), and secretion enhancer strains have been used successfully to boost extracellular expression yield.

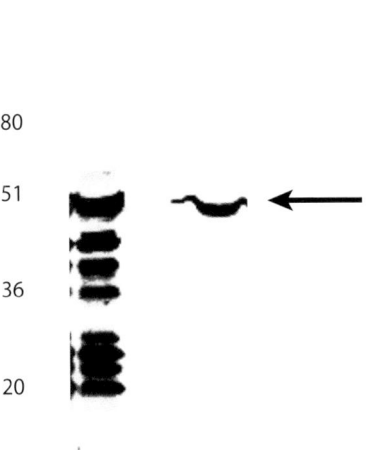

Figure 2. Expression of a 50 kDa human protein (arrow) in a standard *S. cerevisiae* strain (*lane 1*) and in a protease (PrA)-deficient strain (*lane 2*).
Detection of the recombinant human protein was carried out in cleared cell lysates using penta-His antibody.

Wild-type *S. cerevisiae* cells contain a group of proteases localized in the vacuole; these include the aspartic endoproteinase proteinase A (PrA) and a serine protease (protease B or PrB). Further proteases are located in other cellular compartments; however, the vacuolar enzymes are responsible for most of the degradation of endogenous yeast and heterologous proteins (38). These proteases are set free following disruption of the yeast cells during protein expression studies and during purification. The solution to most proteolysis problems arising from protease-containing lysates is to use *S. cerevisiae* strains that are deficient in the major vacuolar proteases (39). Disrupting the *PEP4* gene encoding proteinase A has proved to be sufficient to stabilize the production of many heterologous proteins (40) and significantly reduce the protein degradation of heterologous protein (41).

2.3 Analysis of expression

After having established and confirmed transformed colonies by PCR (see *Protocol 1*), expression of the transformed gene has to be induced and assayed. This is usually done by immunoblotting. Alternatively, enzymatic or functional assays can be done when an appropriate assay is available (which tends to be the exception rather than the rule). Immunoblotting requires an antibody directed against the protein itself – if available – or against a fused 'tag'. This latter approach has been the most common strategy in recent years.

Tags are available as short epitope tags or as fusion proteins. It is widely accepted that using tags may interfere with either the expression yield (positively or negatively) or the activity and correct folding of the expressed protein. They are nevertheless crucial for easy and efficient expression studies. Tags are fused to the N or C terminus, or can be placed at both ends to verify and select for full-length gene products. Tags may also be chosen to work both for detection and for purification by affinity chromatography (6, 42).

Protocols 2 and *3* detail the conditions applied for expressing proteins under the control of the inducible CUP1 promoter and the detection of intracellular proteins in *S. cerevisiae* expressing His-tagged recombinant protein. This protocol describes a method based on an expression vector where the transcription of the cDNA is controlled by the Cu^{2+}-induced CUP1 promoter from the yeast metallothionein gene (33). This strong promoter is induced rapidly by copper sulfate (0.01–1 mM, depending on the copper resistance of the host strain). The intracellular proteins are released from the cells by disrupting the cell walls by mechanical procedures. The proteins are separated by 12.5% sodium dodecyl sulfate polyacrylamide gel electrophoresis (SDS-PAGE) according to Laemmli. Following electrophoresis, proteins in polyacrylamide gel are transferred to polyvinylidene difluoride (PVDF) membranes by semidry electroblotting as described below and characterized by immunodetection.

An efficient procedure to produce and isolate secreted His-tagged proteins from the growth medium has been described (43).

2.4 Recommended protocols

Protocol 1

Yeast transformation and verification of transformants by colony PCR

Equipment and Reagents
- YPD agar (10 g yeast extract, 20 g peptone, 20 g glucose, 15 g agar, dissolved in distilled water to a final volume of 1 l)
- YPD medium (10 g yeast extract, 20 g peptone, 20 g glucose, dissolved in distilled water to a final volume of 1 l)
- 0.1 M Lithium acetate/1 M sorbitol
- PEG 3350 (60%, w/v)/0.1 M lithium acetate
- Single-stranded herring-sperm DNA (10 mg/ml; Fluka/Sigma-Aldrich)
- Dimethyl sulfoxide (DMSO)
- YNB medium (2% glucose, 0.67% yeast nitrogen base (Difco), and amino acids/nucleotides as appropriate for the host strain used (standard concentration: 40 mg/l))
- Shake flasks (100 and 500 ml)
- Distilled water
- Components for PCR: 10× PCR buffer (Qiagen), 5 mM dNTP mix; *Taq* polymerase (2.5 units/µl, ProofStar DNA polymerase; Qiagen), primers (100 pmol/µl)
- Thermocycler
- 1% Agarose gel and equipment and buffers for electrophoresis

Method
1. Streak the host strain on YPD agar and incubate at 28°C for 2 days.
2. Inoculate 20 ml of YPD medium in a 100 ml shake flask with a single colony of *S. cerevisiae* host strain as a pre-culture and grow overnight with shaking (200 r.p.m.) at 28°C.
3. Inoculate 100 ml of YPD medium in a 500 ml shake flask with 3–5 ml from the pre-culture to adjust the culture to an initial OD_{600} of 0.3. Grow with shaking (200 r.p.m.) at 28°C for 4–6 h to reach an OD_{600} of 0.8–1 (exponential phase).
4. Harvest the cells by centrifugation (10 min, 2500 *g*, 4°C).
5. Wash the cell pellet in 20 ml of cold distilled water. Repeat the spin as in step 4, discard the supernatant, and resuspend the cells in 2 ml of 0.1 M lithium acetate/1 M sorbitol.
6. Incubate the cell suspension for 30 min with shaking at 28°C. The cells are now 'competent' for DNA uptake.
7. Take an aliquot of 40 µl of the competent cells (per transformation assay) and add 235 µl of PEG 3350 (60%, w/v)/0.1 M lithium acetate, 5 µl of herring sperm DNA (10 mg/ml), and 5 µl of the plasmid DNA (0.1–1 µg) to be transformed.
8. Mix by pipetting and incubate for 30 min at 28°C.
9. Add 30 µl of DMSO and perform heat shock for 7 min at 42°C.
10. Prepare dilutions (10^{-1} to 10^{-3}) and plate the cells (dilutions as well as undiluted suspension) on YNB medium.
11. Incubate for 4 days at 28°C to select transformants.

12. Pick single colonies (at least three per transformation assay) and restreak on YNB agar (plus appropriate supplements) to obtain pure transformant strains.
13. Select three independent colonies of each restreaked clone to check for the presence of the transformed DNA by colony PCR. The primers preferably used are chosen to amplify the expression cassette.
14. Grow each colony in 1 ml of YNB plus supplements and grow for 24 h at 28°C. Transfer 2 µl into 18 µl of a standard PCR mix. Streak the cells onto selective YNB plates to preserve a copy of the analyzed colony.
15. To perform PCR, add the following:
    - 2 µl of yeast culture
    - 0.6 µl of 100% DSMO
    - 2 µl of 10× PCR buffer
    - 1.6 µl of 5 mM dNTP mix
    - 0.3 µl of *Taq* polymerase (ProofStar DNA polymerase; Qiagen)
    - 0.4 µl of each primer (100 pmol/µl)
    - 12.7 µl of distilled water

    Run the following PCR program in a thermocycler:

Number of cycles	Program
1	95°C for 30 s and 50°C for 60 s
25	72°C for 90 s, 95°C for 30 s, 55°C for 60 s, and 72°C 90 s
1	72°C for 10 min

16. Analyze aliquots of the PCR products by 1% agarose gel electrophoresis.

Protocol 2

Protein expression

Equipment and Reagents
- WM VIII medium according to Lang & Looman (24)
- Erlenmeyer flasks (20 and 100 ml)
- 0.5 M $CuSO_4$
- 10× PBS (84 mM Na_2HPO_4, 16 mM KH_2PO_4, and 1.5 M NaCl)
- 1× PBS
- 1 M Na_2HPO_4
- 1 M NaH_2PO_4
- 3 M NaCl
- 100 mM Phenylmethylsulfonyl fluoride (PMSF)
- Lysis buffer (100 mM $NaPO_4$ pH 8.0, 300 mM NaCl, and 1% Triton X-100)
- Glass beads (0.5 mm diameter, sterile, acid-washed)
- 1 M Tris/HCl (pH 6.8) stock
- 4× SDS-PAGE sample buffer (0.2 M Tris/HCl, pH 6.8, 8% (w/v) SDS, 40% (v/v) glycerol, 5% (v/v) 2-mercaptoethanol, and 0.4% (w/v) bromophenol blue)

Method

1. Transfer a single colony of selected yeast transformant into 5 ml of protein production medium WM VIII (supplemented with amino acids and nucleotides as required by the host strain) in a 20 ml Erlenmeyer flask and incubate for 2 days at 28°C with shaking at 200 r.p.m.

2. Inoculate 30 ml of WM VIII medium in a 100 ml Erlenmeyer flask with 300 µl of the pre-culture and incubate for 16–24 h at 28°C with shaking until the OD_{600} reaches 7–10.

3. Induce protein expression of the culture by adding $CuSO_4$ to a final concentration of 0.5 mM and incubate the culture for another 4–24 h[a,b].

4. Harvest a defined volume of cells (7–10 ml, corresponding to 70 OD_{600}) by centrifugation (10 min, 2500 *g*, 4°C) and resuspend the cell pellet in 5 ml of cold 1× PBS.

5. Repeat the centrifugation and discard the supernatant. The washed cell pellets are either subjected to cell disruption directly or stored at −70°C (stable for at least 1 month).

6. Resuspend the (thawed) cells in 300 µl of ice-cold lysis buffer and add PMSF to a final concentration of 1 mM.

7. Add 1 vol. of sterile, acid-washed glass beads (0.5 mm diameter) and disrupt cells by seven to ten cycles of 1 min full-speed vortexing/1 min cooling on ice.

8. Remove beads and cellular debris by centrifugation (15 min, 10 000 *g*, 4°C) and transfer the supernatant to a fresh tube.

9. Take a 30 µl sample and add 10 µl of 4× SDS-PAGE sample buffer. Heat the sample at 95°C for 5 min and use for SDS-PAGE[c].

10. Use cleared lysate (step 8) for Ni-NTA purification (see *Protocol 4*).

Notes

[a]The optimum time of induction depends on the protein and has to be determined experimentally.
[b]When using a constitutive promoter such as ADH1, the same procedure applies except that no induction is necessary. Cells are harvested in the late exponential phase and disrupted as described above.
[c]The recombinant gene product cannot be discriminated directly in the total yeast proteins after electrophoresis and staining, as the expressed protein is usually present only at low concentration. This requires purification of the recombinant protein from the crude extract or detection of the target protein by immunoblotting.

Protocol 3

Protein analysis by Western blot and immunodetection with anti-His antibody (chemiluminescent method)

Equipment and Reagents
- Whatman 3MM paper
- PVDF membrane
- 100% Methanol (or ethanol)
- Distilled water
- Transfer buffer (25 mM Tris base, 192 mM glycine, and 10% methanol)
- PBST (8.4 mM Na_2HPO_4, 1.6 mM KH_2PO_4, 150 mM NaCl, and 0.1% (v/v) Tween 20)
- Blocking solution (2% (w/v) BSA in PBST) (prepare fresh or keep at $-20°C$)
- Primary antibody (penta-His antibody; Qiagen)
- Secondary antibody (alkaline phosphatase (AP) or horseradish peroxidase (HRP)-conjugated anti-mouse IgG)
- Secondary antibody dilution buffer (5% (w/v) nonfat dried milk in PBST)

Method
1. To transfer proteins in the polyacrylamide gel to PVDF membrane by semi-dry electroblotting, cut eight pieces of Whatman 3MM paper and a piece of membrane to the same size as the gel. To avoid contamination, always handle the filter paper and membrane with gloves.
2. Wet the PVDF membrane in 100% methanol (or ethanol) for 15 s and then transfer it to a container of distilled water for 2 min.
3. Equilibrate the membrane for at least 5 min in the transfer buffer.
4. Soak the filter paper in transfer buffer.
5. Immerse the gel in transfer buffer.
6. Place four sheets of filter paper in the center of the anode electrode plate (positive, usually red), avoiding air bubbles, and place the membrane on top of the filter paper.
7. Place the gel on top of the membrane and place four sheets of filter paper on top of the gel. Air bubbles can be removed by gently rolling a Pasteur pipette over each layer in the sandwich.
8. Place the cathode plate cover (negative, usually black) on the top of the assembled transfer stack.
9. Connect the anode and the cathode lead to their corresponding power supply outputs. Follow the manufacturer's instructions regarding current, voltage, and transfer times. The current density is determined by the size of the gel: 1 mA/cm^2 is recommended (1 h transfer).
10. After transferring, mark the orientation of the membrane and the position of bands of the pre-stained protein standard. You can stain the blot to assess the quality of transfer using Ponceau S red (reversible stain). Incubate the membrane in staining solution (0.5% Ponceau S, 1% acetic acid) with gentle agitation for 2 min. The blot will be destained in distilled water or during the following immunological detection procedure.
11. Incubate the membrane for at least 1 h in blocking solution at room temperature.
12. Incubate with primary antibody (penta-His antibody; Qiagen) diluted 1:1000–1:2000 in blocking solution at room temperature for 1 h[a].
13. Wash the membrane three times for 10 min each in PBST at room temperature.

14. Incubate with secondary antibody solution for 1 h at room temperature. Either AP- or HRP-conjugated anti-mouse IgG may be used. Dilute according to the manufacturer's recommendations. Use 5% nonfat dried milk in PBST for incubation with the secondary antibody when using the chemiluminescent detection method.
15. Wash the filter four times for 10 min each in PBST at room temperature[b].
16. Perform the detection reaction with AP or HRP chemiluminescence reagent and expose to X-ray film or detect by a luminescent imaging system according to the manufacturer's recommendations[c,d] (see *Fig. 3*).

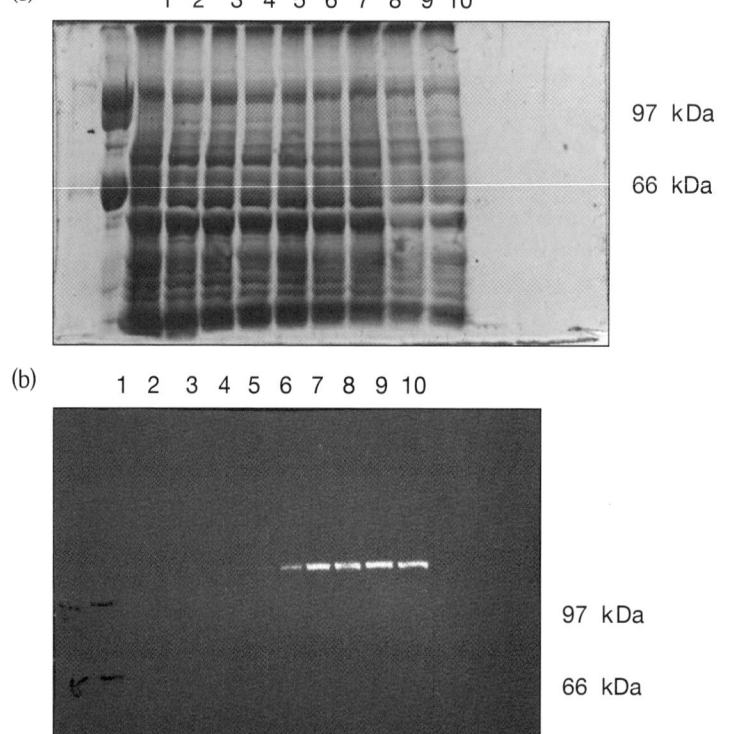

Figure 3. Typical protein gel and immunoblot of a standard yeast expression analysis. (*a*) Protein gel after Coomassie Brilliant Blue staining showing the yeast total soluble proteins. (*b*) Western blot of the same proteins highlighting a 130 kDa heterologous fungal protein. *Lanes 2–10* are cell lysates taken at different times of cultivation. *Lane 2* is the pre-culture, *lane 3* was taken at 14 h of the main culture, *lane 4* at 16 h, and *lane 5* at 20 h. *Lane 6* represents the culture at the time of induction of expression (using copper sulfate) and *lanes 7–10* were taken at 37, 40, 42 and 44 h of fermentation, respectively.

Notes

[a]Do not use dilution buffer containing milk powder for anti-His antibody. This will reduce sensitivity.
[b]Secondary anti-mouse IgG antibodies conjugated either with AP or HRP are available from many suppliers. We have obtained good results with rabbit anti-mouse IgG HRP conjugate from Dako. It is important that the secondary antibodies used recognize mouse IgG1 and that they are used at the highest dilution recommended to avoid nonspecific signals. Detection with anti-His–HRP and Ni-NTA conjugates available from Qiagen does not require the use of secondary antibodies.

cChromogenic or chemiluminescent substrates for detection are applicable. In contrast to chromogenic substrates, chemiluminescent substrates allow multiple exposures so that the results obtained can be optimized and are more sensitive. In our experience, the Western Lightning Chemiluminescence Reagent Plus (Perkin-Elmer) yields good results.

dIt is essential to use suitable negative and positive controls for the blotting and detection procedure. As a negative control, a cleared cell lysate from a yeast clone that carries the empty expression vector without insert is usually applied. As a positive control, a purified His_6-tagged protein, such as GFP (green fluorescence protein), that is easy to express in large amounts and easy to detect may be used.

Protocol 4

Affinity purification of His_6-tagged proteins via Ni-NTA agarose

Equipment and Reagents
- 50% Ni-NTA agarose (Qiagen)
- 1 M Imidazole
- 3 M NaCl
- Wash buffer (50 mM Na_2PO_4, pH 8.0, 300 mM NaCl, and 5 mM imidazole)
- Elution buffer (50 mM $NaPO_4$, pH 8.0, 300 mM NaCl, and 250 mM imidazole)

Method
1. Add 50 µl of a 50% slurry of Ni-NTA resin (100 µl resin has a capacity for 500 mg–1 mg of His_6-tagged protein) to 150 µl of cleared lysate containing whole-cell proteins and mix gently by shaking for 60 min at 4°C.

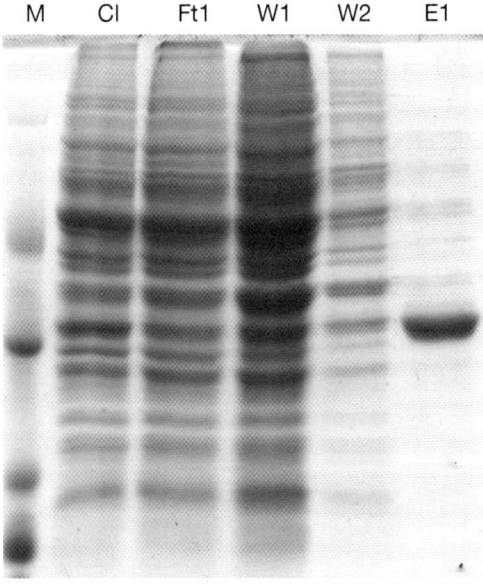

Figure 4. Ni-NTA purification of recombinant His-tagged protein (GFP) from yeast lysate. M, molecular mass marker; Cl, cleared lysate; Ft1, flow-through; W1 and W2, wash fractions; E1, eluate from the Ni-NTA resin.

2. Load the lysate/Ni-NTA mixture onto an unpacked column.
3. Collect the column flow-through. Save the flow-through for SDS–PAGE analysis.
4. Wash three times with 400 µl of wash buffer and collect the wash fractions. Save the wash fractions for SDS–PAGE analysis.
5. Elute the protein three times with 100 µl of elution buffer. Collect the elution fractions and analyze by SDS–PAGE[a] (see *Fig. 4*).

> **Note**
>
> [a]Use 15 µl of each fraction for the protein gel and subsequent immunodetection.

3. TROUBLESHOOTING

- Transformation works best with exponential-phase cells. If a strain is hard to transform, check the growth by following the OD_{600} during the 4–6 h incubation phase and select a time for harvest that is in the exponential-phase growth period. Whilst the method recommended here (using chemically competent cells) is suitable for most strains and gives transformation rates of several hundred transformants per µg of plasmid DNA, electroporation (44) or spheroplast transformation (45) are an alternative for specific mutant strains. When transforming PCR products or linear plasmid DNA for integrative transformation, transformation rates of less than 10 per µg are to be expected.
- Establishing growth and induction conditions for protein expression is best done using an easy-to-express control protein, such as GFP, as a positive control. This is of the utmost importance when initial expression tests with an unknown protein are negative. In many cases, cells are not harvested at the optimal time of expression, which may vary between 24 and 72 h or more depending on the promoter system and the strain used.
- It is advisable to check the growth behavior of the expression strains in comparison with mock transformants during the exponential growth phase by OD_{600} measurements. It is sometimes observed that expression strains show impaired growth. This is due to metabolic burden or toxicity caused by the recombinant protein. If the impaired growth is paralleled by low-level expression, changing the promoter/induction system to a tightly regulated promoter and subsequent separation of the growth and production phases is required.
- If recombinant proteins are required in milligram amounts for functional or structural analyses, the yeast *S. cerevisiae* is a favored expression system. Easy-to-implement fermentation schemes have been described (6), and even if fermentation units are not available, high yields of cells can be achieved in 2–5 l flask batch cultivation (43).
- Isolation of yeast-derived His-tagged proteins via Ni-NTA in some cases requires adaptation of the buffer conditions. In our hands, reducing imidazole concentrations in buffers has yielded good results when the target protein does not bind to the column. Be aware that yeast harbors a number of

endogenous proteins with His patches. These come up as contaminants in His-tag affinity chromatography. If the required purity is not sufficient after the one-step purification, consider incorporating a second purification step, such as gel filtration or ion-exchange chromatography.

4. REFERENCES

1. Romanos MA, Scorer CA & Clare JJ (1992) *Yeast*, **8**, 423–488.
2. Mentesana PE, Dosil M & Konopka JB (2002) *Methods Enzymol.* **344**, 92–111.
3. Pajot-Augy E, Crowe M, Levasseur G, Salesse R & Connerton I (2003) *J. Recept. Signal Transduct. Res.* **23**, 155–171.
★★★ 4. Mendoza-Vega O, Sabatie J & Brown SW (1994) *FEMS Microbiol. Rev.* **15**, 369–410. – *This paper details production processes and media for process scale-up.*
5. Calado CR, Almeida C, Cabral JM & Fonseca LP (2003) *J. Biosci. Bioeng.* **96**, 141–148.
6. Prinz B, Schultchen J, Rydzewski R, *et al.* (2004) *J. Struct. Funct. Genomics*, **5**, 29–44.
7. Wedekind A, O'Malley MA, Niebauer RT & Robinson AS (2006) *Biotechnol. Prog.* **22**, 1249–1255.
8. Bonander N, Hedfalk K, Larsson C, *et al.* (2005) *Protein Sci.* **14**, 1729–1740.
★★★ 9. Fincham JRS (1989) *Microbiol. Rev.* **53**, 148–170. – *This paper presents a comprehensive review of yeast transformation genetics.*
10. Sleep D, Finnis C, Turner A & Evans L (2001) *Yeast*, **18**, 403–421.
11. Sargent PJ, Farnaud S, Cammack R, Zoller HM & Evans RW (2006) *Biometals*, **19**, 513–519.
12. Manivasakam P, Weber SC, McElver J & Schiestl RH (1995) *Nucleic Acids Res.* **23**, 2799–2800.
13. Gray M & Honigberg SM (2001) *Nucleic Acids Res.* **29**, 5156–5162.
14. Lopes TS, Klootwijk J, Veenstra AE, *et al.* (1989) *Gene*, **79**, 199–206.
15. Lee FW & Da Silva NA (1997) *Appl. Microbiol. Biotechnol.* **48**, 339–345.
16. Guerra OG, Rubio IG, da Silva Filho CG, Bertoni RA, Dos Santos Govea RC & Vicente EJ (2006) *J. Microbiol. Methods*, **67**, 437–445.
17. Jansen G, Wu C, Schade B, Thomas DY & Whiteway M (2005) *Gene*, **344**, 43–51.
18. Van Mullem V, Wery M, De Bolle X & Vandenhaute J (2003) *Yeast*, **20**, 739–746.
19. Ugolini S, Tosato V & Bruschi CV (2002) *Plasmid*, **47**, 94–107.
20. Erhart E & Hollenberg CP (1983) *J. Bacteriol.* **156**, 625–635.
21. Okkels JS (1996) *Ann. N.Y. Acad. Sci.* **782**, 202–207.
★ 22. Guldener U, Heck S, Fielder T, Beinhauer J & Hegemann JH (1996) *Nucleic Acids Res.* **24**, 2519–2524. – *This paper describes a widely used and efficient technique of integrating genes into the chromosome with subsequent recovery of the selection marker.*
23. Gueldener U, Heinisch J, Koehler GJ, Voss D & Hegemann JH (2002) *Nucleic Acids Res.* **30**, e23.
24. Wach A, Brachat A, Pohlmann R & Philippsen P (1994) *Yeast*, **10**, 1793–1808.
25. Lang C & Looman AC (1995) *Appl. Microbiol. Biotechnol.* **44**, 147–156.
26. Ruohonen L, Aalto MK & Keranen S (1995) *J. Biotechnol.* **39**, 193–203.
27. Loison G, Findeli A, Bernard S, *et al.* (1988) *Bio/Technology*, **6**, 72–77.
28. Sil AK, Xin P & Hopper JE (2000) *Protein Expr. Purif.* **18**, 202–212.
29. Stagoj MN, Comino A & Komel R (2006) *Biomol. Eng.* **23**, 195–199.
30. Gao CY & Pinkham JL (2000) *Biotechniques*, **29**, 1226–1231.
31. Hottiger T, Kuhla J, Pohlig G, *et al.* (1995) *Yeast*, **11**, 1–14.
32. Labbe S & Thiele DJ (1999) *Methods Enzymol.* **306**, 145–153.
★ 33. Holz C, Hesse O, Bolotina N, Stahl U & Lang C (2002) *Protein Expr. Purif.* **25**, 372–378. – *Original paper describing the method for using yeast in high-throughput expression studies.*
★★ 34. Hitzeman RA, Chen CY, Dowbenko DJ, *et al.* (1990) *Methods Enzymol.* **185**, 421–440. – *A concise review of secretion techniques and signals to be used in* S. cerevisiae.

35. Nuyens F, van Zyl WH, Iserentant D, Verachtert H & Michiels C (2001) *Appl. Microbiol. Biotechnol.* **56**, 431–434.
36. Kjeldsen T (2000) *Appl. Microbiol. Biotechnol.* **54**, 277–286.
37. Parolin C, Corso AD, Alberghina L, Porro D & Branduardi P (2003) *J. Biotechnol.* **120**, 46–58.

★★ 38. Jones EW (1991) *Methods Enzymol.* **194**, 428–453. – *A basic review of yeast proteases.*
39. Van Den Hazel HB, Kielland-Brandt MC & Winther JR (1996) *Yeast* **12**, 1–16.

★★ 40. Harashima S (1994) *Bioprocess Technol.* **19**, 137–158. – *A review of heterologous protein expression in* S. cerevisiae.
41. Holz C, Prinz B, Bolotina N, *et al.* (2003) *J. Struct. Funct. Genomics* **4**, 97–108.
42. Lichty JJ, Malecki JL, Agnew HD, Michelson-Horowitz DJ & Tan S (2005) *Protein Expr. Purif.* **41**, 98–105.
43. Ngamkitidechakul C & Twining SS (2002) *Biotechniques*, **33**, 1296–1300.

★ 44. Manivasakam P & Schiestl RH (1993) *Nucleic Acids Res.* **21**, 4414–4415. – *This paper describes the procedure for transformation of yeast by electroporation.*
45. Burgers PM & Percival KJ (1987) *Anal. Biochem.* **163**, 391–397.

CHAPTER 8
Expression of proteins in *Pichia pastoris*

Geoff P. Lin-Cereghino, Wilson Leung, and Joan Lin-Cereghino

1. INTRODUCTION

Since 1984, the yeast *Pichia pastoris* has been used successfully for the production of over 550 heterologous proteins (see http://faculty.kgi.edu/cregg[8.1]). Once considered an excellent source of single-cell protein, this nontoxic eukaryote was originally developed as a high-protein animal feed but was later shifted to recombinant protein expression. The yield of a given recombinant protein has ranged from a few micrograms to over 10 g/l of *P. pastoris* culture. The increasing popularity of this 'programmable protein machine' is attributed to: (i) the relative ease with which it can be genetically manipulated; (ii) its ability to produce foreign proteins either intracellularly or extracellularly at high levels; (iii) its capability of performing many eukaryotic post-translational modifications, such as proteolytic processing, folding, disulfide bond formation, and glycosylation; (iv) its ability to be scaled up for large-scale shake flask culturing and to grow to extremely high concentrations (>130 g/l dry cell weight) in fermenter conditions; and (v) its availability as a kit from Invitrogen. This chapter describes how to select an expression vector as well as a host strain and then goes over the basic steps of how to synthesize and purify a protein from *P. pastoris*. There is no universally accepted strategy for protein production in this yeast as every protein requires a slightly different protocol, but this chapter is meant to familiarize new users with this expression system so that they can gain a basic understanding of the many options associated with this yeast. More detailed information on the *P. pastoris* expression system may be obtained in one of the many reviews on the system (1-3) and from Invitrogen (http://www.invitrogen.com[8.2]), which offers a wide variety of plasmids, strains, and technical support for using this 'protein machine'. More detailed instructions on some of the procedures discussed in this chapter can be found at http://www.invitrogen.com/content/sfs/manuals/easyselect man.pdf[8.3].

Expression Systems: *Methods Express* (M.R. Dyson and Y. Durocher, eds)
© Scion Publishing Limited, 2007

1.1 Background

P. pastoris is considered to be methylotrophic as it can utilize methanol as its sole carbon source. The conceptual basis for the expression system originated with the observation that many enzymes required for methanol metabolism are present at high levels only when cells are grown on this alcohol (4). The most extreme example of this regulation is displayed by alcohol oxidase (AOX), the first enzyme in the methanol utilization pathway (5). AOX activity is undetectable in cells grown on glucose, glycerol, or ethanol, but is dramatically induced when cells are shifted to methanol (6, 7). In methanol shake-flask cultures, this level is typically ~5% of the total soluble protein, but can rise to greater than 30% in cells fed methanol at growth-limiting rates in fermenters (8). Although two genes, *AOX1* and *AOX2*, encode this protein, *AOX1* is responsible for the majority of the gene product. Approximately 5% of the total poly(A)$^+$ mRNA is from *AOX1* in methanol-grown cells. Based on this observation, it was predicted that the *AOX1* promoter would be an excellent choice for the regulated expression of foreign genes. Fortunately, most techniques needed for the molecular genetic manipulation of *P. pastoris* turned out to be similar to those developed for the baker's yeast, *Saccharomyces cerevisiae*. Therefore, in a fairly short period of time, *P. pastoris* evolved from an obscure methylotroph to a popular host for recombinant protein expression.

1.2 Choosing a plasmid

Expression of any foreign gene in *P. pastoris* requires four basic steps: (i) insertion of the gene into an expression cassette; (ii) introduction of the cassette into the genome; (iii) examination of potential expression strains for the foreign product; and (iv) purification of the heterologous protein in its active form from a superior strain. Therefore, the first task is to select an expression plasmid from the great number of expression vectors available. All expression vectors are *Escherichia coli/P. pastoris* shuttle vectors, containing an origin of replication for plasmid maintenance in *E. coli* and selectable markers that are functional in one or both organisms. Most expression vectors have an expression cassette consisting of a multiple cloning site for insertion of a foreign coding sequence, flanked by a well-characterized promoter, such as that from *AOX1* or *GAP* (glyceraldehyde 3-phosphate dehydrogenase), and termination sequences derived from *AOX1*. The cassette is designed so that the resulting transcription product of cap-5'UTR-ORF-3'UTR-poly(A) is a mature mRNA structure that is familiar to the yeast's cell machinery and does not contain cryptic sequences that may decrease message stability or translational efficiency (9). A list of commonly used expression vectors is shown in *Table 1*. In terms of plasmid features, there are several options to tailor the desired protein's expression to the experimenter's needs, which include:

1. *Intracellular versus extracellular expression.* The recombinant protein can be engineered to be expressed either intracellularly or extracellularly. Proteins are generally expressed in *P. pastoris* in the way that they are produced in the native host. For instance, secreted expression of normally intracellular proteins often

Table 1. Common *P. pastoris* expression vectors

Vector name	Selectable marker	Features	Reference
Vectors for intracellular expression			
pPICZ	Zeo^R	Multiple cloning site for insertion of foreign genes; potential for fusion of foreign protein to His$_6$ and c-myc epitope tags; Zeocin selection for multi-copy strains	Invitrogen
pPIC6	Bsd^R	Similar to pPICZ except that blasticidin resistance used for direct selection of multi-copy strains	Invitrogen
pGAPZ	Zeo^R	Expression controlled by constitutive *GAP* promoter: multiple cloning site for insertion of foreign genes; Zeocin selection for multi-copy strains; potential for fusion of foreign protein to His$_6$ and c-myc epitope tags	Invitrogen,
pFLD	Zeo^R	Similar to pGAPZ except that expression controlled by *FLD1* promoter for inducible expression with methylamine	Invitrogen
pJL-IX	FLD1	*AOX1* promoter with *FLD1* as selectable marker	10
pBLHIS-IX pBLARG-IX pBLADE-IX pBLURA-IX	HIS4 ARG4 ADE1 URA3	*AOX1p* vector series each with one of four biosynthetic gene markers: *HIS4, ARG4, ADE1,* or *URA3*	11
Vectors for secreted expression			
pPICZα	Zeo^R	*AOXI* promoter fused to α-MF prepro signal sequence; multiple cloning site for insertion of foreign genes; potential for fusion of foreign protein to His$_6$ and c-myc epitope tags; Zeocin selection for multi-copy strains	Invitrogen
pPIC6α	Bsd^R	Similar to pPICZα except that blasticidin resistance used for direct selection of multi-copy strains	Invitrogen
pGAPZα	Zeo^R	*GAP* promoter fused to α-MF prepro signal sequence; multiple cloning site for insertion of foreign genes; potential for fusion of foreign protein to His$_6$ and c-myc epitope tags; Zeocin selection for multi-copy strains	Invitrogen, 12
pFLDα	Zeo^R	Similar to pGAPZα except that expression controlled by *FLD1* promoter for inducible expression with methylamine	Invitrogen, 13
pJL1-SX	FLD1	*AOXI* promoter fused to α-MF prepro signal sequence; *FLD1* as selectable marker	10
pBLHIS-SX pBLARG-SX pBLADE-SX pBLURA-SX	HIS4 ARG4 ADE1 URA3	Series of vectors with *AOX1* promoter fused to α-MF prepro signal, each with one of four biosynthetic gene markers: *HIS4, ARG4, ADE1,* or *URA3*	11

leads to misfolding and degradation (2). For intracellular expression, vectors usually lack an initiation codon. It is strongly recommended that the ATG of the heterologous coding sequence be inserted into the first (most 5') restriction site in a multiple cloning site. The *AOX1* 5'UTR encoded by these plasmids is about 115 nt, which is uncharacteristically long for a yeast message. Further lengthening of this mRNA region has been shown to reduce translation efficiency (2). For secretion of engineered proteins, vectors include sequences designed to make in-frame fusions between foreign proteins and various secretion signals, such as the *S. cerevisiae* α-MF prepro-peptide or the *P. pastoris* acid phosphatase secretion signal (*PHO1*) (14). As *P. pastoris* secretes low levels of endogenous proteins, the secreted recombinant peptide often constitutes a majority of the total protein in the medium (7). Therefore, directing a heterologous protein to the culture medium serves as a substantial first step in purification. As the *S. cerevisiae* α-MF-factor prepro-peptide has proved to be the most reliable secretion signal, it is found in most commercially available expression vectors. There have been reports of some successes with other secretion signals; however, these have not been tested with a wide range of recombinant proteins (15–17).

2. *Type of promoter.* The *AOX1* promoter is tightly repressed in glucose medium, but can be induced approximately 1000-fold in methanol medium, making it one of the strongest promoters known (8). The strength of the promoter in methanol makes it ideal for high-level protein production, and the ability to maintain cultures in an 'expression-off' mode in glucose medium is advantageous if the heterologous protein is toxic. Most expression vectors rely on the *AOXI* promoter to drive expression, but there are times where this promoter may not be ideal (18). For example, methanol may not be appropriate for the food industry as the petroleum-related compounds used are health and fire hazards. Alternative promoters include the strong constitutive *P. pastoris GAP* gene promoter, which can be used for expression in glucose and other carbon sources (12). Another option is the glutathione-dependent formaldehyde dehydrogenase (*FLD1*) promoter, which can be induced with either methanol as a sole carbon source (and ammonium sulfate as a nitrogen source) or methylamine as a sole nitrogen source (and glucose as a carbon source) (13). The moderately expressing promoters from the *P. pastoris* genes *PEX8* and *YPT1* have been used in instances where overexpression is problematic; however, their use has been limited (14, 19).

3. *Type of selectable marker.* Two types of marker are utilized in *P. pastoris* expression vectors for the selection of transformants. The first are markers that complement a mutation in a biosynthetic gene of an auxotrophic strain. A total of five biosynthetic markers are readily available in expression vectors for such a purpose, including *ADEI* (PR-amidoimidazolesuccinocarboxamide synthase), *HIS4* (histidinol dehydrogenase), *ARG4* (arginosuccinate lyase), *URA3* (orotidine 5'-phosphate decarboxylase), and *URA5* (orotate phosphoribosyltransferase) from *P. pastoris* (11). In addition, a novel set of expression vectors using the *P. pastoris FLD1* gene as selectable marker has allowed researchers to make expression vectors composed almost entirely of *P. pastoris* DNA (except for the heterologous gene to be expressed) (10). Although these plasmids are constrained for use with *fld1* strains, they are helpful in iso-

lating multi-copy transformants. The majority of expression vectors available, generated mainly by Invitrogen, rely on a dominant resistance marker for selection in both bacteria and yeast. These vectors take advantage of either the *Sh ble* gene from *Streptoalloteichus hindustanus*, which confers resistance to the bleomycin-related drug Zeocin (20) or the blasticidin S deaminase gene (*bsd*) from *Aspergillus terreus*, which confers resistance to the nucleoside antibiotic blasticidin S HCl isolated from *Streptomyces griseochromogenes* (21). These markers have the advantage that they can be used with any strain and can be utilized to isolate transformants with multiple copies of the expression vector. At the present time, vectors containing the Zeocin resistance gene, such as the pPICZ or pPICZα series are much more popular than the Bsd^R vectors. In fact, Zeo^R vectors are the vector of choice for most recombinant expression strategies.

4. *Addition of a peptide tag.* It may be advantageous to use a vector that offers an in-frame fusion to a tag peptide that can be used for immunological detection and/or purification. For instance, certain expression vectors such as the pPICZ and pGAPZ series (Invitrogen) are designed to generate heterologous proteins with C-terminal *c-myc* and His$_6$ tags for visualization and isolation with commercially available antibodies (*c-myc* and His$_6$ tag) or resins (His$_6$ tag). These are particularly useful when an antibody to the heterologous peptide is not available. Usually the tag does not interfere with folding or function of the engineered protein (22). However, if the tag needs to be removed for downstream uses, a protease cleavage site, such as factor Xa or enterokinase, can be engineered between the tag and the recombinant protein.

1.3 Choosing a host strain

Once the expression plasmid has been designed, the strain for expression must be selected. All *P. pastoris* strains are derived from the wild-type strain Y-11430, also known as JC100, from NRRL (Northern Regional Research Laboratories, Peoria, IL) (4). Strain X-33, also wild type, is GS115 (*his4*) transformed with the *P. pastoris* *HIS4* gene. Both JC100 and X-33 strains require no supplementation for growth on minimal media. Either strain is usually the first choice when initially attempting to express a heterologous protein if the expression vector contains a dominant selectable marker. However, depending on the specific traits of the recombinant protein and its expression vector, other strains of the yeast may be better choices. A list of *P. pastoris* strains available for expression is shown in *Table 2*. Some of the most common options include:

1. *Strains for use with biosynthetic markers.* Many expression strains are derived from the *his4* auxotrophic strain GS115 to allow transformation with *HIS4*-based vectors. Several other strains are available that have different combinations of one or more auxotrophic mutations in the *ARG4*, *ADE1*, *HIS4*, *URA3*, and *URA5* biosynthetic genes of *P. pastoris* (11). These strains allow the selection of expression vectors containing the corresponding selectable marker gene, as well as use with dominant markers. The creation of a strain with

Table 2. Commonly used P. pastoris strains

Strain	Genotype	Reference
Wild type strains		
Y-11430	Wild type	Northern Regional Research Laboratories, Peoria, IL, USA
X-33	Wild type (GS115::*HIS4*)	Invitrogen
Auxotrophic strains		
GS115	*his4*	4
GS190	*arg4*	23
JC220	*ade1*	4
JC254	*ura3*	23
GS200	*arg4 his4*	4
JC300	*ade1 arg4 his4*	11
JC301	*ade1 his4 ura3*	11
JC302	*ade1 arg4 ura3*	11
JC303	*arg4 his4 ura3*	11
JC308	*ade1 arg4 his4 ura3*	11
YJN165	*ura5*	24
Methanol utilization strains		
KM71	*his4 arg4 aox1Δ::ScARG4*	7
MC100-3	*his4 arg4 aox1Δ::ScARG4 aox2Δ::Pphis4*	6
Protease-deficient strains		
SMD1168	*his4 Δpep4::URA3 ura3*	25
SMD1165	*prb1 his4*	25
SMD1163	*pep4 prb1 his4*	25
SMD1168 *kex1::SUC2*	*Δpep4::URA3 Δ kex1::SUC2 his4 ura3*	26
Other strains		
GS241	*fld1*	13
MS105	*his4 fld1*	13

different expression cassettes requires sequential transformations; however, the availability of multiple markers allows several engineered peptides to be produced simultaneously in the same cell, which may be required for multi-subunit proteins (27).

2. *Methanol utilization mutants.* Most auxotrophic P. pastoris strains grow on methanol at the same rate as wild-type strains. However, because strains with *AOX* mutations are sometimes capable of producing higher levels of heterologous proteins (28), strains with deletions in one or both *AOX* genes, such as KM71 (*his4 arg4 aox1Δ::ScARG4*) and MC100-3 (*his4 arg4 aox1Δ::ScARG4 aox2Δ::Pphis4*) are available. Although compromised in their ability to metabolize methanol, such strains retain the ability to induce high-level expression from the *AOX1* promoter with reduced amounts of the alcohol present (28).

3. *Protease-deficient strains.* Often, the recombinant protein produced by P. pastoris is found to be partially degraded. Strains deficient in vacuolar proteases are sometimes effective at reducing proteolysis of recombinant proteins (18). The protease-deficient strains SMD1163 (*his4 pep4 prb1*), SMD1165 (*his4*

prb1), and SMD1168 (*his4 pep4*) are particularly useful, as the combination of high cell density and lysis of a small percentage of cells results in a relatively high concentration of vacuolar proteases such as proteinase A (*pep4*) and proteinase B (*prb1*) in the culturing medium. A derivative of SMD1168 in which the gene encoding the carboxypeptidase Kex1, which is involved in C-terminal proteolysis of lysines and arginines, has been disrupted with *S. cerevisiae SUC2* (SMD1168 *kex1::SUC2*) is also available (26). It should be noted, however, that protease-deficient strains typically exhibit slower growth rates and lower transformation efficiencies, whilst requiring greater patience.

4. *Glycosylation mutants.* In comparison with *S. cerevisiae*, *P. pastoris* recombinant proteins tend to be hyperglycosylated less frequently and do not have hyperimmunogenic terminal α-1,3-linked mannosylation (29). However, differences in protein-linked carbohydrate synthesis between *P. pastoris* and humans may affect biological activity and have made the yeast-synthesized proteins inappropriate for human pharmaceutical use because of increased antigenicity (30). Thus, two approaches to engineer human-type *N*-glycans on yeast recombinant proteins have been taken. First, in the GlycoSwitch system available from Research Corporation Technologies (http://www.rctech.com[8.4]), strains have been created to produce uniform, small asparagine-linked glycans on any glycoprotein (31). Any GS115 strain of interest can be converted to a GlycoSwitch strain yielding predominantly $Man_8GlcNAc_2$, $Man_5GlcNAc_2$, or $GlcNAcMan_5GlcNAc_2$ oligosaccharides, all of which have the same structure as processing intermediates of the mammalian *N*-glycosylation pathway. Secondly, in another effort to 'humanize' *P. pastoris*, strains were constructed by eliminating the endogenous yeast glycosylation pathway and by using five eukaryotic proteins (mannosidases I and II, *N*-acetylglucosaminyl transferases I and II, and uridine 5′-diphosphate *N*-acetyl glucosamine transporter) to establish a synthetic glycosylation pathway that generates the human oligosaccharide $GlcNAc_2Man_3GlcNAc_2$ (32, 33). Further advances by GlycoFi Corporation (http://www.glycofi.com[8.5]) include the development of engineered strains blocked in dolichol oligosaccharide assembly. By expressing a given protein in different glycoengineered yeast strains that they have created, GlycoFi can generate a library of glycoproteins, all with an identical peptide backbone, but with different sugar structures attached to them.

5. *Protein disulfide isomerase (PDI) overexpression strains.* The PDI enzyme forms and arranges disulfide bonds during protein folding in the endoplasmic reticulum. Overexpression of the native PDI protein in *P. pastoris* has been found to increase production of disulfide-rich proteins such as Pfs25H, a *Plasmodium falciparum* transmission blocking vaccine candidate (34). Thus, this strain may be an option for secreted proteins that require extensive disulfide formation for efficient folding and export out of the cell.

6. *Strains for use with FLD expression vectors.* Strain MS105 is a *his4* auxotroph with a mutation in the *P. pastoris fld* gene, rendering the strain defective in its ability to grow on methanol as a carbon source or methylamine as a nitrogen source (13). Additionally, *P. pastoris fld1* mutants have increased sensitivity to formaldehyde relative to wild-type cells. Vectors containing the *FLD1*

marker can then be used to select multi-copy expression strains by screening for His⁺-transformed strains of MS015 with high levels of resistance to formaldehyde (10).

2. METHODS AND APPROACHES

2.1 Transformation

Once the plasmid has been constructed and the host strain has been selected, the expression vector must be introduced into the *P. pastoris* genome. Homologous recombination occurs in the methylotroph, allowing the insertion of a desired DNA into a specific site in a yeast chromosome (8). Such strains have been shown to be genetically stable after selection has been removed. There are several methods for *P. pastoris* transformation, including those involving lithium chloride, polyethylene glycol, and electroporation. Although it requires the acquisition of costly equipment, electroporation is usually the method of choice because it provides the highest efficiency, entails relatively few steps, and is the most reliable way to create multi-copy transformants (35). *Protocol 1* describes a method of preparing cells for electroporation with a transformation efficiency of 10^4–10^5/µg for most strain types.

Protocol 1

Preparation of electroporation-competent *P. pastoris* cells

Equipment and Reagents
- 50 ml Sterile baffled flask
- 2.8 l Sterile baffled flask
- Platform shaker/incubator
- Refrigerated floor model centrifuge
- YPD (1% yeast extract, 2% peptone, 2% dextrose)
- 1 M Dithiothreitol (DTT)
- 1 M HEPES (pH 8)
- Sterile 500 ml and 50 ml centrifugation bottles
- Sterilized ice-cold Milli-Q water (1.5 l)
- Sterile plastic pipettes
- Ice-cold sterile sorbitol (1 M)
- Sterile 1.5 ml microfuge tubes
- Spectrophotometer

Method
1. Set up a 10 ml overnight culture of the desired strain in YPD in a 50 ml sterile flask. Grow in a shaking incubator at 30°C overnight.

2. Set up 500 ml of YPD in a large 2.8 l baffled flask (aeration is critical – the YPD must occupy less than 1/5 of the volume of the flask).

3. The next day, inoculate the 500 ml of YPD with enough overnight inoculate to get an OD_{600} of 0.15–0.2.
4. Grow the cells in the shaking incubator at 30°C. Incubate for 6–8 h until the OD_{600} is about 0.8–1.5 (most strains double every 1.5 h in YPD after a lag period).
5. Pour the cells into sterile centrifugation bottles. Pellet the cells by centrifugation at 5000 r.p.m. for 5 min at 4° C.
6. Pour off the supernatant (YPD) and resuspend the pellet in 100 ml of fresh YPD with 2 ml of 1 M sterile HEPES (pH 8.0). Resuspend the cells gently; pipetting is allowed but do not vortex as it will reduce viability.
7. Add 2.5 ml of sterile 1 M DTT to the yeast while stirring. Incubate in a shaker for 15 min at 30°C with mild agitation (100 r.p.m.). DTT will reduce cysteine double bonds in the cell walls and make the cells more permeable. Do not use DTT that has been frozen and thawed several times.
8. Add 400 ml of ice-cold sterile Milli-Q water to the yeast and centrifuge for 5 min at 5000 r.p.m. at 4°C.
9. Pour off the supernatant (water and YPD). Decant on a sterile tissue paper to get rid of the last drop.
10. Resuspend the pellet gently in 500 ml of ice-cold sterile water as in step 6.
11. Centrifuge for 10 min at 5000 r.p.m. at 4°C and pour off the supernatant.
12. Resuspend the pellet in 250 ml of ice-cold sterile water.
13. Centrifuge as in step 11.
14. Pour off the supernatant (decant every last drop) and resuspend the pellet in a total volume of 20 ml of sterile ice-cold 1 M sorbitol. Transfer to 50 ml sterile centrifugation tubes.
15. Centrifuge as in step 11.
16. Resuspend in 1.0 ml of ice-cold 1 M sterile sorbitol.
17. Aliquot the cells into small volumes (45, 90, and 135 μl) in sterile pre-chilled 1.5 ml microfuge tubes on ice.
18. Freeze at –80°C.
19. Thaw a volume of the competent cells on ice as required for electroporation. The cells must be used once they are thawed. They cannot be refrozen.

Plasmid introduction into the *P. pastoris* genome can occur via gene insertion, where one to several copies of the entire plasmid integrates into a specific site, or via gene replacement, where a portion of the expression plasmid usually replaces a section of the yeast chromosome (36). Gene insertion is primarily used because it has higher efficiency and has a greater chance of leading to multi-copy strains. The simplest way is to digest the vector at a unique restriction site in either the biosynthetic marker (for auxotrophic strains) or the expression cassette promoter (for any strain) and then transform it into competent cells, which stimulates the vector to recombine at a homologous locus in the yeast genome via a single crossover event. The vector must be restricted in this region; transformation with nonlinearized plasmids results in a very low efficiency. The free DNA ends stimulate single crossover integration events at pre-determined sequences in the yeast genome. After purification, the DNA is introduced in the electroporation-competent yeast as detailed in *Protocol 2*, which describes how both single-copy and multi-copy strains can be attained. The plating of the transformation reaction on increasingly higher concentrations of Zeocin should allow one to isolate high-copy number strains, as resistance usually correlates with copy number (3, 37). However, for reasons that are not known, some isolates from plates with high Zeocin concentration are not multi-copy. On the other hand, the isolation of multi-copy strains is much more difficult using biosynthetic markers, as there is no easy way to screen for these strains. Thus, our focus will be on using a Zeocin resistance expression plasmid, but *HIS4*-based plasmids, the most popular biosynthetic marker, will also be considered.

Protocol 2

Plasmid preparation and electroporation

Equipment and Reagents
- Agarose gel containing 1× GelStar dye (Cambrex)
- Geneclean II kit (Qbiogene) or QIAquick PCR Purification kit (Qiagen)
- Spectrophotometer
- 2 mm Gap electroporation cuvettes (Harvard Apparatus)
- BTX ECM 630 Electroporator (Harvard Apparatus)
- YPD liquid (see *Protocol 1*)
- YND liquid (0.17% yeast nitrogen base, 0.5% ammonium sulfate, 0.4% glucose)
- Ice-cold sterile sorbitol (1 M)
- Sterile 1.5 ml microfuge tubes
- Microfuge
- YPD plates containing 100, 750, 1000, or 1500 µg/ml Zeocin
- Platform shaker/incubator

Method

1. Digest approximately 5 µg of the expression plasmid with an enzyme that linearizes the plasmid within the *P. pastoris* promoter or the prototrophic marker. The reaction volume should be approximately 50 µl and should contain between 10 and 20 units of restriction enzyme; incubate for at least 2 h to ensure complete digestion.

2. Run about 1/10 of the reaction volume on an agarose gel to confirm efficient linearization.

3. Purify the linearized DNA using a DNA purification kit (i.e. Geneclean II or QIAquick PCR Purification kit) to remove salts and other contaminants. Elute the DNA in as small a volume as possible to obtain a high concentration of linearized plasmid. The DNA concentration should be at least 100 ng/µl, which can be confirmed by measuring A_{260}[a].

4. Obtain electrocompent cells from the –80°C freezer (from *Protocol 1*) and allow them to thaw on ice. Label cuvettes and chill them on ice for at least 5 min.

5. Aliquot 40 µl of cells into the cold, labeled cuvette. Be sure to include a negative control (cells receiving no DNA). Once cells are thawed, they cannot be refrozen because of a huge loss in competence.

6. Add 4 µl of linearized plasmid DNA directly into the sample of cells in the cuvette. Flick the cells and DNA to the bottom of the cuvette and avoid bubbles. Place back on ice for 2–5 min[b].

7. Wipe the cuvette dry with a tissue or paper towel and place in the electroporation chamber. Make sure that it fits snugly.

8. Electroporate the cells by pressing the 'charge' button. The charging parameters for electroporation with the BTX electroporator are: a cuvette gap of 2.00 mm, a charging voltage of 1500 V, a capacitance of 50 µF, and a resistance of 200 Ω.

9. Remove the cuvette from the electroporation chamber and *immediately* add either (a) 1 ml of 1 M sorbitol and mix gently if using a prototrophic marker (i.e. *HIS4*) or (b) add 500 μl of 1 M sorbitol and 500 μl of YPD and mix gently if using a Zeocin resistance marker. Place the cuvette at room temperature and electroporate the other samples. If using a prototrophic marker, go directly to step 10. If using a Zeocin marker, transfer each cell suspension to a sterile 1.5 ml microfuge tube and let them recover in the 30°C shaker for at least 1 h.

10. Plate 100 μl of the suspension on selective medium (YND alone or YND plus the correct amino acids) or on YPD + Zeocin (after the recovery time). Centrifuge the rest of the suspension at 2000 r.p.m. for 60 s, resuspend in a smaller volume, and plate the concentrated cells on another plate(s) of selective medium. If multi-copy strains with a Zeo^R plasmid are needed, then spread 100–200 μl aliquots on selection plates containing increasing concentrations of Zeocin (100–1500 μg/ml).

11. Incubate the selection plates at 30°C. Colonies should start appearing within 2–3 days.

Notes

[a]The plasmid must be relatively free of salt to prevent arcing during electroporation, which will greatly decrease the efficiency.

[b]The volume of cells should always be more than 10× the amount of DNA used, i.e. it is acceptable to use 80 μl of cells as long as the DNA volume does not exceed 8 μl.

2.2 Screening of transformants

Once transformants have been obtained, the next step is to screen a subset of these colonies on a small scale to identify the best expressers. For *HIS4* strains, our laboratory picks approximately ten transformant colonies for screening. However, for Zeo^R plasmids, we usually pick approximately four to five colonies from each Zeocin resistance-level plate. Most of the time, the strains with highest copy number provide the highest expression of recombinant protein; however, there have been several instances where this is not true (38). One explanation for this phenomenon for secreted proteins is that translation above a certain level causes a traffic jam in the secretory pathway, preventing the recombinant peptides from exiting the cell efficiently. Thus, there is a limit to the rate of proteins entering the secretory complex; exceeding this limit causes decreases in secretion efficiency.

As *P. pastoris* cells are extremely sticky, transformants from selection plates must first be single streaked to obtain pure colonies. However, not all such purified colonies will contain the expression cassette because gene conversion events, in which *HIS4* sequences or other marker sequences from the vector integrate into the genome without other vector sequences, occur at a substantial (10–50%) frequency (39). Thus, the presence of a stably integrated expression cassette in transformants should be confirmed prior to small-scale expression assays. One could utilize Southern blotting or real-time PCR, which not only confirm the presence of the expression cassette but also indicate gene copy number. However, these techniques, which require purification of total genomic DNA, are rather time-consuming and not worthwhile at such an initial stage, as the strain with the highest copy number does not always give the highest expression of the heterologous protein. Instead, our laboratory performs a colony PCR analysis (see Protocol 3), which can be completed in a day, to ensure that each colony contains at least one copy of the recombinant gene.

Protocol 3

Colony PCR

Equipment and Reagents
- Plate containing purified colonies
- Sterile yellow pipette tips
- Vortex
- 1.5 ml Microfuge tubes and PCR tubes
- MasterTaq PCR system (Eppendorf)
- Primers that anneal to the recombinant gene[a]
- Waterbath at 95°C
- Thermocycler
- Agarose gel containing 1× GelStar dye (Cambrex)
- Mineral oil

Method

1. Pick a reasonable amount of a fresh colony from a transformation plate using the end of a pipette tip. Colonies should be less than a few days old.

2. Resuspend the cells in a tube containing 500 µl of sterile water. Remember to include a negative control (a yeast colony that is untransformed or contains plasmid without the engineered gene) and a positive control (purified expression vector DNA). Vortex for 30 s.

3. Place 5 µl of the colony resuspension in a thin-walled PCR tube (0.2 or 0.5 ml).

4. To this tube, add master mix containing (per reaction):
   - 5 µl of 10× Eppendorf PCR buffer
   - 1 µl of 10 mM dNTP solution
   - 10 µl of 5× Eppendorf *Taq* master solution
   - 1 µl of Primer 1 (200–300 ng/µl)
   - 1 µl of Primer 2 (200–300 ng/µl)
   - Water to a final volume of 50 µl

5. Overlay with 50 µl of mineral oil (if your thermocycler lacks a hot bonnet).

6. Place the tube in a 95°C waterbath for 5 min and then chill the tube on ice for 2 min.

7. Add the diluted *Taq* polymerase (0.5 units, diluted in 1× PCR buffer).

8. Place on ice.

9. Carry out the PCR with the following steps:

Number of cycles	Program
1	95°C for 5 min
35	95°C for 1 min, 53–60°C for 1 min, and 72°C for 1 min
1	72° C for 10 min

10. Run 10 µl of the PCR on an agarose gel to see whether the expected PCR product has been produced.

> **Note**
>
> ªSelect primers that anneal within the recombinant gene and will produce a product of approximately 200–700 bp. Determine the optimal annealing temperature based on their composition.

2.3 Small-scale expression

After a sufficient number of transformants carrying the expression cassette has been verified, small shake-flask cultures of each strain should be grown to determine whether the foreign protein is synthesized in each and, if so, which strain appears to produce the highest level. For *AOX1* promoter-driven expression, the strains are initially grown in a glucose- or glycerol-based medium where the *AOX1* promoter is tightly repressed and the foreign gene is not expressed, an important quality if the recombinant protein is toxic (2). When the cultures reach the late logarithmic phase and a high cell concentration is attained, the cells are shifted to a methanol medium to initiate foreign protein production. Samples of each culture are removed at selected times and analyzed for the presence of the foreign protein. For intracellular proteins, peak levels are usually reached within 24 h in methanol. For extracellular proteins, peak levels are typically reached anywhere from 24 to 100 h after being shifted to methanol, depending on the characteristics of the foreign protein such as its folding efficiency, its structural stability, and its resistance to degradation by proteases in the medium.

For expression utilizing the *GAP* promoter, where toxicity is not a factor, starter cultures are grown on nonmethanol medium, diluted, and then maintained on the same medium for expression. Because the *GAP* promoter is expressed at a high level on various carbon sources, a range of media can be used, even media containing methanol (12). However, as *GAP* shows best expression in glucose-based medium, it should be used initially in small-scale trials. *Protocols 4* and *5* describe small-scale induction procedures for both *AOX1* and *GAP* promoter-driven expression. In addition, high-throughput formats have been developed using 24- or 96-deep-well plates (40, 41); however, these formats require special equipment accessories for the plates, and the sheer number of samples is probably beyond the needs of most *P. pastoris* users.

Protocol 4

Small-scale *P. pastoris AOX1*-induced expression of recombinant proteins

Equipment and Reagents
- 50 ml Sterile conical tubes
- Shaking temperature-controlled incubator
- Centrifuge
- BMG (buffered minimal medium with 1% glycerol) or BMGY (buffered minimal medium with 1% glycerol and yeast extract)[a]
- BMM (buffered minimal medium with 0.5% methanol) or BMMY (buffered minimal medium with 0.5% methanol and yeast extract)[a]
- Sterile 1.5 ml microfuge tubes
- 100× Protease Arrest (GBiosciences)
- Spectrophotometer

Method

1. On day 1, pick strains from a fresh plate and inoculate 5 ml of BMG medium (plus any required amino acids) in a 50 ml conical tube. Incubate at 30°C overnight on a shaker.

2. On day 2, inoculate sterile 50 ml tubes containing 10 ml BMG medium with approximately 200 µl of the overnight inoculum. This should lead to a starting OD_{600} of 0.10. Certain mutant strains may require more seed culture. Measure the OD of the overnight culture and use enough inoculum to get the desired OD. Incubate at 30°C overnight on a shaker.

3. On day 3, measure the OD_{600} of the cultures and confirm that they are between 2.0 and 5.0. If not, dilute and/or wait as appropriate. Check a few samples under the microscope to confirm that there is no contamination and that the yeasts are budding.

4. Aliquot the necessary volume to provide 5 or 10 OD_{600} units of each culture in a new conical tube and pellet for 5 min at 5000 r.p.m. in a room-temperature centrifuge. Standardize the number of cells going into a new tube so that equal numbers of cells are spun down.

5. Decant off the medium supernatant and invert the tubes on tissues.

6. Resuspend the pellets gently in 5 ml of BMMY or BMM medium with the necessary amino acid supplements.

7. Induce the cultures for the desired number of hours in a 30°C shaker. Remember to add methanol every 24 h to the tubes to account for losses from evaporation and consumption (maintain at 0.5% by adding 25 µl).

8. Take 0.5 ml aliquots of the culture at the desired time points, such as 24, 48, 72, and 96 h post-methanol addition. Determine the OD_{600} at each time point.

9. After the appropriate length of induction time, pellet the culture aliquot in a microfuge at the maximum speed of >10 000 *g* for 1 min and transfer the supernatant to a new microfuge tube. Add Protease Arrest solution to the supernatant fraction.

10. Immediately store the pellets (intracellular fraction) and supernatants (extracellular fraction) at –80°C.

Note

[a] Recipes for all *P.pastoris* growth media can be found at http://www.invitrogen.com/content/sfs/manuals/easyselect_man.pdf[8.3]

Protocol 5

Small-scale *P. pastoris GAP*-driven expression of recombinant proteins

Equipment and Reagents
- 50 ml Sterile conical tubes
- Shaking temperature-controlled incubator
- Centrifuge
- YPD (see *Protocol 1*), BMD (buffered minimal medium with 0.4% glucose) or BMDY (buffered minimal medium with 0.4% glucose and yeast extract)
- Sterile 1.5 ml microfuge tubes
- 100× Protease Arrest (GBiosciences)
- Spectrophotometer

Method
1. On day 1, pick strains from a fresh plate and inoculate 5 ml of YPD (or BMDY) in 50 ml tubes. Incubate at 30°C overnight in a shaking incubator.

2. On day 2, determine the OD_{600} of overnight cultures and dilute to 0.01 in 5 ml of fresh YPD (or other dextrose-based medium). Incubate at 30°C overnight in a shaking incubator.

3. On day 3, take 0.5 ml aliquots of the culture at the desired time points, such as 24, 48, 72, and 96 h post-dilution. Determine the OD_{600} at each time point.

4. After the appropriate length of induction time, pellet the culture aliquot in a microfuge at the maximum speed of >10 000 *g* for 1 min and transfer the supernatant to a new microfuge tube. Add Protease Arrest solution to the supernatant fraction.

5. Immediately store the pellets (intracellular fraction) and supernatants (extracellular fraction) at –80°C.

After the various samples have been collected, the level of heterologous proteins in the sample needs to be analyzed by a functional, immunological, or sodium dodecyl sulfate polyacrylamide gel electrophoresis (SDS-PAGE) assay. For secreted proteins, this is made easier by the fact that most induction media contain few components and *P. pastoris* secretes few endogenous proteins (2). Thus, the detection of proteins can be done by silver or Coomassie blue staining of unprocessed extracellular medium that has been resolved by SDS-PAGE. As the limit of detection by Coomassie blue staining is about 50 ng for a band on a ten-lane 10 cm × 10 cm × 1 mm mini-gel and the volume per lane is approximately 20 µl, if the detectable expression level is 2.5 ng/µl, which is a reasonable expectation, this should produce a visible band. However, if a secreted protein is heterogeneously glycosylated, it may run as a diffuse band, which would make it more difficult to detect.

Analysis of an intracellularly expressed recombinant protein requires that an extract be made from the pellet fraction, as described in *Protocol 6*. In order to reduce proteolysis, it is vital to keep the samples as cold as possible and to utilize chemical inhibitors.

Protocol 6

Production of intracellular extracts from pellet fractions

Equipment and Reagents
- Cell pellets
- Chilled 400–600 μm acid-washed glass beads (Sigma)
- Small scooper
- Multi-tube holder for vortex (Fisher)
- Chilled 1× PBS buffer with 1× Protease Arrest (GBiosciences)
- 1.5 ml Microfuge tubes
- Ice bucket
- Cotton swabs
- Microfuge at 4°C

Method
1. Remove the cell pellets from the freezer and thaw at 37°C until just thawed completely.
2. Add 500 μl of PBS/protease inhibitor cocktail to each tube and vortex to resuspend thoroughly.
3. Add 150 μl of chilled glass beads to each sample.
4. Clean the lips of the tubes with a cotton swab to remove any beads adhering to the lip. This step is vital to ensure a good seal and prevent leakage during the vortexing.
5. Close the tubes well.
6. Place the samples in the multi-tube holder, attach to the vortexer, and shake for 5–7 min at 4°C, preferably in a cold room. Alternatively, if a holder is not available, two to four tubes can be held by hand on the vortexer.
7. Transfer the tubes to a refrigerated microfuge and spin at the maximum speed of >10 000 g for 5 min.
8. Transfer the supernatant (about 400 μl) to a fresh tube. Keep on ice.
9. Use immediately for assays or store at –80°C. As repeated freezing and thawing will diminish enzyme activity, minimize the number of freeze–thaws or freeze each supernatant in several aliquots.

SDS-PAGE cannot be used for proteins expressed intracellularly because of the large number of endogenous proteins. In addition, it is laborious, time-consuming, and expensive. Thus, our laboratory routinely uses a spot Western blot to quantitate the level of heterologous protein in either the extracellular or intracellular fractions of a recombinant strain. Small volumes of extracellular supernatant (25–100 μl) or 10 μg of intracellular cell extract are used. In this method, described in *Protocol 7*, samples are immobilized on a nitrocellulose membrane using a dot blotter. The engineered protein is then detected with the use of a primary antibody against either the protein itself or a tag, followed by a secondary antibody conjugated to the enzyme alkaline phosphatase. This method allows one to screen multiple samples (up to 96) nonradioactively in less than 1 day (see *Fig. 1*).

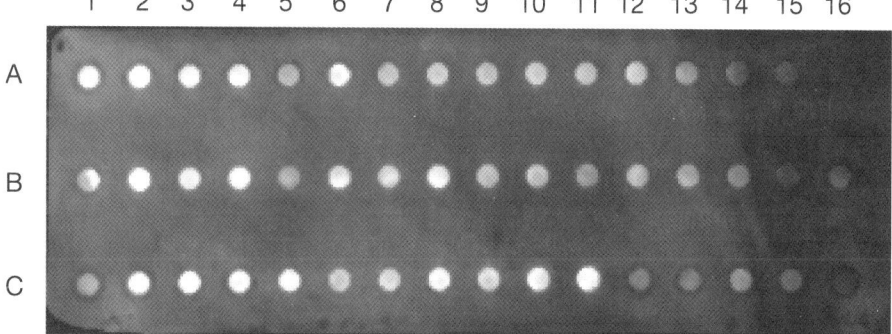

Figure 1. Example of a spot Western blot
Equivalent volumes of extracellular medium of recombinant *P. pastoris* strains were spotted onto a nitrocellulose membrane and the amount of secreted recombinant protein from each strain was quantitated using the spot Western blot protocol. Sample C1 represents a negative control strain harboring only empty vector.

Protocol 7

Spot Western blotting

Equipment and Reagents
- Whatman paper
- Nitrocellulose
- Spot blotter (Topac)
- Aspirator pump
- Western Star System including antibodies (Applied Biosystems)
- 1× PBS
- 50% Tween 20
- Rocker platform
- Film or chemiluminescent imager

Method
1. Remove the desired protein samples from the −80°C freezer.
2. Keep the samples on ice (can be thawed with warm water for a short time).
3. Plan the loading layout; remember to include both positive (purified antigen) and negative controls.
4. Cut out a piece of Whatman paper using the spot Western template provided with the spot blotter apparatus.
5. With gloved hands, cut out a piece of nitrocellulose membrane to the exact size of the Whatman paper.
6. Soak the nitrocellulose membrane in sterile 1× PBS for 5 min.
7. Place the Whatman paper over the wells, followed by the nitrocellulose membrane on top.
8. Secure the lid over the membrane, making sure that there are no bubbles, and screw the apparatus shut. Connect the tubing and apply suction with the pump.

9. Load the desired amount of protein (between 10 and 100 µl of extracellular supernatant or 10 µg of intracellular extract) directly into the wells.
10. Remove the tubing from the apparatus and turn off the suction.
11. Remove the nitrocellulose membrane from the apparatus and set aside to air dry. Mark the blot for orientation.
12. Prepare I-Block solution as follows: add 0.12 g of I-Block (provided in the kit) to 60 ml of 1× PBS, mix together and heat in a microwave, making sure it does not boil over. After the solution has cooled, add 120 µl of 50% Tween 20.
13. Pre-wet the nitrocellulose membrane with 1× PBS in a plastic container.
14. Drain off the PBS and add 20 ml of I-Block solution directly over the membrane.
15. Rock for 1 h to overnight (if overnight, place at 4°C).
16. Drain off the I-Block buffer.
17. Dilute a sufficient amount (see manufacturer instructions) of the appropriate primary antibody (e.g. mouse anti-myc) in 10 ml of I-Block solution and add to the membrane.
18. Incubate for 1 h on a rocker at room temperature.
19. Prepare 200 ml of wash buffer by adding 400 µl of 50% Tween 20 to 200 ml of 1× PBS.
20. Wash the membrane twice for 5 min each using 30 ml of wash buffer.
21. Dilute a sufficient amount (see manufacturer instructions) of the appropriate secondary antibody (e.g. goat anti-mouse) in 10 ml of I-Block.
22. Add to the membrane and incubate for 45 min on a rocker.
23. Wash the membrane three times for 5 min each using 30 ml of wash buffer.
24. Transfer the membrane to a new plastic container.
25. Prepare the assay buffer containing water and 4 ml of 10× assay buffer (provided in the kit)
    Wash the membrane twice for 2 min in 20 ml of assay buffer.
26. Prepare the substrate solution containing 3 ml of CPD-Star and 150 µl of Nitroblock II (provided in the kit).
27. Dry off any excess assay buffer from the membrane by holding a corner against a tissue and place on a flat surface.
28. Pipette the substrate solution directly onto the membrane drop by drop, covering the entire membrane with a thin layer of solution.
29. Allow to incubate for 5 min.
30. Drain off any excess solution from the membrane but do not allow it to dry out.
31. Place the membrane into a clear transfer envelope (provided in the kit).
32. Expose for between 1 and 30 min.

The spot Western blot will identify the most productive strains but not reveal anything about the nature of the engineered peptide. With secreted foreign proteins, particular care must be taken to ensure that the product is of the expected molecular mass, as proteolytic degradation and hyperglycosylation can be a significant problem for some secreted proteins (3, 9). Thus, a standard Western blot, which includes SDS-PAGE prior to immunodetection, should also be performed on samples displaying the highest level of protein expression. The Western blot must contain some of the desired protein isolated from its native source to act as a comparison. In addition, if the product has a biological activity that is measurable, then the proteins from the best isolates should also be assayed for functionality. If desired, the copy number of these selected strains can also be estimated using Southern blotting or real-time PCR.

2.4 Optimization

At this point, one should try to find the growth conditions that give optimal expression by varying parameters such as medium composition, pH, percentage of methanol (or other carbon source), and temperature. Optimizing several culturing factors often gives salient improvements in protein yield (39). Although the optimization phase can be rather extensive, it is dependent on one's time and resources. It is a trial-and-error process with no set end point. However, it is usually worth the effort, as the conditions for highest expression depend on the individual protein. For instance, the best yields are usually found with richer media (http://www.invitrogen.com/content/sfs/manuals/easyselect_man.pdf[8.3].); however, for one protein studied in our laboratory, we found that culturing in BMM (buffered minimal medium with methanol) provided about 50% higher expression than BMMY or YPM (media containing yeast extract and peptone).

2.5 Small-scale purification

At the same time as optimization is being worked out, it is best to try a protein purification strategy on a small scale, as growth medium content may influence the purification options. For instance, yeast extract and peptone in the expression medium may prove to be annoying contaminants. Recombinant proteins can be isolated with size exclusion, affinity, or other types of conventional chromatography. However, for those scientists who are less biochemically inclined, a good option is to express the heterologous protein as a fusion to some type of peptide tag. Proteins expressed from the pPICZ and pGAPZ vectors may be translated in frame with a polyhistidine region that is both an antigen for a commercially available antibody and a moiety for binding to a nickel affinity column (http://www.invitrogen.com[8.2]). *Protocol 8* describes a small-scale, rapid method for purification of His_6-tagged proteins using a commercially available kit. The procedure works better for secreted than intracellular proteins due to reduced competition from endogenous proteins.

Protocol 8

Purification of a His$_6$-tagged protein under native conditions

Equipment and Reagents
- Nickel-NTA kit (Qiagen)
- Lysis buffer (50 mM NaH$_2$PO$_4$, 300 mM NaCl, 10 mM imidazole, pH 8.0)
- 10× Adjustment buffer (500 mM NaH$_2$PO$_4$, 1.5 M NaCl, 50 mM imidazole, pH 8.0)
- Wash buffer (50 mM NaH$_2$PO$_4$, 300 mM NaCl, 20 mM imidazole, pH 8.0)
- Elution buffer (50 mM NaH$_2$PO$_4$, 300 mM NaCl, 250 mM imidazole, pH 8.0)
- Microcentrifuge

Method
1. Start with *P. pastoris* supernatant or intracellular extract containing the recombinant His$_6$-tagged protein. A spot or full Western blot should have been done to confirm its presence and approximate its concentration prior to beginning the purification.
2. Equilibrate the spin column from the kit with 600 µl of lysis buffer. Centrifuge for 2 min at 700 ***g*** (about 2000 r.p.m.) in the microcentrifuge. Columns need to have an open lid.
3. Add 60 µl of adjustment buffer to 540 µl of the *P. pastoris* sample that contains the recombinant His$_6$-tagged protein. Mix well.
4. Load the 600 µl of adjusted sample (step 3) onto the equilibrated column seated in a catcher tube. Spin for 3 min at 700 ***g*** in the microcentrifuge. Columns need to have a closed lid.
5. Wash the column three times with 600 µl of wash buffer. Spin for 3 min at 700 ***g*** in the microcentrifuge with an open cap.
6. Elute the bound protein twice with 100 µl of elution buffer for 3 min at 700 ***g*** with a closed lid and save the elutant.
7. The 200 µl of total elutant should contain the recombinant protein, which can be confirmed by Western blot or Coomassie blue-stained SDS-PAGE.

2.6 Considerations for scaling up expression and purification

After small-scale expression and purification have conditions been worked out, both processes must be scaled up to acquire a reasonable amount of the recombinant product. The highest levels of production are invariably seen in fermenter cultures (43). Compared with shake cultures, fermenters have the potential for much higher cell densities, more uniform culture conditions, and optimal induction of the *AOX1* promoter. Several recent reviews discuss the process with *P. pastoris* (44, 45); unfortunately, fermenter expression is beyond the scope of this chapter. Many researchers do not have access to a fermenter and have no choice but to grow up large volumes of culture in shake flasks to synthesize the necessary amounts of desired proteins. Some scientists also attempt to perform expression in large shake flasks before transitioning to fermenters.

As *P. pastoris* does not like to ferment and prefers respiratory growth, this metabolism requires the cultures to be well aerated. For this reason, the culture volume should never exceed 1/5 of the volume of the flask and these large-scale cultures should be shaken at a significantly high speed (>225 r.p.m.) in an incubator. Certain steps should be adjusted to reflect the optimal conditions for expression of the protein, using information gathered from small-scale expression work. In the same way, the purification regimen for large-scale expression of the desired protein should be adjusted based on work with smaller volumes. The large volume of the extracellular medium usually requires concentration prior to purification; however, components produced by the yeast often cause clogging of concentrators (e.g. Amicon Centricon). Dialysis or ammonium sulfate precipitation prior to concentration may remedy this obstacle (42, 46).

Often the transition from small- to large-scale purification is problematic and these setbacks can only be overcome with much trial and patience. If one is successful at this stage, then one should have a reasonable amount of functional, fairly homogenous recombinant product. In cases with secreted proteins, the recombinant product may be heterogeneously glycosylated and may appear as a diffuse group of bands (1). To remove the carbohydrate groups from the protein, a sample of the peptide may be digested with peptide:*N*-glycosidase F (New England Biolabs). It is important to digest the positive-control protein RNAse B provided by the supplier in parallel with the engineered protein to ensure that the enzyme works under the reaction conditions used. Digested and undigested samples should subsequently be resolved by SDS-PAGE to determine the molecular mass contribution from glycosylation.

3. TROUBLESHOOTING

Despite the success of over 550 recombinant proteins produced in *P. pastoris*, there are instances where the yeast has failed to produce detectable levels of a foreign peptide (47). Even if a reasonable amount of a protein is synthesized, there may be other problems. Some of these difficulties may be addressed by using the vector or strain option discussed at the beginning of this chapter. However, there are several culturing and genetic changes that may serve as appropriate solutions, as follows:

- Low protein expression
  - Select strains with higher copy numbers of expression vectors.
  - Change the sequence of the open reading frame to reflect the codon preference of *P. pastoris* (46). Because of differences in the codon usage between the native host for the gene and *P. pastoris*, the heterologous mRNA transcripts may not be translated efficiently in the methylotroph.
  - Use the most 5′ restriction site in the polylinker of the expression plasmid so that the 5′UTR is not too long.
  - Construct two different expression plasmids, utilizing both the *GAP* and *AOX1* promoters, to drive the expression of the recombinant gene in the same strain (48). It is postulated that a limiting number of transcription factors for a given promoter exist in a cell.

- Inefficient secretion
  - Culture strains at a lower temperature, such as 23°C instead of 30°C (49).
  - Change to a different secretion signal, such as the *P. pastoris* acid phosphatase (*PHO1*) or others that have been used before (14).
  - Switch to a weaker promoter than *GAP* or *AOX1*, as high expression may be saturating the secretion network in the case of the recombinant protein.
- Protein degradation
  - Change the pH or methanol concentration of the growth medium (34).
  - Grow at a lower temperature (3).
  - Add amino acid-rich supplements such as 2% peptone or casamino acids to the growth medium to act as decoys for proteases (37).
  - Decrease the time between induction and harvest.
  - Add protease inhibitors, such as 10 mM EDTA, to the growth medium (50, 51).
- No tangible expression
  - In parallel with the expression of the desired protein, use a positive-control strain that produces a model protein at a high level. Strains for expression of intracellular β-galactosidase (GS115/pPICZ/*lacZ*) and secreted human serum albumin (GS115/Albumin) can be obtained from Invitrogen. The successful expression of either control peptide will confirm that both technique and reagents are good. The experimenter must then determine what modifications are needed for the production of the desired protein – a task that is hopefully not too elusive.

4. REFERENCES

1. Daly R & Hearn MTW (2005) *J. Mol. Recognit.* **18**, 119–138.
2. Lin Cereghino GP, Sunga AJ, Lin Cereghino J & Cregg JM (2001) In *Genetic Engineering: Principles and Methods*, pp. 343–352. Edited by JK Setlow. Kluwer Academic/Plenum Publishers, New York.
★★ 3. Macauley-Patrick S, Fazenda ML, McNeil B & Harvey LM (2005) *Yeast*, **22**, 249–270. – Excellent discussion of several strategies to decrease protein degradation.
★ 4. Cregg JM, Barringer KJ, Hessler AY & Madden KR (1985) *Mol. Cell. Biol.* **5**, 3376–3385. – First paper describing the use of P. pastoris *for heterologous expression.*
5. Couderc R & Baratti J (1980) *Agric. Biol. Chem.* **44**, 2279–2289.
6. Cregg JM, Madden KR, Barringer KJ, Thill GP & Stillman CA (1989) *Mol. Cell. Biol.* **9**, 1316–1323.
7. Tschopp JF, Brust PF, Cregg JM, Stillman CA & Gingeras TR (1987) *Nucleic Acids Res.* **15**, 3859–3876.
8. Cregg JM & Madden KR (1988) *Dev. Ind. Microbiol.* **29**, 33–41.
9. Romanos MA, Scorer CA & Clare JJ (1992) *Yeast*, **8**, 423–488.
10. Sunga AJ & Cregg JM (2004) *Gene*, **330**, 39–47.
11. Lin Cereghino GP, Lin Cereghino J, Sunga AJ, *et al.* (2001) *Gene*, **263**, 159–169.
12. Waterham HR, Digan ME, Koutz PJ, Lair SV & Cregg JM (1997) *Gene*, **186**, 37–44.
13. Shen S, Sulter G, Jeffries TW & Cregg JM (1998) *Gene*, **216**, 93–102.
★ 14. Lin Cereghino J & Cregg JM (2001) *FEMS Microbiol. Rev.* **24**, 45–66. – A review that covers all of the major points of P. pastoris *expression with a strong emphasis on designing vectors.*
15. Brocca S, Schmidt-Dannert C, Lotti M, Alberghina L & Schmid RD (1998) *Protein Sci.* **7**, 1415–1422.

16. Martinez-Ruiz A, Martinez del Pozo A, Lacadena J, et al. (1998) *Protein Expr. Purif.* **12**, 315–322.
17. Raemaeker RJM, deMuro L, Gatehouse JA & Fordham-Skelton AP (1999) *Eur. J. Biochem.* **65**, 394–403.
18. Brierley RA (1998) *Methods Mol. Biol.* **103**, 149–177.
19. Sears IB, O'Connor J, Rossanese OW & Glick BS (1998) *Yeast*, **14**, 783–790.
20. Higgins DR, Busser K, Comiskey J, Whittier PS, Purcell TJ & Hoeffler JP (1998) *Methods Mol. Biol.* **103**, 41–53.
21. Kimura M, Takatsuki A & Yamaguchi I (1994) *Biochim. Biophys. Acta*, **1219**, 653–659.
22. Arnau J, Lauritzen C, Petersen GE & Pedersen J (2006) *Prot. Exp. Pur.* **48**, 1–13.
23. Cregg JM, Shen S, Johnson M & Waterham HR (1998) *Methods Mol. Biol.* **103**, 17–26.
24. Nett JH & Gerngross TU (2003) *Yeast*, **20**, 1279–1290.
25. Gleeson MAG, White CE, Meninger DP & Komives EA (1998) *Methods Mol. Biol.* **103**, 81–94.
26. Boehm T, Pirie-Shepard S, Trinh LB, Shiloach J & Folkman J (1999) *Yeast*, **15**, 563–567.
27. Vuorela A, Myllyharju J, Nissi R, Pihlajaniemi T & Kivirikko KI (1997) *EMBO J.* **16**, 6702–6712.
28. Chiruvolu V, Cregg JM & Meagher MM (1997) *Enzyme Microb. Technol.* **21**, 277–283.
★★ 29. Bretthauer RK & Castellino FJ (1999) *Biotechnol. Appl.Biochem.* **30**, 193–200. – *A complete and clear description of glycosylation in* P. pastoris.
30. Verostek MF & Trimble RB (1995) *Glycobiology*, **5**, 671–681.
31. Callewaert N, Laroy W, Cadirgi H, et al. (2001) *FEBS Letters* **503**, 173–178.
32. Bobrowicz P, Davidson RC, Li H, et al. (2004) *Glycobiology*, **14**, 757–766.
33. Hamilton SR, Bobrowicz P, Bobrowica B, et al. (2003) *Science*, **301**(5637), 1244–1246.
34. Tsai CW, Duggan PF, Shimp RL, Miller LH & Narum DL (2006) *J. Biotechnol.* **121**, 458–470.
35. Cregg JM, Cereghino L, Shi J & Higgins DR (2000) *Mol. Biotech.* **16**, 23–52.
36. Cregg JM & Madden KR (1989) *Mol. Gen. Genet.* **219**, 320–323.
37. Clare JJ, Rayment FB, Ballantine SP, Sreekrishna K & Romanos MA (1991) *BioTechnology*, **9**, 455–460.
38. Hohenblum H, Gasser B, Maurer M, Borth N & Mattanovich D (2004) *Biotechnol. Bioeng.* **85**, 367–375.
39. Cregg JM & Madden KR (1987) In *Biological Research on Industrial Yeasts*, pp. 1–18. Edited by GG Stewart, I Russell, RD Klein & RR Hiebsch. CRC Press, Boca Raton, FL.
★★ 40. Bottner M & Lang C (2005) *Methods Mol. Biol.* **267**, 277–286. – *A concise overview of the steps required for high-throughput expression from transformation to protein purification.*
★★ 41. Weis R, Luiten R, Skranc W, Schwab H, Wubbolts M & Glieder A (2004) *FEMS Yeast Research* **5**, 179–189. – *An excellent description of how to optimize growth parameters to give reproducible, high-level expression on a high-throughput scale.*
42. Triches Damaso MC, Almeida MS, et al. (2003) *Appl. Environ. Microbiol.* **69**, 6064–6072.
43. Higgins DR & Cregg JM (1998) *Methods Mol. Biol.* **103**, 1–15.
44. Cos O, Ramon R, Montesinos JL & Valero F (2006) *Microb. Cell Fact.* **5**, 1–20.
45. Zhang W, Bevins MA, Plantz BA, Smith LA & Meagher MM (2000) *Biotechnol. Bioeng.* **70**, 1–8.
46. Yadava A & Ockenhouse CF (2003) *Infect. Immun.* **71**, 4961–4969.
47. Burrowes OJ, Diamond G & Lee TC (2005) *J. Biomed. Biotechnol.* **4**, 374–384.
48. Wu JM, Lin JC, Chieng LL, Lee CK & Hsu TA (2003) *Enzyme Microb. Technol.* **33**, 453–459.
49. Li A, Xiong F, Lin Q, et al. (2001) *Protein Expr. Purif.* **21**, 438–445.
50. Shi X, Karkut T & Chamankhah M (2003) *Protein Exp. Purif.* **28**, 321–330.
★★ 51. Vad R, Nafstad E, Dahl A & Gabrielsen O (2005) *J. Biotechnol.* **116**, 251–260. – *A good exploration of different approaches to optimize expression.*

CHAPTER 9
Improved baculovirus expression vectors

Richard B. Hitchman, Robert D. Possee, and Linda A. King

1. INTRODUCTION

In recent years, the baculovirus expression vector system has become a well-established and popular method for producing high yields of structurally, functionally, and antigenically authentic foreign proteins in insect cells (1). Traditionally, when making a recombinant baculovirus, the foreign gene is cloned into a transfer vector, which contains sequences that flank the polyhedrin (*polh*) locus in the virus genome. The virus genome (generally *Autographa californica* multinucleopolyhedrovirus (AcMNPV)) and the transfer vector are introduced into the host insect cell, and homologous recombination between the flanking sequences common to both DNA molecules effects insertion of the foreign gene into the virus genome, resulting in a recombinant virus genome. The genome then replicates to produce recombinant budded virus (BV) that can be harvested from the culture medium. A major disadvantage of this method is that it generates a mixture of recombinant and original parental virus after the initial round of replication. Consequently, the recombinant virus has to be separated from the parental virus by plaque purification, a process that is labor-intensive, technically demanding, and time-consuming. More recently, bacterial artificial chromosomes (BACs) have been incorporated into the baculovirus genome to allow recombinant virus construction in *Escherichia coli* (2). Several commercial bacmid-based systems are now available (e.g. Bac-to-Bac and BaculoDirect; Invitrogen), but all still produce a mixed pool of both parental and recombinant viruses, which then need to be separated from each other by antibiotic selection or nucleoside analogs.

2. METHODS AND APPROACHES

2.1 Principles of *flash*BAC

The *flash*BAC system is a new platform technology for the production of recombinant baculoviruses, and has been designed specifically to remove the need to

148 ■ CHAPTER 9: IMPROVED BACULOVIRUS EXPRESSION VECTORS

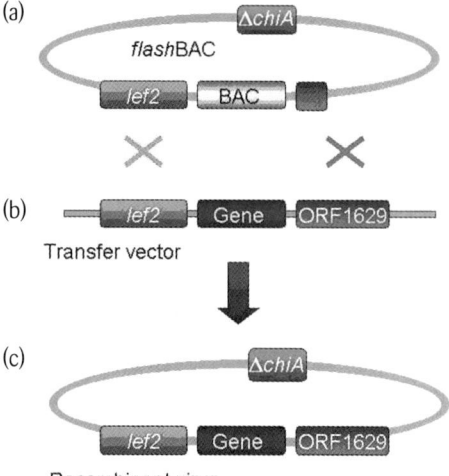

Figure 1. Schematic illustrating construction of a recombinant baculovirus using *flash*BAC.
*flash*BAC DNA (*a*) is mixed with transfer vector DNA (*b*) containing the gene of interest to be inserted into the viral genome. Homologous recombination in insect cells removes the BAC and restores ORF1629 (*c*), allowing the production of infectious recombinant virus.

separate recombinant virus from parental virus by plaque purification or any other means. It reduces the production of recombinant virus to a one-step procedure in insect cells and is thus fully amenable to high-throughput and automated production systems. *flash*BAC technology is based on an AcMNPV genome that lacks part of an essential gene (ORF1629) and contains a BAC at the *polh* locus, replacing the polyhedrin-coding region (see Fig. 1). The essential gene deletion prevents virus replication within insect cells, but the BAC allows the viral DNA to be maintained and propagated within bacterial cells. Viral DNA is then isolated from the bacterial cells by lysis and purified on cesium chloride gradients, ready for use in co-transfections. A recombinant baculovirus is produced by simply co-transfecting insect cells with *flash*BAC DNA and a transfer vector containing the foreign gene of interest. Recombination within the insect cells (see Fig. 1a, b) restores the function of the essential gene, allowing the virus DNA to replicate and produce BV, simultaneously inserting the foreign gene of interest under the control of the *polh* gene promoter and removing the BAC sequence (see Fig. 1c). The recombinant BV can then be harvested from the culture medium of the transfected insect cells and forms a seed stock of recombinant virus. As it is not possible for nonrecombinant virus to replicate, there is no need for any selection system.

*flash*BAC also contains modifications to its genome to maximize protein secretion and membrane protein targeting. Chitinase (*chiA*) is an auxiliary gene that encodes an enzyme with exo- and endochitinase activity (3). In an infected insect, chitinase (together with cathepsin) facilitates host cuticle breakdown and tissue liquefaction at the very late stages of infection, thus releasing the virus to infect more hosts (4). Confocal and electron microscopy observations of insect cells infected with AcMNPV have shown that chitinase is targeted to the endoplasmic reticulum where it is densely packed in a paracrystalline array, severely compromising the function and efficacy of the secretory pathway (5, 6). ChiA is nonessential for replication in insect cell culture and its deletion from *flash*BAC has improved the efficacy of the secretory pathway and resulted in a greatly enhanced (up to 60-fold in some instances) yield of recombinant proteins that are secreted or membrane targeted (in comparison with recombinant viruses that synthesize chitinase) (7).

2.2 Insect cell culture requirements

When working with insect cells, aseptic technique must be used at all times and culturing should be carried out in a laminar flow cabinet or around a flame. The insect cells most commonly used are Sf21 cells (8), originally derived from the pupal ovarian cells of *Spodoptera frugiperda* (fall army worm), Sf9 cells (9), which are a clonal isolate of Sf21, and High Five cells (BTI-TN-5B1-4; Invitrogen), a clonal isolate derived from the parental *Trichoplusia ni* (cabbage looper) ovarian cell line TN-368 (10–14). Generally, Sf21 or Sf9 cells are used for co-transfections, virus amplification, and plaque assays. High Five cells should not be used to produce or amplify virus because of the increased possibility of generating virus mutants (15). However, they may produce higher yields of secreted recombinant proteins with more authentic post-translational modifications than Sf9 cells (16). Most insect cells can be cultivated over a temperature range of 25–30°C, but the optimal temperature for cell growth and infection of Sf21 and Sf9 cells is considered to be 27–28°C. Insect cell culture medium utilizes a phosphate buffering system, rather than the carbonate-based buffers that are commonly used for mammalian cells. This means that CO_2 incubators are not required. Serum is required for the maintenance of certain cell lines, but many have now been adapted to serum-free conditions. We recommend carrying out any cell culture work prior to handling virus and only handling one cell line at a time. It is extremely important that the insect cells used for the production of recombinant viruses and proteins are of the highest quality. This is achieved by subculturing cells before they become overgrown, i.e. reach stationary phase, and by using cells that have been subcultured no more than approximately 30 times. *Protocol 1* describes the method we routinely use for counting and assessing cell viability.

Protocol 1

Counting insect cells and determining cell viability

Equipment and Reagents
- Insect cells
- Serum-supplemented or serum-free insect cell culture media
- Tissue culture hood
- Phase-contrast microscope
- Tissue culture dishes (35 mm^2)
- Neubauer counting chamber
- 0.4% Trypan blue (Invitrogen)

Method
1. Observe 1 ml of cells under the phase-contrast microscope using ×10 and ×40 objectives[a].
2. Pipette a sample of cells into a Neubauer counting chamber and allow them to fill the chamber by capillary action.
3. Count all of the cells within the 5 × 5 square grid on the counting chamber, using the phase-contrast microscope (×10 objective). Count cells touching the triple lines on the top and left of the squares. Do not count cells touching the triple lines on the bottom or right side of the squares[b].
4. Repeat the count to give a more statistically correct estimate of cell density.
5. Calculate the number of cells in 1 ml of culture by multiplying the average number of cells from the 5 × 5 square by 10^4. If the cells were diluted before counting, also multiply by the dilution factor.
6. To assess cell viability, mix 0.9 ml vols of cell suspension and 0.1 ml of 0.4% trypan blue[c] (2%, w/v) (1:2 dilution of cells).
7. Load the cells into a counting chamber and count the cells as described above. Calculate the percentage viability of a culture (diluted 1:2) as follows:

    % dead cells = (total blue cells counted × 2 × 100)/(total cells counted × 2)
    % viable cells = 100 − % dead cells

Notes
[a]Healthy cells look bright, round, and refractile, and many should be in the process of dividing into daughter cells.
[b]The 5 × 5 square gives the number of cells present in 0.1 µl of culture.
[c]Nonviable cells absorb the dye and appear blue, whilst live and healthy cells exclude the blue-colored dye. Healthy, exponential-phase cultures should contain more than 95% unstained, viable cells. However, viable cells will adsorb trypan blue over time and this can affect counting and viability results.

2.2.1 Maintaining insect cells in suspension culture

Insect cell lines can be maintained as either suspension cultures, in shake or stirred flasks, or in monolayer cultures in T-flasks or dishes. Generally, insect cells adapted to serum-free medium are cultivated in shake cultures with the aid of an orbital shaker platform, whilst cells adapted to serum-supplemented media are cultivated in stirrer flasks or monolayer cultures (to prevent excessive foaming and subsequent cell damage). A method for growing cells in suspension culture is described in *Protocol 2*. Typical densities for suspension culture of different cell lines are shown in *Table 1*. We recommend maintaining stock cultures without antibiotics; otherwise, you may be maintaining a low-level contaminant that can cause inefficient virus replication or protein production. However, the addition of antibiotics to 'dead-end' procedures such as plaque assays (see *Protocol 6*) may be useful in preventing contamination (penicillin and streptomycin prepared with 5 units/ml penicillin G sodium and 5 µl/ml streptomycin sulfate in 0.85% saline). More recently WAVE technology has become available and is becoming increasingly popular for use with baculovirus (17). WAVE uses plastic bags, which are inflated with oxygen and then half filled with medium. The cells (and virus) are introduced by a syringe using ports on the bag, which is placed on a rocker platform. The bag is gently rocked at the required rate. Oxygen and fresh medium can be introduced at the required rate using a controller and pump, and cells can be removed by using separate ports. The system caters for different volumes of cells from 100 ml up to 20 l of cells.

Table 1. Insect cell seeding and maximum growth densities in suspension culture

Cell line	Medium	Seeding cell density ($\times 10^6$ cells/ml)	Maximum cell density ($\times 10^6$ cells/ml)
Sf9	Serum-free medium	0.3–0.5	6.0
Sf21	Serum-containing medium	0.1–0.2	2.0–2.5
High Five	Serum-containing medium	0.7–1.0	2.0–2.5

Protocol 2

Passaging suspension cultures of insect cells

Equipment and Reagents
- Insect cells
- Serum-supplemented or serum-free insect cell culture media
- Sterile glass or disposable shake flasks
- Tissue culture hood
- Selection of sterile pipettes (1–10 ml)
- 5% (w/v) Sodium dodecyl sulphate (SDS)

Method
1. Calculate how much of the existing culture needs to be diluted in fresh medium to obtain the required cell density and volume for the new culture (see *Table 1*).
2. Pipette the medium into a fresh flask and add the required volume of cells[a].
3. Record the following information on the flask that the cell culture is maintained in:
   - Cell line
   - Passage number
   - Date of each passage
   - Density to which the culture has been split
   - Medium used
4. Also keep a separate record of liquid nitrogen batch and the date and passage when cells were first raised from liquid nitrogen.
5. Discard any unused cells into SDS or an autoclavable discard container.

Note
[a] Each time a cell stock is transferred in this way, the passage number of the culture increases by one.

2.3 Maintaining insect cells in monolayer culture

The maintenance of cells in monolayer culture does not normally require counting of cell densities. Cells are observed under the phase-contrast microscope until they become confluent. When confluence is observed under the microscope, the cells are subcultured as described in *Protocol 3* (see *Table 2*). The exception to this is when setting up a monolayer culture flask (T-flask) from cells growing in a suspension culture (when a known number of cells are used to seed a T-flask; see *Table 3*).

Table 2. Typical ratios employed to subculture insect cells maintained in T-flasks

Cell Line	Estimated ratio of existing culture to fresh medium
Sf9	1:5
Sf21	1:10
High Five	1:5

Table 3. Seeding densities of insect cells in T-flasks

Cell Line	T-flask size (cm^2)	Seeding density ($\times 10^6$)
Sf9 or Sf21	T25	1.0–1.5
Sf9 or Sf21	T75	3.0–5.0
High Five	T25	0.5–0.9
High Five	T75	2.0–3.0

Protocol 3

Subculturing cells in monolayer culture

Equipment and Reagents
- Insect cells
- Serum-supplemented or serum-free insect cell culture media
- 25 or 75 cm^2 T-flasks
- Selection of sterile pipettes (1–10 ml)
- Tissue culture hood
- 5% (w/v) SDS

Method
1. Observe the health and confluency of the cells using an inverted phase-contrast microscope (using ×10, ×20, and ×40 objectives)[a].
2. When the cells are fully confluent, dislodge them from the surface of the flask by banging the flask down firmly on a solid surface[b]. Do not hold the neck of the flask when carrying out this operation and minimize foaming, particularly when using serum-free medium.
3. Using a graduated pipette, gently pipette the culture up and down to break up clumps of cells into a single-cell suspension. Avoid causing bubbles and frothing in the culture.
4. Split the culture into a new T-flask using the appropriate medium and ratios shown in Table 2[c].
5. The following information should be recorded on each T-flask:
   - Cell line
   - Passage number
   - Date of each passage
   - Density to which the culture has been split to (i.e. 1:5 or 1:10)
   - Medium used
6. A separate record should also be kept detailing;
   - Liquid nitrogen batch
   - Date and passage when cells were raised from liquid nitrogen
7. Discard any unused cells into SDS or an autoclavable discard container.

Notes

[a]Healthy cells attach well to the bottom of a T-flask forming a monolayer and double every 18–24 h. Loosely attached cells or cells floating in the medium are frequent in cultures that are overgrown, whereas in healthy cultures only a few will be visible.
[b]Trypsin and other enzymes are not recommended to dislodge insect cells as they may damage the cells. Cell scrapers can also cause cell damage and should be used only if absolutely necessary.
[c]Each time a culture is split, its passage number increases by one.

2.4 Construction of recombinant baculoviruses using *flash*BAC

Once the target gene of the protein to be expressed has been cloned into a suitable transfer vector, it is ready to be inserted into the virus genome. Any transfer vector based on homologous recombination with native viral sequences at the *polh* locus may be used with *flash*BAC, including the *polh* promoter, dual, triple, and quadruple expression vectors, and those that use other gene promoters such as *p10* and *ie1*. Recombinant viruses can then easily be made using the one-step process described in Protocol 4. Generally, after incubation for 5 days, the co-transfection mixture will yield a virus titer in the region of 1×10^7 plaque-forming units (p.f.u.)/ml, which can then be used directly for scaling up to a high-titer stock. After co-transfection and harvesting of the seed stock of virus for further amplification, you may also harvest the remaining cells from the dish to use for SDS-polyacrylamide gel electrophoresis (PAGE)/Western blotting. This will give a quick check for gene expression, although in our experience the levels of expression in these cells will vary depending on the quality of the transfer vector DNA used.

Protocol 4

Production of recombinant baculovirus using *flash*BAC

Equipment and Reagents
- *flash*BAC DNA (100 ng per co-transfection)
- Sterile transfer vector DNA (500 ng per co-transfection)[a]
- Sf9 cells (0.9×10^6 cells/dish) from a culture in exponential-phase growth
- Serum-free insect cell culture medium
- 35 mm² Tissue culture treated dishes (one per recombinant virus and an extra dish for a mock-transfected control)
- Transfection reagent[b]
- Incubator set at 28°C
- 1% Virkon (Amtec) or other suitable disinfectant
- Inverted phase-contrast microscope
- Plastic box to house dishes in the incubator
- Sterile bijoux or similar[c]
- Selection of sterile pipettes (1–10 ml)

Method
1. One hour before use, dispense 2 ml of insect cells at the correct density (1.5×10^6 Sf21 or 1×10^6 Sf9 cells/dish) into 35 mm² dishes on a flat surface.

2. Prepare a co-transfection mix by adding 1 ml of serum-free, antibiotic-free medium into a sterile, disposable polystyrene container (7 ml bijoux are convenient).

3. Add 5 µl of Lipofectin or other transfection reagent and mix.

4. Add 100 ng of *flash*BAC DNA (5 µl from the *flash*BAC kit) and 500 ng of transfer vector DNA (containing the gene under investigation or the control provided in the kit) and mix with gentle agitation. In the mock-transfection control, omit the DNA from the medium.

5. Incubate at room temperature for 15–30 min to allow the liposome–DNA complexes to form.
6. Remove the culture medium from the 35 mm² dishes of cells using a sterile pipette, ensuring that the cell monolayer is not disrupted[d].
7. If using cells maintained in serum-supplemented medium, wash the monolayer twice with 1 ml of serum-free medium[e].
8. Immediately add the 1 ml of DNA + liposome complex dropwise to the center of each dish and incubate in a plastic sandwich box at 28°C for a minimum of 5 h or overnight.
9. After the incubation period, add a further 1 ml of the appropriate insect cell culture medium to each dish and continue the incubation for 5 days in total.
10. Following the 5 day incubation period, harvest the medium containing the recombinant virus into a sterile bijoux and store in the dark at 4°C[f].

Notes

[a] Any vector designed for double-crossover, homologous recombination with baculovirus DNA at the *polh* locus is suitable (see http://expressiontechnologies.com[9.1] for a detailed list of suitable vectors).

[b] We routinely use Lipofectin (Invitrogen), but other reagents have been tested and found to work equally well including FuGENE 6 (Roche), GeneJuice (Novagen), Tfx-20 (Promega), and CELLFECTIN (Invitrogen).

[c] Plasticware used to prepare the transfection mixture must be made from polystyrene and not from polypropylene, as the complexes will bind to polypropylene.

[d] When removing liquid from a dish of cells, tip the dish at a 30–60° angle so the liquid pools towards one side the dish. Do not allow the cell monolayer to dry out at this point.

[e] Washing the cells removes any residual serum, which inhibits liposome-mediated transfection of DNA into cells.

[f] If the control transfer vector supplied with the *flash*BAC kit has been used to make recombinant virus, the *lac*Z-positive infected cells can be stained using X-gal. Add 1 ml of appropriate insect cell culture medium containing 15 μl (2%, w/v) X-gal (in *N,N*-dimethylformamide) and incubate at 28°C. After 5 h, the cells and culture medium will appear blue in color, confirming the production of recombinant virus.

2.4.1 Amplification of recombinant baculovirus

Following the co-transfection, a portion of the viral seed stock generated in *Protocol 4* is used to infect insect cells at a low multiplicity of infection (m.o.i.) (<1 p.f.u./cell) to prepare a high-titer passage 1 stock of recombinant virus. Using a low m.o.i. ensures that only a few cells are infected initially; the virus replicates to release BV, which then infects more cells and so on. In this way, multiple rounds of replication occur and high virus titers are obtained; it also reduces the chances of any defective virus particles occurring. If the cells are infected at a high m.o.i. (>1 p.f.u./cell), all of the cells will be infected initially and only one round of replication will occur, giving a poor virus amplification and low-titer virus stock. Insect cells also have a relatively high dissolved oxygen content requirement, particularly when infected with virus. Maintaining the appropriate dissolved oxygen content is important for cell growth and virus replication, and this can be achieved in shake, spinner, and tissue culture flasks by using vented caps and by not over-tightening lids. Flasks must not be filled to more than 50% of their total volume to maximize the surface area exposed to the air.

METHODS AND APPROACHES · 157

Protocol 5

Amplification of recombinant *flash*BAC virus

Equipment and Reagents
- Seed stock of recombinant virus prepared in *Protocol 4*
- 50 ml Culture of insect cells (Sf21 or Sf9) in exponential-phase growth
- Serum-supplemented or serum-free insect cell culture media
- Suitable shake/spinner culture flask
- Selection of sterile pipettes (1–10 ml)
- Incubator set at 28°C
- Inverted phase-contrast microscope
- Sterile pipettes
- 1% Virkon (Amtec) or other suitable disinfectant

Method
1. Observe the health and viability of cells under an inverted microscope[a].
2. Set up a 50 ml culture at low density, e.g. 0.5×10^6 Sf9/Sf21 cells/ml and allow the cells to grow to the required density (e.g. for approximately 48 h). Monitor cell growth using an inverted microscope and hemocytometer as described in *Protocol 1*.
3. When the cells reach the correct density (1×10^6 Sf21 cells/ml or 2×10^6 Sf9 cells/ml), add 0.5 ml of the recombinant virus seed stock (from *Protocol 4*) and incubate with shaking or stirring (as appropriate) until the cells appear to be well infected (normally 4–5 days)[b].
4. Harvest the culture medium by centrifugation at low speed (e.g. 1500 g) at 4°C for 15 min.
5. Decant aseptically and store the recombinant virus in the dark at 4°C[c].

Notes
[a]It is important that the cells are healthy and have good viability to ensure that virus replication occurs efficiently to amplify high-titer stocks of virus for subsequent use in expression studies.
[b]Virus-infected cells will appear uniformly rounded and grainy with distinct enlarged nuclei.
[c]The virus inoculum may be stored for 6–12 months or longer in the dark at 4°C. The titer of the virus will start to decrease over time, but the addition of serum (2–5%) may help to prevent this.

2.4.2 Quantification of recombinant virus by plaque assay

It is important that the titer of the recombinant virus is known so that, in expression studies, cells can be infected with a known, high m.o.i. This will ensure that all cells are infected simultaneously to produce a synchronous culture for accurate optimization studies. Titration also allows reproducibility among protein production runs and maximizes the chance of detecting the expressed protein, especially where levels of expression are lower than expected. It is also important to have an estimate of the virus titer when preparing stocks of virus prior to the protein production stage. Amplification of virus using an excessively high m.o.i. may result in the generation of defective particles (18) and waste valuable virus stocks. Traditionally, baculoviruses have been titrated by plaque assay (19–21), as described in *Protocol 6*, or by end-point dilution (22). Although the plaque assay method of titration is still considered the most accurate, this claim is being increasingly challenged by new technologies (see *Protocol 8*).

Protocol 6

Plaque assay for titration of recombinant baculovirus

Equipment and Reagents
- 35 mm^2 Tissue culture treated dishes (eight dishes per virus to be titrated and two control dishes)
- Insect cells (Sf9 or Sf21)[a]
- Inverted phase-contrast microscope
- Virus to be titrated (from *Protocol 5*)
- Appropriate serum-free or serum-supplemented culture medium
- Low-gelling-temperature agarose for cell culture (2%, w/v, in sterile dH$_2$O, sterilized by autoclaving)
- Antibiotics (optional) (penicillin and streptomycin prepared with 5 units/ml penicillin G sodium and 5 µl/ml streptomycin sulfate in 0.85% saline; 1:100 final dilution)
- Incubator at 28°C and a plastic sandwich box
- Phosphate-buffered saline (PBS, pH 6.2, sterilized by autoclaving)
- Neutral red (e.g. from Sigma). Prepare a stock solution at 5 mg/ml in water, filter through a 0.2 µm filter and store at room temperature. For use, dilute 1:20 in PBS (do not store after dilution)
- Sterile pipettes, disposable tips, and bijoux to prepare serial dilutions
- 1% Virkon (Amtec) or other suitable disinfectant

Method
1. Confirm the health and viability of the insect cells to be used under an inverted microscope[b].
2. Seed ten 35 mm^2 dishes with an appropriate number of cells to form a subconfluent monolayer (1.5 × 10^6 Sf21 or 0.9 × 10^6 Sf9 cells/dish) and leave for 1 h at room temperature on a level surface.
3. Add 0.45 ml of appropriate medium into each of seven sterile bijoux (or microcentrifuge tubes). To the first bijoux, add 50 µl of the recombinant virus to be titrated (this will be a 10^{-1} dilution) and mix thoroughly by vortexing or inversion. Using a fresh pipette tip, remove 50 µl from the 10^{-1} bijoux and transfer it to the next one (this will be a 10^{-2} dilution) and vortex. Continue diluting the virus in this way to a dilution of 10^{-7}.
4. One hour after seeding the dishes, use the microscope to confirm that the cells have formed a subconfluent monolayer and then remove the culture medium and discard into 1% Virkon or other disinfectant.
5. Add 100 µl volumes (using fresh pipette tips) from each of the 10^{-4} to 10^{-7} dilutions to duplicate dishes (eight dishes in total, four dilutions to be plated) dropwise to the center of each dish. Add 100 µl of suitable medium to the remaining two control dishes.
6. Incubate the dishes at room temperature for 40–60 min on a level surface to allow virus adsorption[c].
7. Meanwhile, completely melt the solidified agarose and, after cooling to about 50°C, mix with an equal volume of appropriate insect cell culture medium. If using antibiotics, add these to the culture medium before preparing the overlay. You will need 2 ml of agarose/medium overlay for each dish.
8. After the virus adsorption period, carefully aspirate the virus inoculum from each of the 35 mm^2 dishes using a sterile pipette by tipping the dish to one side[d] and discard into 1% Virkon or other disinfectant[e].

9. Gently pipette 2 ml of the agarose overlay down the side of each dish, allowing it to roll over the cells so as not to disturb the monolayer and incubate at room temperature until the agarose has solidified[f].

10. Add 1 ml of appropriate insect cell culture medium to each dish, as a liquid feed overlay (with antibiotics if required).

11. Place the dishes into a secure container (e.g. a sandwich box) and incubate at 28°C for 3 days (Sf21 cells) or 3–4 days (Sf9 cells).

12. After 3–4 days, confirm that the cells have formed a confluent monolayer under the microscope. Remove the liquid overlay from the dishes and replace with 1 ml of diluted neutral red stain and incubate for 3–4 h at 28°C.

13. After 3–4 h, pour off the stain (into disinfectant) and invert the dishes (place on tissue paper, which can then be discarded by autoclaving), before replacing the lids. Leave the dishes in the dark, in the inverted position, for the plaques to clear[g].

14. Plaques are visible as clear areas against a red background (only live cells take up the stain). Select one set of duplicate dishes with between 10 and 30 plaques (ideally) and count them. Calculate the average number of plaques for that dilution. To determine the virus titer, use the following calculation:

    Titer of virus (p.f.u./ml) = average plaque count × dilution factor × 10 (as 0.1 ml of virus was added)

15. If you observe no plaques, it may be because the titer of the virus is too low to be detected using the dilutions that you have plated out (also try plating out the lower dilutions, from 10^{-1} to 10^{-7}). Alternatively, the virus titer may be so high that it has lysed all of the cells and plaques have merged into each other (try plating out higher dilutions).

Notes

[a]We routinely use Sf21 cells in serum-supplemented medium as they produce distinct, large plaques in 3 days, compared with smaller, less-distinct plaques in 4 days for Sf9 or Sf21 cells grown in serum-free medium.

[b]It is important that the cells are healthy and have good viability to ensure the formation of distinct plaques. Seeding cells too densely will result in poor virus growth and small plaques, whilst seeding cells too thinly will require longer incubation periods resulting in large, ill-defined, and diffuse plaques.

[c]Ensure that the cell monolayers do not dry out by removing them to the bench if working inside a cabinet.

[d]A common error is to forget to change tips when preparing the virus dilutions, resulting in virus carry-over and a false, high virus titer. Also, take care not to disturb the cell monolayer or allow it to dry out during this process.

[e]If the virus inoculum is not completely removed from the cells before adding the agarose overlay, it will interfere with the gelling process and produce cracks, causing the overlay to fall away when the dishes are inverted. It may also cause the plaques to look smeared by allowing the virus to spread randomly, rather than being contained within foci of cells.

[f]Ensure that the temperature of the agarose overlay is not too high when poured over the cells and that the cells have not been allowed to dry out at any stage. The latter is characterized by large areas of bright pink stain with a glassy appearance.

[g]This may take a few hours or may occur very rapidly, depending on the strength of the neutral red.

2.4.3 Analysis of recombinant protein expression

With the exception of the quick analysis of cells used in the transfection, as described in *Protocol 1*, most analyses of gene expression will require infecting a dish of fresh cells with the recombinant virus. *Protocol 7* describes a quick and easy method to check that your virus is producing recombinant protein at satisfactory levels, before proceeding to more-lengthy optimization steps. The best time to harvest the recombinant protein can then be examined by taking samples at different times after infection. The most commonly used times are at 24, 48, 72 and 96 h post-infection (p.i.). Some proteins may be very stable and accumulate to high levels by 96 h p.i., whilst others may start to degrade and thus need to be harvested much earlier. Protein expression can then be analyzed by SDS-PAGE (see *Fig. 2*). To optimize the m.o.i., set up dishes of cells or shake cultures and infect with different m.o.i. (2, 5, and 10 are recommended as a guide). Harvest samples at different times after infection (24, 48, 72, and 96 h p.i.) and examine protein synthesis. You may achieve good levels of protein synthesis at 48 h p.i. with a very high m.o.i. (e.g. 10), but if you wait until 72 h p.i., you may achieve the same levels with an m.o.i. of only 2 or 5, thus using less virus inoculum. Protein expression may also be compared between Sf9 and High Five cell lines (see section 2.2).

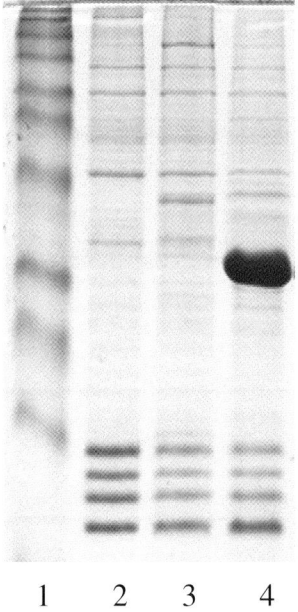

Figure 2. Analysis of recombinant protein expression by SDS-PAGE.
Lane 1, molecular mass marker; *lane 2*, uninfected cells; *lane 3*, mock-infected cells; *lane 4*, virus-infected cells. Cells were infected at a high m.o.i. and harvested at 48 h p.i. Image supplied by D. Carpentier (Oxford Brookes University, UK).

Protocol 7

Analysis of recombinant protein expression

Equipment and Reagents
- 35 mm^2 Tissue culture treated dishes
- Insect cells (Sf9 or High Five) and appropriate medium
- Inverted phase-contrast microscope
- Recombinant virus stock (from *Protocol 5*)
- Incubator at 28°C and a plastic sandwich box
- Selection of sterile pipettes (1–10 ml)
- Microfuge tubes (1.5–2 ml)
- PBS
- 1% Virkon (Amtec) or other suitable disinfectant
- 4× SDS-PAGE sample buffer (50% 0.5 M Tris/HCl, pH 6.8, 40% glycerol, 8% 2-mercaptoethanol, 0.04 g bromophenol blue, 0.8 g SDS, prepared in a final volume of 25 ml and stored frozen in 1 ml aliquots)

Method
1. Seed 35 mm^2 dishes (one per virus to be tested and a mock-infected control) with 1×10^6 cells per dish and leave to settle for 1 h.
2. Remove the medium and add 200 µl of the recombinant virus stock to be tested and incubate at room temperature for a further hour.
3. Remove the virus inoculum and add 1.5 ml medium and incubate at 28°C for 48 h.
4. Use a sterile pipette tip to scrape the cells from the dishes into suspension and remove into sterile microfuge tubes.
5. Pellet the cells at 1500 *g* for 10 min. Remove the supernatant fraction and discard[a].
6. Add 80 µl of PBS to the pellet and vortex.
7. Next add 20 µl of 4× SDS-PAGE sample buffer and mix. These samples can then be stored at −20°C until needed for SDS-PAGE analysis.
8. Prior to electrophoresis, boil the samples at 100°C for 5 min.

Note
[a] For secreted protein expression, the supernatant would be analyzed.

2.5 Automated production of recombinant baculoviruses

The simplicity of the *flash*BAC system and its ability to produce recombinant virus without any parental background make it ideal for use with automated liquid handlers (23). As each robotic system will be different, *Protocol 8* is not detailed but is simply a guide to illustrate the ease with which 24 recombinant viruses can be made simultaneously in a 24-well plate format. The robotic system used must be capable of pipetting under sterile conditions; smaller robots, for example, may be conveniently placed in a Class II hood (see *Fig. 3*).

2.5.1 Automated screening of recombinant proteins

High-throughput production of recombinant viruses requires a rapid and sensitive method for testing the presence of the expression product during the early stages of the process. This is easily carried out using *flash*BAC, where the pure and relatively high-titer co-transfection mixture can be used to infect cells directly in a 24-well plate format. The cells and supernatant can then be harvested and assayed for protein production as described in *Protocol 9*. This method can also be scaled up to a 24-deep-well block format (24).

Figure 3. A BaculoWorkstation (NextGen Sciences Ltd).
The workstation is located within a Class II hood, which provides a sterile environment for working with insect cells and viruses.

Protocol 8

Automated production of multiple recombinant baculoviruses

Equipment and Reagents
- BaculoWorkstation (NextGen Sciences Ltd) or similar robot, situated within a sterile environment, e.g. a Class II microbiological cabinet
- Filtered tips (ideally carbon coated for liquid level sensing)
- Tissue culture plates (24- and 96-well)
- flashBAC DNA (100 ng per co-transfection)
- Sterile transfer vector DNA (500 ng per co-transfection) with suitable tag for expression testing, e.g. His tag (see Protocol 4)
- Sf9 cells from a culture in exponential-phase growth (2×10^5 cells/well are required)
- Serum-free insect cell culture medium
- Transfection reagent, e.g. Lipofectin (Invitrogen)
- Parafilm
- Incubator set at 28°C
- 1% Virkon (Amtec) or other suitable disinfectant
- Inverted phase-contrast microscope
- Sterile bijoux or similar[a]

Method
1. Confirm the health and viability of the insect cells to be used under an inverted microscope and prepare a stock of Sf9 cells in serum-free medium at a density of 5×10^5 cells/ml.
2. Program the robot to aliquot 400 µl (2×10^5 cells/well) into each well of a 24-well tissue culture plate. Allow the cells to settle and attach for 1 h before use.
3. During the 1 h incubation period, program the robot to prepare 24 co-transfection mixes in the wells of a 96-well plate (made from polystyrene and with U- or V-shaped wells).
4. Dispense the following to each of 24 wells of a 96-well plate in the following order to give a final volume of 20 µl:
   - 8 µl of serum-free medium
   - 2 µl of Lipofectin or other transfection reagent
   - 5 µl of flashBAC DNA (100 ng)
   - 5 µl of transfer vector DNA (500 ng)
5. Mix the reagents by gentle pipetting (dispense and aspirate three times at slow speed).
6. Dispense the 20 µl co-transfection mix to the appropriate wells of the 24-well plate containing the cell monolayers[b].
7. Replace the lid and seal with Parafilm to prevent evaporation. Incubate at 28°C for 5 days.
8. Harvest the culture medium containing recombinant virus from each well and store at 4°C in the dark.

Notes
[a]Plasticware used to prepare the transfection mixture must be made from polystyrene and not from polypropylene.
[b]There is no need to change the medium; simply add the co-transfection mix to the medium and mix by gentle pipetting.

Protocol 9

Automated expression screening of recombinant viruses

Equipment and Reagents
- BaculoWorkstation (NextGen Sciences Ltd) or similar robot, situated within a sterile environment, e.g. a Class II microbiological cabinet
- Filtered tips (ideally carbon coated for liquid level sensing)
- 24-Well tissue culture plates
- Multiple recombinant viruses, e.g. produced in *Protocol 7*
- Sf9 cells from a culture in exponential-phase growth (4×10^5 cells/well are required)
- Serum-free insect cell culture medium
- Parafilm
- Incubator set at 28°C

Method
1. Prepare a master mix of Sf9 cells in serum-free medium at a cell density of 5×10^5 cell/ml.
2. Use the robot to aliquot 400 µl (4×10^5 cells/well) into each well of a 24-well tissue culture plate. Allow the cells to settle and attach for 1 h before use.
3. During this incubation period, program the robotic system to dispense the recombinant viruses (generated in *Protocol 8*) directly onto these cells using standard liquid handling protocols.
4. Add 200 µl of each virus into each of the wells of the 24-well plate, allowing a well for an uninfected control.
5. Replace the lid and seal with Parafilm to prevent evaporation. Incubate at 28°C for 48–72 h.
6. Harvest the culture medium and pellet from each well into individual sterile containers and use for expression testing, e.g. by SDS-PAGE and Western blotting.

2.5.2 Rapid quantification of recombinant baculovirus

A rapid and highly reproducible titration method based on quantitative PCR (QPCR) has been developed in our laboratory (25), which offers the versatility to titrate not only single viruses, but also multiple viruses in a semi-automated, high-throughput format. During QPCR, a target-specific probe anneals downstream from one of the primer sites. The probe is tagged with a reporter fluorescent dye (FAM) at its 5′ end and a quencher dye (TAMRA) at its 3′ end. Whilst the probe is intact, the proximity of the two dyes results in fluorescent resonance energy transfer, with the quencher greatly reducing the fluorescence of the reporter dye. During PCR amplification, the 5′→3′ exonuclease activity of *Taq* polymerase cleaves the dual-labeled probe as the primer extends, resulting in release of the reporter dye and interruption of the fluorescent resonance energy transfer, therefore increasing the reporter dye fluorescence signal. This cleavage removes the probe from the target DNA strand, permitting primer extension. Subsequent rounds of PCR amplification result in additional cleavage of reporter dye molecules from their respective dual-labeled probes, culminating in a successive increase in fluorescence intensity proportional to the amount of PCR product produced. This increase in fluorescence can be quantified in real time and permits quantification during the exponential phase of the reaction as opposed to end-point accumulation of a product by conventional PCR. Quantitative analysis is represented by means of an amplification plot (see *Fig. 4*, also available in the color section), illustrating the fluorescence signal versus the cycle threshold (C_t) of the reaction. The C_t value is the point at which the fluorescence intensity rises above the background-noise baseline and is determined at the most exponential phase of the PCR. This value is inversely proportional to the amount of target DNA molecules and therefore permits quantification of the template DNA using a standard curve. The *baculo*Quant titration method relies on being able to obtain consistently reproducible quantities of viral DNA from known volumes of inoculum. Having tested other commercially available kits, we recommend using the High Pure Viral Nucleic Acid kit (Roche Molecular Diagnostics). The greater the number of virus particles within a sample, the higher the concentration of virus DNA that will be present and, correspondingly, the lower their C_t value. By comparing this C_t value to a known standard curve, it is possible to establish an accurate titer, which can be converted to equivalent p.f.u./ml (Qp.f.u./ml), using an equation derived from linear regression analysis.

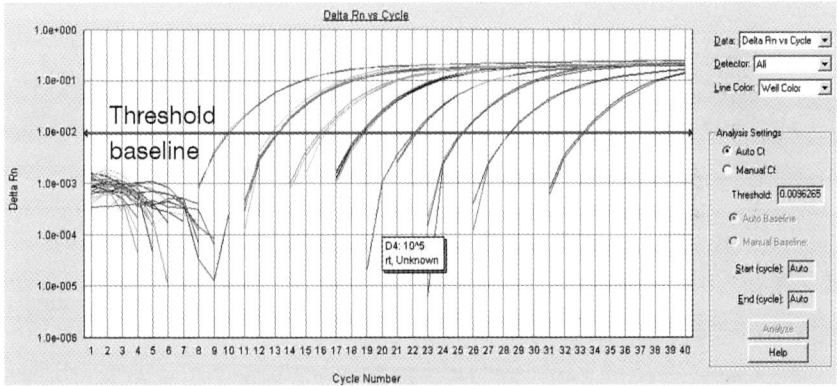

Figure 4. Typical amplification plot generated using the Applied Biosystems Sequence Detection System 7500, showing the threshold baseline (see page xxiii for color version).
A clear distinction between different virus titers can be seen from the plot as represented by cycle number.

Protocol 10

Rapid baculovirus titration using *baculo*Quant

Equipment and Reagents
- Stock of BV to be titrated
- *baculo*Quant kit (Oxford Expression Technologies Ltd) containing:
  - 2.5 µM forward primer (5'-CGGCGTGAGTATGATTCTCAAA-3')
  - 2.5 µM reverse primer (5'-ATGAGCAGACACGCAGCTTTT-3')
  - 2.5 µM dual-labeled probe (5'-FAM-AAAAGTCTACGTTCACCACGCGCCAAA-TAMRA-3')
  - DNA standard (1000 ng at 25 ng/µl)
  - BaculoQuant manual
- High Pure Viral Nucleic Acid kit (Roche Molecular Diagnostics)
- QPCR master mix, i.e. ABsolute Blue QPCR ROX Mix (ABgene)
- Applied Biosystems 7500 Real-Time PCR Sequence Detection System (or other QPCR system)
- Optical 96-well plates and plate seals (Applied Biosystems)
- Filter pipette tips

Method

1. Remove 200 µl of BV from the recombinant baculovirus stock and isolate the viral DNA using the High Pure Viral Nucleic Acid kit, following the manufacturer's instructions[a].

2. Prepare a log series of tenfold dilutions of the supplied DNA standard in final volumes of 10 µl (25, 2.5, 0.25, 0.025, and 0.0025 ng/µl). Two microlitres of each of the diluted standards will then be used in each DNA standard reaction (see step 5).

3. Prepare QPCRs as a master mix as follows according to how many viruses you need to titrate:

Component	Volume (final conc.) per reaction
ABsolute Blue QPCR ROX mix	12.5 µl
Water	7.5 µl
Forward primer	1 µl (100 mM)
Reverse primer	1 µl (100 mM)
Probe	1 µl (100 mM)

Each virus sample, along with the control sample and each of the five standard DNA samples (step 2), should be analyzed in triplicate. Control reactions contain water in the place of DNA.

4. Aliquot 23 µl of master mix into the appropriate number of wells in a 96-well plate. Each reaction will be carried out in triplicate, i.e. three lots of five wells for standards, three wells for the water control, and three wells per unknown virus.

5. Add 2 µl of your purified viral DNA from step 1 (or 2 µl of DNA standards or water), giving a final volume of 25 µl per reaction.

6. Dispense each 25 µl reaction into a 96-well plate, making a note of their location on the plate.

7. Enter the position of each reaction into the Sequence Detection System software along with the fluorescent dyes used and the standard DNA dilutions.

8. Carry out DNA amplification following the manufacturer's instructions, e.g. ABI Quantitative Analysis, using the following cycling conditions:

   - UNG activation: 1 cycle of 2 min at 50°C
   - *Taq* activation/UNG deactivation: 1 cycle of 15 min at 95°C
   - Annealing/extension: 40 cycles of 15 s at 95°C and 1 min at 60°C

9. At the end of the QPCR cycle, export the C_t values into an EXCEL spreadsheet. Calculate the mean C_t value for each virus and enter this value into the equation provided in the *baculo*Quant manual (OET Ltd). This will convert the QPCR-derived titer to its plaque assay equivalent (Qp.f.u./ml).

> **Note**
> ªElute the viral DNA from the column in a final volume of 50 µl.

3. TROUBLESHOOTING

- It is important not to let your cells overgrow, i.e. to subculture them on a regular basis, and to avoid using cells that have undergone more than 30 continuous passages since being raised from liquid nitrogen. Check the health and viability of your cells prior to each experiment.
- Poor viral amplification is generally caused by poor-quality cells or infection at the wrong m.o.i. (i.e. too little or too much virus). We recommend always titrating your virus stock to allow calculation of an exact m.o.i.
- We recommend always checking your construct by DNA sequencing. For example, is the coding region of the gene downstream of the *polh* gene promoter inserted in such a way that the gene's AUG start codon is the first AUG after the promoter sequences? This is important, as translation occurs at the first AUG in the mRNA.
- Poor expression levels may be improved by optimization, for example, using different cell lines and time points for harvesting (see section 2.4.3).

4. REFERENCES

1. Kost TA, Condreay JP & Jarvis DL (2005) *Nat. Biotechnol.* **23**, 567–575.
2. Luckow VA, Lee SC, Barry GF & Olins PO (1993) *J. Virol.* **67**, 4566–4579.
3. Hawtin RE, Arnold K, Ayres MD, *et al.* (1995) *Virology*, **12**, 673–685.
4. Hawtin RE, Thomas CA, Gooday GW, Kuzio JA, King LA & Possee RD (1997) *Virology*, **238**, 243–253.
★★★ 5. Thomas CA, Hawes CR, Lee BY, Min MK, King LA & Possee RD (1998) *J. Virol.* **72**, 10207–10212. – *Publication describing the importance of viral chitinase.*
6. Saville GP, Patmanidi AL., Possee RD & King LA (2004) *J. Gen. Virol.* **85**, 821–831.
7. Possee RD, Saville GP, Thomas CJ, Patminidi AL & King LA (2001) *Proceedings of a Joint International Symposium of Insect COE Research Program and Insect Factory Research Project*, 22–23 October, Tsukuba, Japan.
8. O'Reilly DR, Miller L & Luckow VA (1992) *Baculovirus Expression Vectors – a Laboratory Manual*. WH Freeman, New York.
9. Vaughn JL, Goodwin RH, Tompkins GJ & McCawley P (1977) *In Vitro*, **13**, 213–217.
10. Hink WF (1970) *Nature* **226**, 466–467.
11. Davis TR, Wickham TJ, McKenna KA, Granados RR, Shuler ML & Wood HA (1993) *In Vitro Cell Dev Biol Anim.* **29**, 388–390.
12. Granados RR, Guoxun L, Derksen ACG & McKenna KA (1994) *Invertebr. Pathol.* **64**, 260–266.

13. **Wickham TJ, Davis T, Granados RR, Shuler ML & Wood HA** (1992) *Biotechnol. Prog.* **8**, 391–396.
14. **Wickham TJ & Nemerow GR** (1993) *Biotechnol. Prog.* **9**, 25–30.
15. **Friesen PD & Nissen MS** (1990) *Mol. Cell Biol.* **10**, 3067–77.
16. **Davis TR, Trotter KM, Granados RR & Wood HA** (1992) *Biotechnology*, **10**, 1148–1150.
17. **Weber W, Weber E, Giesse S & Memmert K** (2002) *Cytotechnology*, **38**, 77–85.
18. **Kool M, Voncken JW, van Lier FL, Tramper J & Vlak JM** (1991) *Virology*, **183**, 739–746.
19. **Hink WF & Vail PV** (1973) *J. Invertebr. Pathol.* **22**, 168–174.

★★ 20. **King LA & Possee RD** (1992) *The Baculovirus Expression System: a Laboratory Guide.* Chapman & Hall, London. – *A complete guide to baculovirus methods and techniques.*

21. **O'Reilly DR, Miller LK & Luckow VA** (1994) *Baculovirus Expression Vectors: a Laboratory Manual.* Oxford University Press: New York.
22. **Lynn DE** (1992) *BioTechniques*, **13**, 282–285.
23. **Hunt I** (2005) *Protein Expr.Purif.* **40**, 1–22.
24. **Bahia D, Cheung R, Buchs M, Geisse S & Hunt I** (2005) *Protein Expr. Purif.* **39**, 61–70.

★ 25. **Hitchman RB, Siaterli EA, Nixon CP, King LA** (2007) *Biotechnol Bioeng.* **96**, 810–814. – *The original publication first describing baculovirus titration by Taqman QPCR.*

CHAPTER 10

Transient transfection of insect cells for rapid expression screening and protein production

Kathryn H. Loomis, Courtney R. Rockwell, Heather D. Sternard, Keith W. Yaeger, and Robert E. Novy Jr

1. INTRODUCTION

Expressing recombinant proteins in insect cells is a proven method for producing active, soluble, and in some cases appropriately post-translationally modified recombinant proteins. Baculovirus-mediated expression is the most commonly used method (1). An alternative approach is the generation of stable insect cell lines through the use of appropriately designed plasmids (2, 3). In both cases, the time required to obtain expression results is substantial, 1.5–3 weeks for baculoviruses and numerous weeks for stable cell lines. In addition, baculovirus infection can potentially cause undesired alterations in cell physiology that may perturb the desired post-translational processing pathways (4). Transient transfection represents a significantly faster approach for protein expression in insect cells. Although this technique is commonly used in mammalian tissue culture applications, its use in insect cells is not as widespread and is underappreciated. The method is relatively simple and involves mixing an appropriate expression vector with a transfection reagent, applying the mixture to cultured insect cells, growing the transfected cells for 48–72 h, and harvesting and lysing the transfected cells, followed by purification of the recombinant protein. Critical factors influencing expression are the enhancer and promoter elements driving transcription from the plasmid and the efficiency of the transfection reaction. Although transient transfection has limitations for large-scale applications (cultures greater than 10 l), it does represent a useful alternative for volumes up to a few liters. The method can also be utilized on a small scale for rapid screening of expression in insect cells prior to devoting resources to generate baculovirus recombinants for larger-scale applications.

Expression Systems: *Methods Express* (M.R. Dyson and Y. Durocher, eds)
© Scion Publishing Limited, 2007

2. METHODS AND APPROACHES

To accelerate the process of recombinant protein expression and purification in insect cells, EMD developed a start-to-finish approach based on transient transfection, the InsectDirect System. This system employs an optimal combination of expression vector, high-efficiency transfection reagent, insect cell lysis reagents, and immobilized metal affinity chromatography (IMAC) resin for affinity purification of His•Tag fusion proteins (5-7) (see *Fig. 1*). Although yields commonly fall within the range of 0.5-5 mg of target protein per liter of transfected culture, we have produced several proteins at 10-80 mg/l of culture (5, and unpublished data). In addition, the approach has proved to be directly scaleable in the 1 ml to 1 l (4 × 250 ml) range of culture/transfection volumes (5).

2.1 Media and insect cells

The majority of results generated with the InsectDirect System have utilized *Spodoptera frugiperda* Sf9 insect cells grown in serum-free medium. Medium supplemented with serum for cell growth is also compatible with this method.

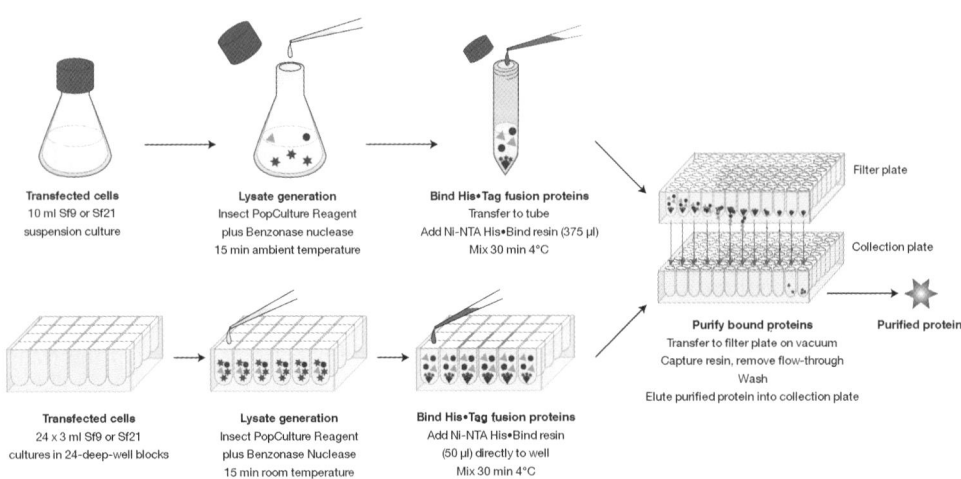

Figure 1. InsectDirect System.
The diagram depicts the process for shake flask and 24-well plate cultures for the InsectDirect System. Transfected cultures are lysed by the addition of Insect PopCulture Reagent and the viscosity resulting from released nucleic acids is decreased by the addition of Benzonase nuclease. Ni-NTA His•Bind resin is to added to capture the fusion proteins bearing a His•Tag coding sequence. Lysate/resin mixtures from multiple shake flask cultures or 24-well plate-based cultures are then processed using a filter-plate-based procedure to yield purified protein. The purification steps for the shake flask cultures and the lysis and purification steps in the 24-well plate-based cultures are compatible with robotic liquid handling automation (8).

2.2 Transient expression vectors

The pIEx (plasmid for immediate-early expression) vectors are derivatives of the pIE1 vectors originally generated for the creation of stable insect cell lines (6). The pIEx vectors feature the *hr5* (homologous region 5) enhancer and the *ie1* (immediate-early 1) promoter from *Autographa californica* multiple nucleopolyhedrovirus (AcMNPV) (7). This promoter/enhancer combination uses endogenous insect cell transcriptional machinery, avoiding baculovirus infection and the associated cytopathic effects (8–11). When used with a high-efficiency, low-toxicity, liposome-based transfection reagent, transient transfection with these vectors can give recombinant protein yields similar to that obtained from baculovirus infection (12, 13). In other cases, however, a significantly higher level of protein can be obtained using the baculovirus system, indicating that yield is a protein-dependent phenomenon (14). In general, a relative correlation is observed between the two systems. Although the absolute yield may differ, proteins expressing at high or low levels in the transient system also express at relative high or low levels in the baculovirus system.

Any vector carrying transcription elements that are active in insect cells is technically compatible with transient transfection. Therefore, vectors originally designed for the selection of stable insect cell lines can also be used in a transient mode. Examples include vectors based on the OpIE2 promoter (*ie2* promoter derived from *Orgyia pseudotsugata*) (3) and vectors based on a combination of the insect actin promoter and *Bombyx mori* nucleopolyhedrovirus-derived *hr3* enhancer (2). These vectors carry additional elements on the plasmid backbone that facilitate cell line selection. Accordingly, these types of vector are significantly larger than pIEx. Given that transfection efficiency is inversely correlated to plasmid size (15), the relatively small size of the pIEx vectors is likely to be another factor contributing to their improved performance compared with the cell line selection vectors (6, and unpublished data).

A new alternative vector, pIEx/Bac, is compatible with both transient transfection and baculovirus generation. This vector features the *hr5* enhancer/*ie1* promoter along with the late/very late AcMNPV-derived *p10* promoter on a baculovirus transfer vector backbone (14). The AcMNPV-derived sequences (*lef2*/ORF603 upstream and partial ORF1629 downstream) facilitate homologous recombination into the *polH* locus of AcMNPV. pIEx/Bac is compatible with traditional *Bsu*36I-linearized baculovirus DNA (BacVector, EMD; BacPAK, Clontech; and BaculoGold, BD Biosciences) and with a recent innovation designed for faster baculovirus generation (BacMagic DNA, EMD; *flash*BAC DNA, Oxford Expression Technologies).

2.3 Transfection reagent

Liposome-based formulations are widely used for the transfection of insect cells and numerous formulations are available commercially. The reagent used in this approach, Insect GeneJuice Transfection Reagent, is a liposome formulation with low toxicity and high transfection efficiency, which is optimized for the transfection of plasmid DNA into insect cells.

2.4 Insect cell lysis

Transfected cells can be concentrated by centrifugation prior to generating a lysate. Alternatively, cell lysis can be achieved in a high-throughput 'in-media' mode without harvesting suspended cells by centrifugation. In both cases, detergent-based, buffered lysis reagents are used to create a total cell extract. The lysis reagents are formulated to generate extracts while avoiding protein denaturation.

2.5. IMAC purification of His•Tag fusion proteins

Recombinant proteins generated with this approach typically have a 6× or 10× His•Tag sequence and are purified using IMAC (5). When purifying His•Tag fusion proteins from eukaryotic cell extracts, it is typically necessary to use more resin than would theoretically be needed based on manufacturers' stated binding capacity. Presumably this is due to the presence of metals and/or compounds in the extracts and/or growth medium.

The following section provides protocols for 24-well plate to medium-scale (250 ml) applications of the InsectDirect technique. Section 2.7 demonstrates applications of the technique using pIEx and pIEx/Bac recombinants and offers comparisons with the baculovirus technique.

2.6 Recommended protocols

Protocol 1

Transfection of Sf9 cells

Equipment and Reagents
- BacVector or TriEx medium (EMD)
- Sf9 insect cells (EMD)
- Insect GeneJuice Transfection Reagent (EMD)
- Recombinant pIEx plasmid DNA (endotoxin-free)
- Sterile tissue culture flasks or plates (Nalgene or Corning)
- Laminar flow hood
- Incubator at 28°C
- Temperature-controlled orbital shaker at 28°C
- Inverted phase-contrast microscope

Method
The following procedures describe the method for introducing plasmid DNA into Sf9 cell cultures in a 24-well plate adherent culture and 10 ml and 3 ml suspension cultures. Alternative formats require adjustment to the amount of DNA, Insect GeneJuice Transfection Reagent and cell seeding densities as indicated in *Table 1*.

Twenty-four-well plate adherent culture
1. One hour prior to transfection, seed 2×10^5 Sf9 cells[a] in each well of a 24-well plate in 0.5 ml of medium[b]. Do not swirl the plates as cells will cluster in the middle.

Table 1. Transfection mixtures

	24-Well plate	Six-well plate	10 ml (125 ml flask)	100 ml (1 l flask)	1 l (4 × 1 l flask)
Number of cells	2×10^5	1×10^6	1×10^7	1×10^8	1×10^9
Culture volume prior to transfection	0.5 ml	2.5 ml	9 ml	90 ml	950 ml
Amount of plasmid DNA	0.3 µg	1.5 µg	15 µg	150 µg	1.5 mg
Volume of Insect GeneJuice Transfection Reagent	1.5 µl	7.5 µl	75 µl	750 µl	7.5 ml
Volume of serum-free medium added to both plasmid and Insect GeneJuice Transfection Reagent for dilution	Up to 20 µl	Up to 100 µl	Up to 0.5 ml	Up to 5 ml	Up to 25 ml
Volume of serum-free medium added to Insect GeneJuice Transfection Reagent/DNA complexes after 15 min incubation	160 µl	0.8 ml	NA	NA	NA

NA, Not applicable.

2. For each well to be transfected, dilute 0.3 µg of DNA[c] with 15 µl of serum-free medium. Also, dilute 1.5 µl of Insect GeneJuice Transfection Reagent[d] with 20 µl of serum-free medium[e].
3. Slowly add the diluted DNA *dropwise* to the diluted Insect GeneJuice Transfection Reagent. Mix immediately by gentle vortexing or pipetting to avoid precipitation.
4. Incubate the Insect GeneJuice Transfection Reagent/DNA mixture at room temperature for 15 min.
5. After the 15 min incubation, add 160 µl of serum-free medium to the Insect GeneJuice Transfection Reagent/DNA transfection mixture.
6. Aspirate the medium from the cells and add the transfection mixture to the cells.
7. Cover the plate and carefully transfer to a flat-bottomed, covered storage container containing a damp paper towel for moisture. Incubate the cells at 28°C for 48–72 h. Optional: Remove transfection mixture after 4 h and replace with complete growth medium.
8. Harvest cells for characterization or reporter assays (see *Protocol 2*).

Suspension culture – 10 ml

1. Seed 1×10^7 Sf9 cells[a] in 9 ml of serum-free medium[b] in a 125 ml Erlenmeyer shake flask.
2. In a sterile tube, dilute 15 µg of plasmid DNA[c] with 0.5 ml of serum-free medium. Also, dilute 75 µl of Insect GeneJuice Transfection Reagent[d] with 0.425 ml of serum-free medium[e].
3. Add the DNA *dropwise* to the Insect GeneJuice Transfection Reagent and mix immediately by gentle vortexing or pipetting to avoid precipitation.
4. Incubate at room temperature for 15 min.
5. Add the transfection mixture to the cells.
6. Incubate the cells at 28°C, shaking at 150 r.p.m. for 48–72 h.
7. Harvest the cells for characterization or reporter assays (see *Protocol 2*).

Suspension culture (3 ml) in 24-deep-well blocks – high-throughput application

1. Seed 3×10^6 Sf9 cells[a] in 2.7 ml of serum-free medium[b] per well of a 24-deep-well block.
2. In a sterile tube, dilute 4.5 µg of plasmid DNA[c] with 0.15 ml of serum-free medium. Also, dilute 22.5 µl of Insect GeneJuice Transfection Reagent[d] with 0.123 ml of serum-free medium[e].
3. Add the DNA *dropwise* to the Insect GeneJuice Transfection Reagent and mix immediately by gentle vortexing or pipetting to avoid precipitation.
4. Incubate at room temperature 15 min.
5. Add the transfection mixture to the cells.
6. Incubate the cells at 28°C, shaking at 300 r.p.m.[f] for 48–72 h.
7. Harvest cells for characterization or reporter assays (see *Protocol 2*).

Notes

[a] Passage cells regularly (e.g. every 2–3 days) and avoid confluent growth. For transfection, use only rapidly proliferating cells. Conditions for cell growth and density should be consistent for optimum reproducibility.

[b] Transfection has been optimized using BacVector Insect Cell Medium. Other media have not been extensively tested but are likely to be compatible. Some medium formulations may cause formation of precipitate upon addition of Insect GeneJuice Transfection Reagent. This precipitate can significantly decrease transfection efficiency.

[c] Use only high-quality, supercoiled, endotoxin-free DNA (e.g. purified using Novagen's Mobius or UltraMobius plasmid kit).

[d] Insect GeneJuice Transfection Reagent is compatible with both serum-containing and serum-free media. Note that serum must not be present during the formation of the DNA/Insect GeneJuice Transfection Reagent complex.

[e] The ratio of Insect GeneJuice Transfection Reagent to DNA is a crucial factor for transfection optimization. We recommend 5 µl of Insect GeneJuice Transfection Reagent per 1 µg of DNA as a starting point. For optimization, vary between 4 and 8 µl of Insect GeneJuice Transfection Reagent per 1 µg of DNA.

[f] The orbital shaker must be set at 250–300 r.p.m. for 24-deep-well blocks to ensure adequate aeration.

Protocol 2

Harvesting transfected cells

Equipment and Reagents

- Insect PopCulture Reagent (EMD)
- CytoBuster Protein Extraction Reagent (EMD)
- Benzonase nuclease (EMD)
- Phosphate-buffered saline (PBS: 43 mM Na_2HPO_4, 15 mM KH_2PO_4, 137 mM NaCl, 27 mM KCl, pH 7.4) or Hanks' buffered salts solution (HBSS: 0.137 M NaCl, 5.4 mM KCl, 0.25 mM Na_2HPO_4, 0.44 mM KH_2PO_4, 1.3 mM $CaCl_2$, 1.0 mM $MgSO_4$, 4.2 mM $NaHCO_3$)
- Microcentrifuge
- Tabletop centrifuge

Method
Insect PopCulture cell extract preparation
1. Add 0.05 culture volumes of Insect PopCulture Reagent to the culture[a,b].
2. Mix by inverting gently several times and incubate for 15 min at room temperature[c,d,e].
3. His•Tag fusion proteins can be purified from total culture extracts using either Ni-NTA His•Bind resin or NiMAC His•Bind Fractogel resin. Other affinity purification methods are likely to be compatible; however, His•Bind resin (IDA agarose) and GST•Bind resin are not suitable for use with Insect PopCulture Reagent extracts.

CytoBuster cell extract preparation
Extraction of adherent cells
1. Aspirate the culture medium from the cells.
2. Wash the cells once with PBS or HBSS.
3. Add 100–300 µl of CytoBuster and incubate at room temperature for 5–15 min.
4. Transfer the extract to a suitably sized tube and centrifuge for 5 min at 16 000 g (4°C).
5. Transfer the supernatant to a new tube and proceed with analysis and/or purification[f].

Extraction of suspension cells
1. Pellet the cells by low-speed centrifugation (e.g. 5 min at 2500 g).
2. Resuspend the cells in CytoBuster Reagent using 150 µl per 10^6 cells (the optimal amount of CytoBuster Reagent may vary based on cell size).
3. Incubate at room temperature for 5–15 min.
4. Transfer the extract to a suitably sized tube and centrifuge for 5 min at 15 000 g (4°C).
5. Transfer the supernatant to a new tube and proceed with analysis and/or purification[f].

Notes
[a]Sf9 cells and TriEx Sf9 insect cells have been used successfully with Insect PopCulture Reagent. Other insect lines should be compatible, although it is possible that line-dependent differences could occur.

[b]TriEx Insect Cell Medium and BacVector Insect Cell Medium have been used successfully with Insect PopCulture Reagent. Other serum-free media are likely to be compatible. Serum may interfere with downstream applications such as protein assays and purification.

[c]Cell extracts may become viscous from nucleic acids released during cell lysis. These nucleic acids can interfere with effective protein purification. Benzonase nuclease degrades all forms of DNA and RNA (single-stranded, double-stranded, linear, and circular) to 5'-monophosphate-terminated oligonucleotides 2–5 bases long (2, 3), reducing sample viscosity. Benzonase nuclease treatment generally is not recommended for purification of proteins that must be nuclease-free. However, depending on the processing methods, Benzonase nuclease may be removed during purification by anion exchange chromatography.

[d]Insect PopCulture Reagent extraction along with Benzonase nuclease treatment can be performed at room temperature or at 4°C. However, at 4°C, incubation times may need to be increased as Benzonase nuclease activity decreases at lower temperatures.

[e]Acidic pH (<5.0) can degrade components of Insect PopCulture Reagent.

[f]Extracts prepared with CytoBuster Reagent can be used immediately or may be stored for extended periods of time at −20 or −80°C.

Protocol 3

Production of recombinant baculoviruses using pIEx/Bac recombinant and BacMagic DNA

Equipment and Reagents
- BacVector Medium (EMD)
- Sf9 Insect Cells (EMD)
- Insect GeneJuice Transfection Reagent (EMD)
- BacMagic DNA (EMD)
- pIEx/Bac Vector (EMD) DNA recombinant
- 35 mm tissue culture dishes (Nalgene or Corning)
- Sterile Erlenmeyer shake flasks (Nalgene or Corning)
- Laminar flow hood
- Incubator operating at 28°C
- Temperature-controlled orbital shaker operating at 28°C
- Inverted phase-contrast microscope

Method

Preparation of cell cultures for transfection

For each co-transfection, prepare one 35 mm plate. We also recommend including plates for positive and negative transfection controls.

1. Seed dishes with insect cells at least 1 h before use. Use 1×10^6 Sf9 cells in 2 ml of BacVector Insect Cell Medium per dish. Gently rock the plates in a side-to-side and back-and-forth pattern to ensure an even monolayer. *Do not swirl plates* as cells will cluster in the center.

2. During the 1 h incubation, prepare a co-transfection mix of DNA and Insect GeneJuice Transfection Reagent (see below).

Preparation of transfection mixture

1. For each transfection, assemble the following components in the order listed in a sterile 6 ml polystyrene tube to give a final volume of 1.015 ml. Do not substitute with a polypropylene or polycarbonate tube:
   - 1 ml of BacVector Insect Cell Medium
   - 5 µl of Insect GeneJuice Transfection Reagent
   - 5 µl of BacMagic DNA (100 ng total)
   - 5 µl of transfer vector DNA (500 ng total)

2. Mix with gentle agitation or vortexing. Incubate at room temperature for 15–30 min to allow complexes to form.

3. Just prior to the end of the transfection mixture incubation period, remove the culture medium from a 35 mm plate using a sterile pipette. Do not disturb the cell monolayer. When removing the liquid from a dish of cells, tip the dish at a 30–60° angle, so the liquid pools to one side of the dish. Do not let the monolayer dry out[a].

4. Immediately after the medium has been removed from the cells, add 1 ml of transfection mixture dropwise to the center of the dish. Incubate in a humidified container at 28°C overnight (minimum of 5 h).

5. After the initial incubation period, add 1 ml of BacVector Insect Cell Medium to each dish. Continue the incubation for a total of 5 days. Serum can be added to the medium at this point, if desired.

6. After 5 days of incubation, harvest the medium containing the seed stock of recombinant baculovirus. The expected titer is generally 1×10^7 plaque-forming units (p.f.u.)/ml. Negative-control cells will have formed a confluent monolayer. Virus-infected cells will appear grainy with enlarged nuclei and will not have formed a confluent monolayer.

Amplification of recombinant virus

Amplification of the recombinant virus is necessary before proceeding with experimental work. The following provides a method for amplification of virus in cells grown in a 100–200 ml suspension culture, but can be scaled down proportionately if desired.

1. Prepare a 100–200 ml culture of Sf9 cells at $1-2 \times 10^6$ cells/ml. Cells should be infected at a low multiplicity of infection (m.o.i.) of <1 p.f.u./cell[b].

2. Add 0.5 ml of recombinant virus seed stock to the cell culture. Incubate with shaking until the cells are well infected (usually 4–5 days).

3. When the cells appear to be well infected[c], harvest the cell culture medium by centrifugation at 1000 *g* for 20 min at 4°C. Remove the supernatant aseptically. Store the supernatant (recombinant virus) in the dark at 4°C[d].

4. A plaque assay to determine accurate titer is strongly recommended before using the virus in subsequent experiments.

Notes

[a]For cells maintained in serum-supplemented medium, wash the monolayer twice, each time with 1 ml of serum-free medium, before proceeding with co-transfection.
[b]The total culture volume should not comprise more than 20% of the flask volume. Shake flasks should be shaken at speeds to maximize aeration, e.g. 150 r.p.m.
[c]Under a phase-contrast inverted microscope, cells infected with virus appear grainy when compared with healthy cells. The infected cells become uniformly rounded and enlarged, with distinct enlarged nuclei.
[d]The virus inoculum can be stored in the dark at 4°C for 6–12 months, but the titer will begin to drop after 3–4 months. Titer the virus before use and reamplify if necessary. The addition of 2–5% serum when using serum-free medium can help avoid a drop in titer. Virus may be frozen at –80°C for longer periods. Avoid multiple freeze–thaw cycles.

2.7 Examples of results

2.7.1 Scaleable nature of the InsectDirect System

Yields obtained from the InsectDirect method have proved to be scaleable over the range of volumes tested to date. Similar yields per volume of culture are often obtained after scaling up a transfection from a 10 ml volume to a 100 ml or 1 l volumes. As *Fig. 2* demonstrates, yields per volume of culture often improve at a larger scale.

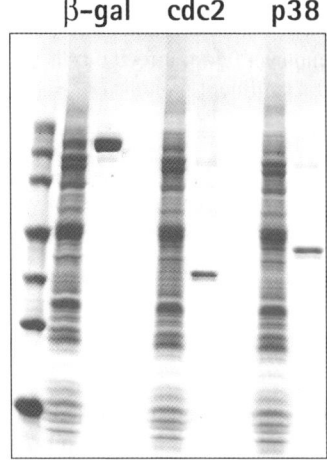

Figure 2. Scaleable nature of the InsectDirect System.
Suspension cultures of 10 ml were seeded with 1×10^6 Sf9 cells/ml grown in BacVector Insect Cell Medium. DNA (15 μg) was diluted with 0.5 ml of BacVector Medium, and 75 μl of Insect GeneJuice Transfection Reagent was diluted with 425 μl of BacVector Medium. The diluted DNA and Insect GeneJuice Reagent were combined and incubated at room temperature for 15 min. The transfection mixture was added slowly to the cells and incubated at 28°C with shaking for 72 h. Total culture extracts were prepared by the addition of Insect PopCulture Reagent (0.5 ml) followed by the addition of Benzonase Nuclease (5 μl). Ni-NTA His·Bind resin (0.25 ml per culture) was added to the extracts. After mixing on a rocking platform for 30 min at 4°C, the mixture was centrifuged and the retained resin was washed with 5 ml of 1× Ni-NTA Wash Buffer. The target protein was eluted with 0.75 ml of 1× Ni-NTA elution buffer. The protein concentration of the eluate was determined by bicinchoninic acid assay.

2.7.2 InsectDirect system and baculovirus comparison

The absolute yield of purified protein from any expression system is often dependent on the target protein. We have observed several examples in which the yield obtained from the InsectDirect System approach was similar to that obtained from the baculovirus approach when similar culture volumes were transfected or infected, respectively. *Fig. 3* provides four examples in which this was the case.

2.7.3 Dual-purpose pIEx/Bac for transient transfection and baculovirus-mediated expression

The pIEx/Bac dual-purpose vector was compared with pIEx in the transient transfection mode and with pTriEx in the baculovirus mode. *Renilla* luciferase (Rluc) reporter enzyme was used for the comparison. *Fig. 4(a)* shows the enhancer and/or promoter elements driving transcription from each vector and also provides the comparative results. Although the pIEx/Bac recombinant provides less Rluc activity than pIEx in the transient mode, it performs significantly better in the baculovirus mode relative to the *p10* promoter only-based pTriEx construct (see *Fig. 4b*).

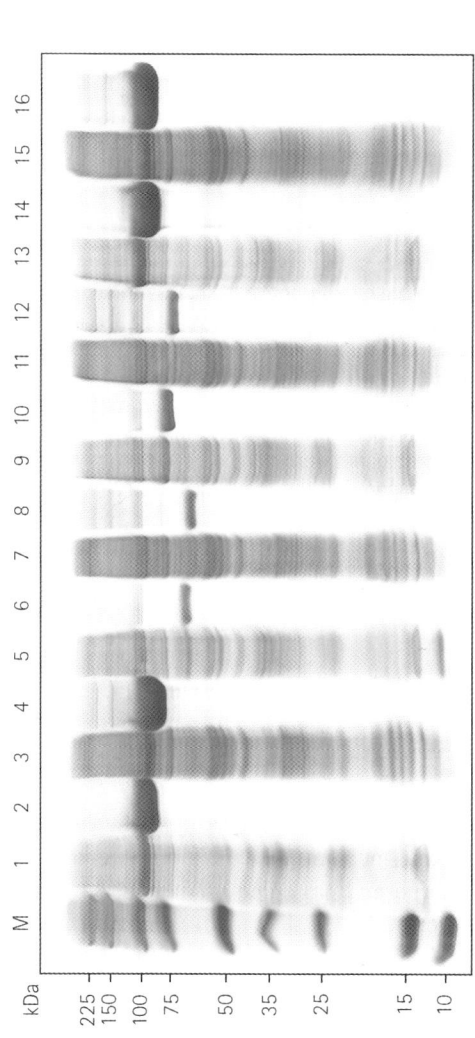

Figure 3. Comparison of the baculovirus and InsectDirect System for expression and purification.
For the InsectDirect System, Sf9 insect cells were grown in BacVector Insect Cell Medium. Suspension cultures of Sf9 cells (1×10^7 cells seeded in 9 ml of medium) were transfected with 15 μg of recombinant plasmid using Insect GeneJuice Transfection Reagent. For the transfections, a modified protocol was used: 15 μg of DNA was diluted with 0.5 ml of BacVector Medium, and 75 μl of Insect GeneJuice Transfection Reagent was diluted with 425 μl of BacVector Medium; the diluted DNA and Insect GeneJuice Transfection Reagent were combined and incubated at room temperature for 15 min before slowly adding the transfection mixture to the cells. For baculovirus, TriEx Sf9 cells were grown in TriEx Insect Cell Medium. Suspension cultures of TriEx Sf9 cells (1×10^7 cells seeded in 9 ml of medium) were infected at an m.o.i. of 5. Both the transfected and baculovirus-infected cultures were incubated at 28°C with shaking for 72 h. Total culture extracts were prepared by the addition of 600 μl of Insect PopCulture Reagent, followed by the addition of 5 μl of Benzonase Nuclease. Samples were removed at this point to analyze the crude fraction. Ni-NTA His•Bind Resin (375 μl per culture) was added to the extracts. After mixing on a rocking platform for 30 min at 4°C, the mixture was centrifuged and the retained resin was washed with 5 ml of 1× Ni-NTA wash buffer. The target protein was eluted with 750 μl of 1× Ni-NTA elution buffer. After the addition of SDS-sample buffer, samples representing 15 μl of crude and purified fractions were added to adjacent lanes and analyzed by 4–20% gradient SDS-PAGE. The gel was stained with RAPIDstain reagent (GBiosciences). HSPA, heat-shock protein α; HSPB, heat-shock protein β; PP, protein phoshatase; PK, protein kinase.

Lane	Sample
M	Perfect Protein markers, 10–25 kDa
1	vBAC-2cp - HSPA, crude
2	vBAC-2cp - HSPA, purified
3	pIEx-7 - HSPA, crude
4	pIEx-7 - HSPA, purified
5	vBAC-2cp - PP, crude
6	vBAC-2cp - PP, purified
7	pIEx-7 - PP, crude
8	pIEx-7 - PP, purified
9	vBAC-2cp - PK, crude
10	vBAC-2cp - PK, purified
11	pIEx-7 - PK, crude
12	pIEx-7 - PK, purified
13	vBAC-2cp - HSPB, crude
14	vBAC-2cp - HSPB, purified
15	pIEx-7 - HSPB, crude
16	pIEx-7 - HSPB, purified

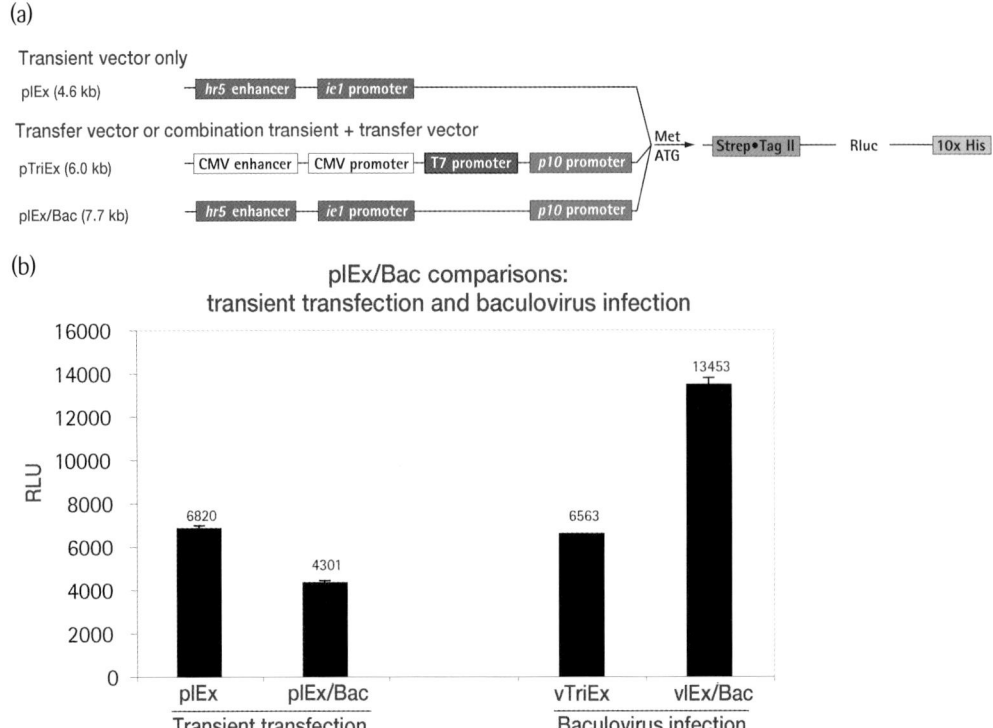

Figure 4. pIEx/Bac vector performance in transient transfection and baculovirus infection modes. (a) The *hr5* enhancer/*ie1* promoter elements drive transcription from pIEx and pIEx/Bac in the transient transfection mode. The *p10* promoter drives late/very late transcription in baculovirus derived from pIEx/Bac and pTriEx recombinants. In the case of pIEx/Bac, the *hr5* enhancer/*ie1* promoter would also direct early transcription in the baculovirus mode. In all cases, the open reading frame for *Renilla* luciferase (Rluc) was cloned into the Radiance Ek/LIC (http://www.emdbiosciences.com/radiance[10.1]) cloning site of each vector. (b) Rluc recombinants were compared for the production of Rluc activity. pIEx and pIEx/Bac recombinants were compared in the transient transfection mode. pTriEx and pIEx/Bac recombinants were compared in the baculovirus mode. Transfected or infected cultures were harvested after 72 h and Rluc activity was assayed using the MightyLight RLuc Assay kit and Dynex Technologies Microtiter Plate Luminometer. RLU, relative light units.

3. TROUBLESHOOTING

- Growth of Sf9 insect cells in culture
  - In order to prevent contamination, use sterile technique when working with uninfected Sf9 insect cells and medium. Whenever possible, dedicate one laminar flow hood to culturing and passaging uninfected Sf9 insect cells and a second for all virus work.
  - To maximize viability and protein production, maintain cells at a concentration between 0.2×10^6 and 5×10^6 cells/ml. Overdilution will result in cell death. Be especially careful about overdilution during passages 1–5.

- Split cells every 2–4 days.
- For best results, use cells that are between passage 5 and 25.
- To ensure proper aeration during cell growth, the total culture volume should not comprise more than 20% of the flask volume and threaded caps should be kept slightly loose.

• Transfection of Sf9 insect cells
- Transfection was optimized for Sf9 insect cells grown in BacVector Insect Cell Medium. Other media and cell types have not been tested extensively and may or may not work in a similar manner. If another medium is used, monitor for possible precipitation during addition of the transfection mixture. The presence of precipitate can significantly reduce transfection efficiency.
- A ratio of 5 µl of Insect GeneJuice Transfection Reagent per 1 µg of DNA is optimal for Sf9 cells grown in BacVector Insect Cell Medium. An optimization experiment for alternative cells and/or media may need to be done for best results. Test between 3 and 8 µl of Insect GeneJuice Transfection Reagent per 1 µg DNA.
- A precipitate may form if the DNA and Insect GeneJuice are not diluted in serum-free medium prior to mixing. Thoroughly mixing (by gentle vortexing or pipetting) of the DNA and Insect GeneJuice Transfection Reagent dilutions is crucial.
- Best results are obtained using supercoiled, endotoxin-free DNA.
- Use only serum-free medium during formation of the DNA/Insect GeneJuice complex. Serum may be present in the cell culture to which the complexes are added.
- Use only healthy, rapidly proliferating cells for transfection.

• Production of recombinant baculoviruses using BacMagic DNA
- The following transfection reactions are optional, but highly recommended:
  • Negative transfection control: instead of the recombinant transfer plasmid, use a corresponding amount of medium or Tris/EDTA.
  • Positive transfection control: instead of the recombinant transfer plasmid, use 500 ng of the supplied transfection control plasmid.
- Use only healthy, rapidly proliferating cells for transfection and virus amplification.
- The amount of seed stock virus added to the amplification culture can be adjusted up or down if the titer of the high-titer stock is low.
- The time of infection for the amplification culture may need to be longer than 5 days if the virus titer from the co-transfection is too low for the volume of cells being infected.

Acknowledgements

The authors would like to thank Scott Monsma and Kath Kramer for graphics and editing.

4. REFERENCES

1. **Summers MD** (2006) In: *Advances in Virus Research: Insect Viruses: Biotechnological Applications*, pp. 3–73, Edited by BC Bonning, K Maramorosch, and AJ Shatkin. Elsevier Academic Press, New York.
2. **Farrell PJ, Lu M, Prevost J, Brown C, Behie LA & Iatrou K** (1998) *Biotechnol. Bioeng.* **60**, 656–663.
3. **Pfeifer TA, Hegedus DD, Grigliatti TA & Theilmann DA** (1997) *Gene*, **188**, 183–190.
4. **Jarvis DL, Fleming JGW, Kovacs GR, Summers MD & Guarino LA** (1990) *BioTechnology*, **8**, 950–955.
★★★ 5. **Loomis KH, Yaeger KW, Batenjaney MM, et al.** (2005) *J. Struct. Funct. Gen.* **6**, 189–194. – *Provides examples of high yields using the InsectDirect System.*
★ 6 **Jarvis DL, Weinkauf C & Guarino LA** (1996) *Protein Expr. Purif.* **8**, 191–203. – *Original description of the pIE1 vectors.*
★ 7. **Loomis K, Novy R & Yaeger K** (2002) *Innovations*, **16**, 7–10. – *Original description of the pIEx vectors and comparison with a vector designed for stable cell line selection.*
★ 8. **Loomis K, Batenjany M, Bibikov S, Novy R, Bartnicki D & Wong S** (2004) *Innovations*, **19**, 10–12. – *Original description of the InsectDirect System.*
★★★ 9. **Guarino LA & Dong W** (1994) *Virology* **200**, 328–335. – *Characterization of the AcMNPV hr5 enhancer.*
★★★ 10. **Pullen SS & Friesen PD** (1995) *J. Virol.* **69**, 156–165. – *This paper describes the regulation of the ie1 promoter by DNA sequences within the 5′ noncoding leader region.*
★★★ 11. **Rodems SM & Friesen PD** (1993) *J. Virol.* **67**, 5776–5785. – *Characterization of the AcMNPV hr5 enhancer.*
12. **Loomis K, Rockwell C, Yaeger K & Novy R** (2005) *Innovations*, **21**, 12–14.
13. **Loomis K, Rockwell C & Novy R** (2006) *Innovations*, **24**, 4–7.
★★ 14. **Loomis K, Rockwell C, Sternard H & Novy R** (2007) *Innovations*, **25**, 14–17. – *Provides a detailed description of the new dual-purpose pIEx/Bac expression vector.*
15. **Yin W, Xiang P & Li Q** (2005) *Anal. Biochem.* **346**, 289–294.

CHAPTER 11
Generation of stable CHO cell lines for protein expression

Zhijian Lu, Haley Laken, Jimin Zhang, Xiaotian Zhong, and Richard Zollner

1. INTRODUCTION

The modern biopharmaceutical industry owes its establishment and success largely to the economical industrial scale-up of recombinant protein production. Among different methods employed for making recombinant biotherapeutics, mammalian cell expression systems have become the workhorse that is used in the manufacturing of the majority of marketed protein drugs (1, 2). In fact, about 60–70% of biotherapeutics are produced in mammalian cells with Chinese hamster ovary (CHO) cells as the predominant host (3). Industry-wide accumulation of experience with CHO cells over the past 20 years is significant in several aspects. First, it has demonstrated the suitability of this host cell for proper folding, glycosylation, and secretion of therapeutic proteins. Secondly, the continuing improvement of this system has achieved grams per liter levels of expression productivity, thus effectively addressing the cost-of-goods issue for this class of biotech product. Thirdly, dozens of FDA-approved protein drugs produced in CHO cells have paved a clear regulatory path for biotherapeutic products generated from this system.

The CHO cell expression system has attained prominence not only in biotherapeutic manufacturing but also in different fields of life sciences research, especially for protein structural studies. In the past decade, stable CHO expression has proved highly effective for a number of novel structural determinations. More and more structural biologists have been realizing that, for many important mammalian proteins, the mammalian system is the only route forward. In particular, CHO expression is a powerful tool for the production of cell-surface receptors and secreted proteins. With protein materials generated by stable CHO cell lines, a number of crystal structures have been solved, including CD2 (4), CD4 (5), CD46 (6), CD58 (7), CTLA4 and B7-1 (8), Gb1pα (9), ICAM2 (10), IL12 (11), Lingo-1 (12), P-Selectin (13), and SEMA4D (14). Some cytosolic enzymes expressed in stable CHO lines also provide high-quality protein materials for crystal structure determination. For instance, the X-ray crystal structure of cytosolic phospholipase A2 pro-

Expression Systems: *Methods Express* (M.R. Dyson and Y. Durocher, eds)
© Scion Publishing Limited, 2007

duced by a stable CHO line has been resolved (15). The CHO expression system has become an essential tool for structural studies and a wider use of the system by structural laboratories is expected in the near future.

With increasing familiarity to researchers in the life science fields, the CHO expression system has found wide application in developing assays for studying biological processes and drug discovery screens (16, 17). It is common to use the CHO expression system to derive cell lines stably expressing target proteins to construct a functional or a reporter gene assay. In this chapter, we will describe some protocols that are generally applicable to the expression of secreted recombinant proteins as well as host cell engineering. These protocols can be adapted for different purposes with appropriate modifications. We will assume that the readers are familiar with the general concepts and techniques of cell culture. We also provide references to the seminal original research articles on key CHO technologies for readers to grasp the ideas behind these protocols.

1.1 Cell lineage

The commonly used CHO cells are the descendents of CHO-K1, which originated from the extended culturing of primary ovary cells from Chinese hamster by Puck (18). The clone arose from this culture without application of exogenous immortalization agents. The karyotype of CHO-K1 is 20 ($n=22$) (19), indicating some spontaneous transformation, which apparently contributed to the immortality. When the CHO-K1 cell line was submitted to the American Type and Culture Collection (ATCC) in 1970, it had been passaged more than 400 times from the original culture. One of the significant developments that made CHO cells very useful in recombinant protein technologies was the establishment of lineages that lacked dihydrofolate reductase (DHFR) (20). Using [^{3}H]deoxyuridine to kill cells with intact DHFR, the authors were able to select DHFR-deficient lines (DUKX) after mutagenesis steps by UV irradiation or mutagen treatment.

The establishment of DHFR-deficient CHO DUKX cell lines enabled the utilization of methotrexate (MTX) as a selection agent for clones transfected with vectors containing a modular DHFR cDNA (21, 22). When clones from the initial DHFR selection were subjected to increasing concentrations of MTX, new clones arose with many more copies of the transforming gene, which formed the foundation of the widely practiced amplification of heterologous protein expression in CHO cells (23–26).

The CHO-K1 lineage of cells started with anchorage-dependent and serum-dependent growth characteristics. Individual cell lines established for expressing recombinant proteins were adapted to grow in defined serum-free media in suspension suitable for large-scale manufacturing purposes (24, 27). However, this could be a long and unpredictable process. In order to accelerate the time line for obtaining CHO cell lines for research and large-scale industrial production, sublines such as CHO-SSF3 and PA-DUKX were developed with a better tendency to grow in serum-free suspension (28, 29). Invitrogen also provide their own versions of CHO cells that can grow in serum-free media in suspension. *Table 1* lists the commonly available CHO lines and their sources.

Table 1. Commonly used CHO cells

Name	Vendor	Catalog no.
CHO-K1	ATCC	CCL-61
CHO/*dhfr*⁻	ATCC	CRL-9096
CHO-S	Invitrogen	11619-012
DG-44 (*dhfr*⁻)	Invitrogen	12613-014

1.2 Selection markers

The development of CHO DUKX made use of DHFR as the logical choice for a selection marker for single-vector introduction, with the added benefit of possible gene expression amplification (22). There are instances, however, when multiple selection markers are required for successful recombinant protein expression, such as proteins consisting of multiple subunits or proteins requiring additional post-translational processing enzymes. Nowadays, there are many selectable markers to choose from and *Table 2* lists a few representative ones and the vendor information for the appropriate expression vector.

Table 2. Selection markers and antibiotics

Drug-resistant gene	Selection agent	Vendor
Neomycin phosphotransferase	G418	Invitrogen
Streptoalloteichus hindustanus bleomycin gene	Zeocin	Invitrogen
Hygromycin phosphotransferase	Hygromycin	Clontech
Puromycin *N*-acetyl transferase	Puromycin	Clontech
Dihydrofolate reductase	Methotrexate	ATCC[a]

[a]ATCC catalog no. 37146 for the pSV2-dhfr vector (21).

2. METHODS AND APPROACHES

2.1 Cell line maintenance

CHO cells are a robust and adaptable cell line. Different lineages of CHO cells are capable of growing as an attached monolayer culture or in suspension. This allows for ease of transfection, selection, and expression amplification of CHO cells in attached culture followed by transition to suspension culture for large-scale expression of heterologous proteins. All of the basic manipulations of CHO cells should be carried out in a tissue culture hood. CHO cells are grown at 37°C in a CO_2 incubator with humidified air. The concentration of CO_2 is determined based on the amount of bicarbonate in the medium being used.

CHO cells are normally split two or three times per week. Although CHO cells can be passaged based on pre-determined split ratios, counting cell numbers at each passage (e.g. using a hemocytometer) will help to achieve target seed densities more reliably. It is good practice to observe the cells at each passage, noting the appearance of cells and, in the case of attached cells, the degree of confluency and morphology, and to check for contamination. Attached cells can be passaged using enzymatic or nonenzymatic methods. Enzymatic methods are generally quicker, but if cell-surface receptors are to be preserved or the complete removal of animal-derived components is preferred, then one can choose nonenzymatic detachment methods.

Protocol 1

Passaging of anchorage-dependent CHO cells

Equipment and Reagents
- Phosphate-buffered saline, calcium- and magnesium-free (PBS-CMF)
- Trypsin/EDTA solution (0.1% trypsin[a], 1 mM EDTA)
- Dissociation buffer (0.5–1.0 mM EDTA in PBS-CMF)
- Fresh growth medium
- Hemocytometer or other cell counting device

Methods
1. Remove spent growth medium using a sterile pipette or decant into a clean waste container. Rinse the flask with 0.1–0.2 ml PBS-CMF per cm^2 (refer to *Table 3* for the surface areas of culture vessels). For a T-75 flask, 5–10 ml of PBS-CMF is sufficient. If dissociation buffer is to be used, a second rinse with PBS-CMF will help to ensure quick and complete detachment of the cells[b].

2. Add 0.01–0.04 ml of trypsin/EDTA or dissociation buffer per cm^2 of growth area. For a T-75 flask, 1–3 ml is sufficient. Gently tilt and rotate the flask to ensure coverage of the trypsin or dissociation buffer over all of the monolayer. The flask can be replaced in a 37°C incubator, but in most cases cells will detach in 5–10 min if left in a tissue culture hood at room temperature. Detachment can be encouraged by sharp rapping of the flask against the palm of your hand.

3. Once the cells have detached, add fresh growth medium to stop trypsinization or to dilute the EDTA. The volume added depends in part on the amount of trypsin/dissociation buffer added and a convenient volume to subculture from. For a T-75 flask, bring the volume in the flask to 10 ml. Pipette the cells up and down to break up any clumps.

4. Take a sample and count the cells using a hemocytometer or any other cell counting device. The cells may or may not need to be centrifuged depending on how the cells are to be split or reseeded. If the split ratio is low (1:2 or 1:3), then centrifugation is recommended. Remove the cells to a sterile centrifuge tube and centrifuge for 10 min at 1000 r.p.m. Cells are split in part depending on growth characteristics (slow or fast) and what purpose the cells will be used for and when. Alternatively, new flasks are seeded at a known cell number. CHO cells are generally seeded at $1.5–3.0 \times 10^4$ cells/cm^2, i.e. approximately $1–2 \times 10^6$ cells for a T-75 flask.

Notes

[a] Irradiated trypsin or recombinant trypsin is preferred to reduce the risk of contamination by mycoplasma and viruses.
[b] Cells will detach quickly from flasks if the trypsin or dissociation buffer is at 37°C.

Table 3. Empirical conditions for culture of attached CHO cells

Vessel type	Surface area (cm^2)	Growth medium volume	Cell dissociating agent volume	Approximate seed density (total no. of cells per flask)	Approximate total cell number at confluence
96-Well plate	0.3	200 μl	100 μl	$4.5–9 \times 10^3$	$3–4 \times 10^4$
48-Well plate	0.7	500 μl	100 μl	$1–2 \times 10^4$	$0.7–1.4 \times 10^5$
24-Well plate	2	2 ml	0.5 ml	$3–6 \times 10^4$	$2–4 \times 10^5$
12-Well plate	4	3 ml	0.5 ml	$0.6–1.2 \times 10^5$	$4–8 \times 10^5$
Six-well plate	9.6	5 ml	0.5 ml	$1.4–2.8 \times 10^5$	$1–2 \times 10^6$
60 mm Dish	20	5–6 ml	1 ml	$3–6 \times 10^6$	$2–4 \times 10^6$
100 mm Dish	60	10–15 ml	1–3 ml	$0.9–1.8 \times 10^6$	$0.6–1.2 \times 10^7$
T-25 flask	25	5–6 ml	0.5–1 ml	$3.8–7.5 \times 10^5$	$2.5–5 \times 10^6$
T-75 flask	75	10–15 ml	1–3 ml	$1–2 \times 10^6$	$0.75–1.5 \times 10^7$
T-175 flask	175	25–30 ml	2–4 ml	$2.6–5.2 \times 10^6$	$1.75–3.5 \times 10^7$
Roller bottle	850	100–200 ml	10–20 ml	$1.3–2.6 \times 10^7$	$0.85–1.7 \times 10^7$
Expanded roller bottle	1700	300–400 ml	25–50 ml	$2.5–5 \times 10^7$	$1.7–3.4 \times 10^8$
Ten-layer cell factory or cell stacker	~6360	1000–2000 ml	100–200 ml	$0.9–1.9 \times 10^8$	$0.64–1.3 \times 10^9$

Protocol 2

Passaging CHO cells in suspension culture

Equipment and Reagents
- Disposable or reusable Erlenmeyer plastic or glass shake flasks
- Hemocytometer or other cell counting device
- Humidified incubator controlled at 37°C with 5% CO_2
- Orbital shaker

Methods
1. In a tissue culture hood, take a sample of the culture to be passaged and count using a hemocytometer or other cell counting device.

2. Based on the target seed density, pipette an appropriate volume into the new vessel[a] and add fresh growth medium to the appropriate volume. For example, a 100 ml shaker at 1×10^6 cells/ml can be passaged into a new 100 ml shaker at 2×10^5 cells/ml by pipetting 20 ml of the culture into the new shaker and adding 80 ml of fresh growth media.

3. New flasks are generally seeded in the $1–3 \times 10^5$ cells/ml range. This should be lower if the cells are fast growers and higher if the cells are slower or higher final cell densities are required or if the cells are being changed to new conditions or medium. If the volume of spent medium is greater than one-third of the fresh medium, then the cells should be centrifuged and resuspended in fresh medium. This avoids seeding the cells with a high percentage of medium that has been depleted of nutrients and has a high level of toxic/inhibitory metabolites, e.g. lactate and ammonia.

Notes
[a]Refer to Chapter 12, *Protocol 1*, notes f–h, for the volume of culture medium to use and advice for suspension culture.

2.2 Cryopreservation of CHO cells

Due to the risk of contamination in cell culture and the risk of expression instability, it is important that cell banks be made on a fairly routine basis. In addition to a cell bank of the untransfected CHO host, cell banks made at various stages post-transfection will ensure that, if contamination occurs or for other reasons cells from earlier time points are needed, a cell source is available without having to remake the CHO lineage.

Protocol 3

Freezing CHO cells

Equipment and Reagents
- Growth medium
- Tissue culture hood
- DMSO, tissue culture grade
- Trypsin or dissociation buffer
- Small sterilizing 0.2 µm disposable filters
- Sterile centrifuge tubes and pipettes
- Sterile cryovials (e.g. Corning)

Methods

1. Prepare the freezing medium. Freezing medium is growth medium with 10% DMSO added. This is then filter sterilized. The presence of selection agents is not necessary, nor is it harmful to the cells during this process. They can be left out. DMSO concentrations in freezing medium in cell cultures generally vary from 5 to 15%. For CHO cells, 10% DMSO works well.

2. It can be helpful to prepare your cryovials before you start counting cells. Label the number of vials expected to be frozen. This number is based on the use of the cell bank. Label the vials with cell line name, notebook number, and date.

3. If the cells to be frozen are attached, follow *Protocol 1* to step 4 when the cells have detached[a]. If more than one flask of cells is to be used, combine the detached cells in a centrifuge tube. Take a sample for counting. If the cells are from a suspension culture, sample all flasks that are to be combined and count the cells.

4. Centrifuge the cells at 1000 *g* for 5–10 min. Discard the supernatant.

5. Resuspend cells in calculated volume of freezing media.

6. Aliquot cells into the labeled cryovials[b].

7. Freeze the cells[c,d].

Notes

[a] It is best to freeze CHO cells when they are harvested in exponential phase, usually 2 or 3 days post-subculture. CHO cells can be frozen from attached or suspension culture and in medium with or without fetal bovine serum. CHO cells can be frozen at a cell density ranging from 1×10^6 to 2×10^7 cells/ml depending on the available cells and the need for quick expansion post-thaw.

[b] Based on the cell count above, the target cell density per vial, and the number of vials to be frozen, resuspend the pellet in the appropriate amount of freezing medium. An example follows: two T-175 tissue culture flasks are trypsinized and the cells combined in a sterile centrifuge tube. The cell count shows that there is a total of 5×10^7 cells in the centrifuge tube. If 10 ml of freezing medium is added, then ten 1 ml cryovials can be made with cells at 5×10^6 cells/ml/cryovial.

[c] Cells can be frozen in various ways. There are expensive controlled cooling apparatus that do an excellent job of cooling cells at a rate of –1°C/min. Alternatively, simply putting the cells in a styrofoam container, e.g. the type that 15 ml centrifuge tubes come in, covered with paper towels and placing in a –80°C freezer will freeze cells perfectly adequately. Once cells are frozen, they can then be moved into a mechanical –135°C freezer or a liquid nitrogen freezer for long-term storage.

[d] It is important that at least one vial from each bank be thawed to evaluate cell viability and sterility. If cells thaw at low viability or become contaminated, then a new cell bank should be made.

2.3 Cell line construction

2.3.1 Transfection methods for cell line construction

A number of nonviral methods for introducing an expression construct into a CHO host have demonstrated suitability. No published studies have provided a direct comparison of these methods. Selection of a particular method may depend on the considerations listed in *Table 4*. Any method selected may require optimization for the particular CHO host to be transfected. Points to consider include: DNA amount, ratio of carrier to DNA, cell density, and medium composition. Medium components, in particular serum, can adversely impact some of the methods. Stable transfection and selection can be carried out either in attached or suspension format. Attached culture allows the convenient removal of dead cells that are not stably transfected. Transfection of adherent cells can be carried out in culture vessels of all sizes. The amount of DNA and reagents used should be adjusted accordingly based on the surface area of the vessel (see *Table 3*).

Table 4. Transfection methods for CHO cells

Method	Reference	Considerations
Electroporation	30	Special equipment required
Calcium phosphate	31	Economical
Lipofection	32	A number of different reagents are available commercially
Polymer mediated (polyethylenimine)	33	Economical

Protocol 4

Liposome-mediated transfection

Equipment and Reagents
- CHO cell host and culture medium
- Appropriate culture vessels
- Lipofectamine 2000 (Invitrogen) or another commercially available lipid-based reagent
- Incubator

Method[a]

1. One day prior to transfection, seed the cell stock so that the appropriate confluence (>90%) will be achieved at the time of transfection
2. Determine the appropriate amount of DNA[b] and lipid reagent to be used depending on cell type and culture vessel size (see manufacturers' instructions).
3. On the day of transfection, form the DNA–lipid complex as follows:
   - Combine the culture medium + DNA.
   - Combine the culture medium plus lipid reagent and incubate for 5 min at room temperature.
   - Combine the DNA and lipid reagent and incubate for 20 min at room temperature.
4. Apply the DNA–lipid complexes to the cells to be transfected.
5. Return to the incubator and allow the transfection to proceed for 6 h.
6. After 6 h, remove the transfection mixture from the cells and replace with growth medium.
7. At 24 h post-transfection, expand the cells by splitting 1:10 in a new culture vessel.
8. Add an appropriate selective agent 24 h later.
9. Cell death should be apparent within the next 3 days.
10. Between 5 and 10 days after the addition of the selective agent, growth of stably transfected cells should become apparent, as small colonies form and increase in size in the culture vessel.
11. At this point, there are two options[c]: (i) individual colonies can be selected for expansion and productivity evaluation (see *Protocols 6–8*); or (ii) all stably transfected cells can be combined to form a pool of cells expressing the desired molecule.

Notes

[a]Steps 2–6 may be specific to the lipid reagents manufactured by other companies and may need a minor amount of optimization depending on the specific CHO host used. Step 7 and beyond apply to any of the transfection methods listed in *Table 1*.
[b]The total amount of DNA introduced during the transfection should be within the range specified by the protocol. If more than one expression plasmid is introduced, the total amount of DNA must remain the same and be divided among the plasmids to be transfected. The ratio of DNA for each plasmid can vary and should be determined based on the requirements of the experiment (if more expression of one gene is needed, that gene should make up a larger proportion of the total).
[c]Cryopreservation shortly after selection provides a back-up stock of cells.

2.3.2 Selectable markers for selection of stable cell lines

A gene conferring resistance to a culture additive needs to be included in transfections to generate stable cell lines. A subset of resistance markers and the concentration range of the corresponding selection agents commonly used in CHO cells are listed in *Table 5*.

Determination of the appropriate concentration of drug to be used for selection depends on the cell type and the vectors used to express the selectable marker. The concentration needs to be determined experimentally for each cell type/vector combination.

Different mechanisms are employed with drug selection and cells typically need to be actively dividing for the drug to effectively kill the cells that are not stably transfected.

Table 5. Ranges for common selection agents

Selection agent(s)	Typical concentration range	Reference
Neomycin/G418	0.4–1 mg/ml	34
Zeocin	50–300 µg/ml	35
Hygromycin	0.05–1 mg/ml	36
Puromycin	1–10 µg/ml	37
MTX[a]	0–100 µM	38

[a]Amplifiable marker: usually requires the use of a DHFR-deficient host.

Protocol 5

Determination of selective agent concentration

Equipment and Reagents
- CHO cell host and culture medium
- Selective agent stock solution (100–1000×)
- Six-well plates

Method
1. Plate out the attached cells to 25–50% confluence in a six-well plate.
2. Prepare the medium containing a range of drug concentrations.
3. After cells have attached, add medium containing a different concentration of drug to each well. Be sure to include one well with no drug.
4. Monitor the health of the cells for 3–7 days:
   - Check and record the culture confluence.
   - Split any cultures that become confluent during the monitoring period.
   - If cells do not survive this procedure, a lower drug concentration may need to be selected.
5. Choose the minimum concentration at which no cells survive.
6. Test this selected concentration on cells that have been transfected with an expression plasmid harboring a selectable marker – growth of stably transfected cells should be apparent after cell death ceases.

2.3.3 Clonal cell line isolation

There are several methods, involving various levels of complexity, of isolating cell colonies from cell culture dishes. We describe below three different protocols for manual operation.

Protocol 6

Ring cloning method[a]

Equipment and Reagents
- Growth medium
- PBS-CMF
- Sterile glass cloning cylinders (e.g. Corning)
- Autoclaved Vaseline (or Dow Corning 976V high vacuum silicone)
- Trypsin or trypsin/EDTA solution (see *Protocol 1*)
- Dissociation buffer (see *Protocol 1*)
- Sterile 24-, 48-, or 96-well plates

Method
1. Mark the cell colony location on the bottom of the cell culture dish.
2. Remove the medium and wash cells with PBS-CMF.
3. Apply a sterile glass column on top of the cell colony and sealed with autoclaved Vaseline.
4. Detach the cells by adding one drop of trypsin or trypsin/EDTA or dissociation buffer for 1–2 min.
5. Add 50–100 µl of selection medium and transfer the cells to 24-, 48-, or 96-well plates and continue the drug selection.

Note

[a]This method requires the colonies in the dish to be far enough apart to allow only one colony in a glass column.

Protocol 7

Trypsin filter paper method

Equipment and Reagents
- Growth medium
- Trypsin or trypsin/EDTA solution (see *Protocol 1*)
- PBS-CMF
- Sterile 3M filter paper disks (2 mm in diameter)
- Sterile 24-, 48-, or 96-well plates

Method
1. Mark the cell colony location on the bottom of the cell culture dish.
2. Remove the medium and wash the cells with PBS-CMF.
3. Apply the 3M filter paper disks (soaked in trypsin solution) directly on top of individual colonies for 2 min.
4. Swab the colony with the trypsin filter paper and transfer the cells to 24-, 48-, or 96-well plates with fresh selection medium to continue the selection.

Protocol 8

Pipette tip method

Equipment and Reagents
- Growth medium
- Trypsin or trypsin/EDTA solution (see *Protocol 1*)
- PBS-CMF
- Sterile pipette tips
- Sterile 24-, 48-, or 96-well plates

Method
1. Mark the cell colony location on the bottom of the cell culture dish.
2. Remove the medium and wash the cells with PBS-CMF.
3. For each 10 cm dish, add 1 ml of trypsin and monitor the cell shape under a microscope. When cells become rounded, slowly remove the trypsin and add growth medium back to the culture dish to cover all of the cells.
4. Scrape and aspirate the cells of individual colonies into pipette tip, and then transfer them to 24-, 48-, or 96-well plates with fresh selection medium to continue the selection.

2.3.4 Screening stable cell lines (expression analysis)

To identify high-level expressing stable cell lines, efficient analytical methods are necessary. Several methods are commonly used for this purpose, such as enzyme-linked immunosorbent assay (ELISA), sodium dodecyl sulfate polyacrylamide gel electrophoresis (SDS-PAGE), Western blots, and dot blots. To carry out expression analysis, it is common to replace the serum-containing selection medium with serum-free medium when cells have grown to confluence. The cells are then left in serum-free medium at 37°C for 24–48 h before harvesting the conditioned medium for analysis.

1. ELISA is a powerful method for detecting and quantitating protein expressed into a conditioned medium. Common enzymes used for ELISA are horseradish peroxidase, alkaline phosphatase, and β-D-galactosidase. ELISA can be performed in 96-well (or 384-well) polystyrene plates. ELISA requires a pair of target-specific antibodies. For detailed protocols, see (39).
2. SDS-PAGE can be used to determine the purity and relative molecular mass of the target protein in conditioned medium after the initial screening. If the cell has a high level of expression of the target protein, Coomassie blue staining should be able to visualize the protein band without having to concentrate the conditioned medium. The minimum amount of protein required for Coomassie blue staining is approximately 0.2 µg per lane.
3. Western blotting. After SDS-PAGE separation, Western blotting can be used if (i) the expression level of the protein is relatively low; or (ii) the conditioned medium contains serum, which makes Coomassie blue staining difficult. A

specific antibody that recognizes the denatured target protein is required. The method has been described in detail in (40).

4. Dot blotting provides a convenient alternative to ELISA and can be used for a quick initial screening. This method is not quantitative and can only be used for serum-free conditioned medium. It can be done in 2–3 h. Generally, 5 µl of conditioned medium is needed to spot directly on nitrocellulose membrane. The rest of the procedure is as for Western blotting after protein transfer from the polyacrylamide gel to nitrocellulose.

2.4 Scaling up of CHO cells for recombinant protein production

In order to produce sufficient recombinant protein, the established CHO cell line has to be scaled up. This section describes procedures for large-scale growth of both attached and suspension CHO cells.

2.4.1 Adherent cells

This section describes procedures for growing CHO cells at scales sufficient to produce enough protein for research needs. For adherent CHO cells, fetal bovine serum is commonly present as it is required for their growth and attachment. However, fetal bovine serum presents purification issues. Fortuitously, CHO cells can remain adherent and quite healthy for several days after serum has been removed from the confluent culture. This allows the production of conditioned medium from adherent cells in serum-free conditions. Serum-free expression medium can be harvested from the attached culture at 24, 48, or even 96 h, and additional serum-free expression medium can be added. The best time for harvesting and refeeding needs to be tested empirically. Harvest time is based on expression levels and the quality of protein expressed, as well as cell health, e.g. levels of viability and attachment. It is sometimes better to carry out repeated harvests rather than one extended production phase. Unlike hybridomas, CHO cells should be kept as viable as possible during expression periods to ensure consistent protein quality.

Protocol 9

Scaling up of adherent cells for expression

Equipment and Reagents
- Complete growth medium and growth medium without serum
- Culture vessels with a large surface area (e.g. T-175, roller bottles, cell factories)

Method
1. Determine the doubling time of established CHO cell lines by seeding at the desired density, growing them under the cloning conditions, harvesting cells at the time of confluence, and counting the harvested cells.
2. Confirm secretion of the recombinant protein using appropriate analytical methods, such as a Western blot or protein capture or activity assay.
3. Determine the ultimate scale of expression based on the expression level and the amount of protein required. Then choose the type, size, and number of tissue culture containers required.
4. Continue splitting confluent CHO cells in ratios no higher than 1:8 until the culture reaches the target size. During this scale-up, always use the medium that the cell line was cloned in.
5. When the final subculture reaches confluency, remove the medium, rinse the culture once with PBS-CMF, and then add serum-free medium for the eventual protein production[a].
6. Harvest the conditioned medium at a predetermined time, usually 24, 48, or 72 h after the medium switch. Add fresh serum-free medium to culture the cells again for the next harvest. A second, third, or even forth serum-free harvest can be obtained based on the state of the cells, i.e. if they are still well attached and healthy-looking.
7. Removed cell debris by centrifugation or filtration prior to purification.

Notes
[a] A convenient scenario is to seed flasks with cells late in the week, e.g. Thursday or Friday, in growth medium and allow the cells to grow to confluency over the weekend. Three to four days after seeding, the growth medium can be removed and serum-free medium added. The first harvest of serum-free conditioned medium is then obtained 24–72 h after the medium switch.

2.4.2 Scaling up of suspension cells for expression

Once CHO cells have been adapted to growth in serum-free suspension, the eventual volume of the harvest is limited simply by the size of the equipment used. The basic strategy for scaling up is diluting/splitting cultures at high density to low density in culture vessels such as shakers, spinners, or bioreactors of various degrees of sophistication. Cells are maintained on their normal split regimen, splitting every 3 or 4 days. Seed densities remain the same during scaling up[a]. Once the cells are at the final volume, the expression duration can be lengthened and various strategies implemented to increase titers. These strategies can include medium concentrate additions, temperature shifts, and/or inducers such as sodium butyrate.

Note

ᵃCells should remain under selective pressure for as long as possible during scaling up. The timing of selection removal is based mostly on cost. Stable CHO cells will continue to secrete protein for many generations after the selection agent has been removed. There is generally plenty of time from removal of selection pressure and the end of the production period before expression decreases to unproductive levels.

2.5 Specialized application: host engineering

CHO cells have been broadly used for mutant cell line engineering, as they possess a pseudodiploid and stable karyotype, can grow in suspension as well as adherent cultures, and have a high colony-forming efficiency. In addition, CHO cells express a functional haploidy that has allowed the isolation of recessive mutants at frequencies of 10^{-5} to 10^{-7} (41). By using a number of different lectins that specifically recognize certain sugar conformations, many of which are cytotoxic, a wide range of subclones has been isolated that express altered carbohydrates at the cell surface (42). These so-called lectin-resistant cells are almost all glycosylation-defective CHO mutants and are described using the broad generic symbols 'Lec' for recessive mutants and 'LEC' for dominant mutants.

The Lec3.2.8.1 cell used for making proteins in structural studies is one of the glycosylation-defective mutants expressing the most drastically modified carbohydrate structures. It carries four glycosylation mutations (42). Lec2 and Lec8 mutants are defective in transporting one of the nucleotide sugars into the Golgi compartment (43, 44). Lec3 lacks a cytosolic UDP-*N*-acetylglucosamine 2-epimerase activity, which is required for the biosynthesis of CMP-sialic acid used in Golgi modifications (45). The Lec1 mutation is epistatic to the mutations of Lec2, Lec3, and Lec8. It is defective in UDP-*N*-acetylglucosamine:α-3-D-mannoside β-1,2-*N*-acetylglucosaminyltransferase I (GlcNAc-TI) activity, which initiates complex and hybrid *N*-linked carbohydrate synthesis in the medial Golgi (46, 47). The substrate of GlcNAc-TI, $Man_5GlcNAc_2Asn$ (a product of mannose trimming from high-mannose oligosaccharides by mannosidases), accumulates in the absence of GlcNAc-TI. Therefore, the combined mutations in Lec3.2.8.1 cause this CHO cell line to express uniform glycoproteins with all *N*-linked carbohydrates in the Man_5 oligomannosyl form, which is sensitive to treatments with endoglycosidase H and *N*-glycopeptidase F (48). The following protocol describes a basic method for deriving lectin-resistant CHO cells.

Protocol 10

Isolation of Lec1 mutants in CHO cells

Equipment and Reagents
- Growth medium
- 10 cm dishes
- Six-well plates
- *Phaseolus vulgaris* L lectin (L-PHA) (2 mg/ml stock in PBS, pH 7.2; EY Laboratories)
- Concanavalin A
- *Lens culinaris* lectin (LCH)
- Ethylmethane sulfonate (Sigma)[a]
- Methylene blue solution (0.2%, w/v, in 50% methanol)

Method

1. Plate CHO cells (1×10^7) into 10 cm dishes in growth medium supplemented with 10% fetal calf serum. Incubate cells overnight in the absence (control) or presence of 200 or 500 µg/ml ethylmethane sulfonate.

2. The next day, remove the ethylmethane sulfonate by washing with culture medium and allow the cells to grow for an additional 4–5 days. Then replate them at 10^5 cells/plate in 10 cm dishes and treat them with L-PHA at a concentration of 50 µg/ml[b].

3. Pick 24 colonies (two to three colonies per plate) after an 8 day selection and expand the cell clones in six-well plates.

4. Plate L-PHA-resistant clones as well as wild-type CHO DUKX cells into three 96-well plates at a density of 2×10^3 cells/well to perform a phenotype test. Treat one plate of cells with L-PHA at concentrations of 200, 400, 600, and 1000 µg/ml, or without L-PHA (control). Treat another plate of cells with concanavalin A at concentrations of 1, 2, 5, and 10 µg/ml, or without concanavalin A (control). Test a third plate with LCH at concentrations of 10, 40, 80, and 500 µg/ml, or without LCH (control).

5. Remove the medium from the plates after selection for 4 days and stain the cells with methylene blue solution. Examine the cells by microscopy. Colonies that are resistant to L-PHA at >250 µg/ml, to ConA at ~10–20 µg/ml, and to LCH at >500 µg/ml exhibit a Lec1 phenotype (47).

Notes

[a] A 10 mg/ml stock solution is made by adding 81 µl of ethylmethane sulfonate to 10 ml of medium and immediately vortexing for 1 min. An aliquot is taken immediately for addition to the cell culture.

[b] The correct lectin concentrations used for selection can be determined by a survival curve experiment. Cells are plated at 10^5 cells/plate, selected with L-PHA at increasing concentrations for 8 days, and colonies are counted. A concentration of 50 µg/ml L-PHA normally gives one to two colonies per plate for nonmutagen-treated cells.

3. REFERENCES

1. Harrison S, Adamson S, Bonam D, et al. (1998) Semin. Hematol. **35**, 4-10.
2. Eriksson RK, Fenge C, Lindner-Olsson E, et al. (2001) Semin. Hematol. **38**, 24-31.
★★★ 3. Wurm FM (2004) Nat. Biotechnol. **22**, 1393-1397. - A good review of mammalian cell expression for biotherapeutics production.
4. Jones EY, Davis SJ, Willaims AF, Harlos K & Stuart DI (1992) Nature, **360**, 232-239.
5. Wu, H, Kwong PD & Hendrickson WA (1997) Nature, **387**, 527-530.
6. Casasnovas JM, Larvie M & Stehle T (1999) EMBO J. **18**, 2911-2922.
7. Butters TD, Sparks LM, Harlos K, et al. (1999) Protein Sci. **8**, 1696-1701.
8. Stamper CC, Zhang Y, Tobin JF, et al. (2001) Nature, **410**, 608-611.
9. Dumas JJ, Kumar R, Seehra J, Somers WS & Mosyak L (2003) Science, **301**, 222-226.
10. Casasnovas JM, Springer TA, Liu JH, Harrison SC & Wang JH (1997) Nature, **387**, 312-315.
11. Yoon C, Johnston SC, Tang J, Stahl M, Tobin JF & Somers WS (2000) EMBO J. **19**, 3530-3541.
12. Mosyak L, Wood A, Dwyer B, et al. (2006) J. Biol. Chem. **281**, 36378-36390.
13. Somers, WS, Tang J, Shaw GD & Camphausen RT (2000) Cell, **103**, 467-479.
14. Love; CA, Harlos K, Mavaddat N, et al. (2003) Nat. Struct. Biol. **10**, 843-848.
15. Dessen A, Tang J, Schmidt H, et al. (1999) Cell, **97**, 349-360.
16. Satoh K, Ohyama K, Aoki N, Iida M & Nagai F (2004) Food Chem. Toxicol. **42**, 983-993.
17. Tang Y, Luo J, Fleming CR, et al. (2004) Assay Drug Dev. Technol. **2**, 281-289.
18. Puck TT, Cieciura SJ & Robinson A (1958) J. Exp. Med. **108**, 945-956
19. Kao FT & Puck TT (1968) Proc. Natl. Acad. Sci. U.S.A. **60**, 1275-1281.
20. Urlaub G & Chasin LA (1980) Proc. Natl. Acad. Sci. U.S.A. **77**, 4216-4220.
21. Subramani S, Mulligan R & Berg P (1981) Mol. Cell. Biol. **1**, 854-864.
★★★ 22. Kaufman RJ & Sharp PA (1982) J. Mol. Biol. **159**, 601-621. - Concept development of gene amplification in recombinant protein expression.
23. McCormick F, Trahey M, Innis M, Dieckmann B & Ringold G (1984) Mol. Cell. Biol. **4**, 166-172.
24. Kaufman RJ, Wasley LC, Spiliotes AJ, et al. (1985) Mol. Cell. Biol. **5**, 1750-1759.
25. Page MJ (1985) Gene, **37**, 139-144.
★★ 26. Jun SC, Kim MS, Baik JY, Hwang SO & Lee GM (2005) Appl. Microbiol. Biotechnol. **69**, 162-169. - A good illustration of CHO line construction with expression amplification.
27. Israel DI, Nove J, Kerns KM, Moutsatsos IK & Kaufman RJ (1992) Growth Factors, **7**, 139-150.
28. Gandor C, Leist C, Fiechter A & Asselbergs FA (1995) FEBS Lett. **377**, 290-294.
★★ 29. Sinacore MS, Drapeau D & Adamson SR (2000) Mol. Biotechnol. **15**, 249-257. - Methodology for deriving cell lines growing in serum-free medium.
30. Baum C, Forster P, Hegewisch-Becker S & Harbers K (1994) Biotechniques, **17**, 1058-1062.
31. Batard P, Jordan M & Wurm F (2001) Gene, **270**, 61-68.
32. Hodgson CP & Solaiman F (1996) Nat. Biotechnol. **14**, 339-342.
33. Galbraith DJ, Tait AS, Racher AJ, Birch JR & James DC (2006) Biotechnol. Prog. **22**, 753-762.
34. Southern PJ & Berg P (1982) J. Mol. Appl. Genet. **1**, 327-341.
35. Mulsant P, Tiraby G, Kallerhoff J & Perret J (1988) Somat. Cell Mol. Genet. **14**, 243-252.
36. Gritz L & Davies, J (1983) Gene, **25**, 179-188.
37. de la Luna S & Ortin J (1992) Methods Enzymol. **216**, 376-385.
38. Simonsen CC & Levinson, AD (1983) Proc. Natl. Acad. Sci. U.S.A. **80**, 2495-2499.
39. Page M & Thorp R (1996) In The Protein Protocols Handbook, pp. 1117-1118. Edited by JM Walker. Humana Press, Totowa, NJ.
40. Page M & Thorp R (1996) In The Protein Protocols Handbook, pp. 317-320. Edited by JM Walker. Humana Press, Totowa, NJ.
41. Siminovitch L (1976) Cell, **7**, 1-11.
★★ 42. Stanley P (1984) Annu. Rev. Genet. **18**, 525-552. - An excellent review of glycosylation mutants of animal cells.

43. Deutscher SL, Nuwayhid N, Stanley P, Briles EI & Hirschberg CB (1984) *Cell*, **39**, 295–299.
44. Deutscher SL & Hirschberg CB (1986) *J. Biol. Chem.* **261**, 96–100.
45. Hong Y & Stanley P (2003) *J. Biol. Chem.* **278**, 53045–53054.
46. Kumar R, Yang Y, Larsen RD & Stanley P (1990) *Proc. Natl. Acad. Sci. U.S.A.* **87**, 9948–9952.
47. Sarkar M, Hull E, Nishikawa Y, *et al.* (1991) *Proc. Natl. Acad. Sci. U.S.A.* **88**, 234–248.
★★ 48. Stanley P (1989) *Mol. Cell. Biol.* **9**, 377–383. – *A detailed characterization of a multiple-glycosylation-defective CHO mutant.*

CHAPTER 12
Transient expression in HEK293-EBNA1 cells

Roseanne Tom, Louis Bisson, and Yves Durocher

1. INTRODUCTION

Fast and efficient production of recombinant proteins (r-proteins) remains a major challenge for the academic and biopharmaceutical communities. Pure r-proteins are often required in large amounts (hundreds of milligrams to gram quantities) when being developed as biotherapeutics, or in smaller quantities (milligrams) for high-throughput screening campaigns and structural or functional studies. Mammalian cells are often preferred over prokaryotic systems when expressing cDNAs of mammalian origin due to their superior capability to conduct elaborate post-translational modifications. The conventional way to produce large quantities of r-proteins in mammalian cells is to transfect a small volume of cells and then select for clones that have stably integrated one or multiple copies of the gene of interest into their chromosomes (1). This process is tedious and generally requires many months to isolate stable clones that express adequate levels of the r-protein of interest. Whilst transfection of mammalian cells is a well-known technique that has been widely used on a small scale for many decades, only recently has this technology become scaleable through key developments in transfection-compatible culture media, highly efficient expression vectors, serum-free and suspension-growing cell lines, and cost-effective transfection reagents (2, 3). Large-scale transfection of mammalian cells is now establishing itself as a 'must-have' technology in the scientific community, as it allows the production of milligram to gram quantities of r-proteins within a few days after cDNA cloning into the appropriate expression vector (4–7) (see Fig. 1a–c). Thus far, only Chinese hamster ovary (CHO), African green monkey (CV-1, COS-1, and COS-7) and human embryonic kidney (HEK) 293 cells have been successfully used for large-scale transfection (2). More recently, efficient insect cell protocols have been elaborated (see Chapter 10) (8). Many gene carriers can be used for the large-scale transfection of mammalian cells, including commercially available cationic lipid mixtures, but so far only two carry out the process in a cost-

Expression Systems: *Methods Express* (M.R. Dyson and Y. Durocher, eds)
© Scion Publishing Limited, 2007

(a)

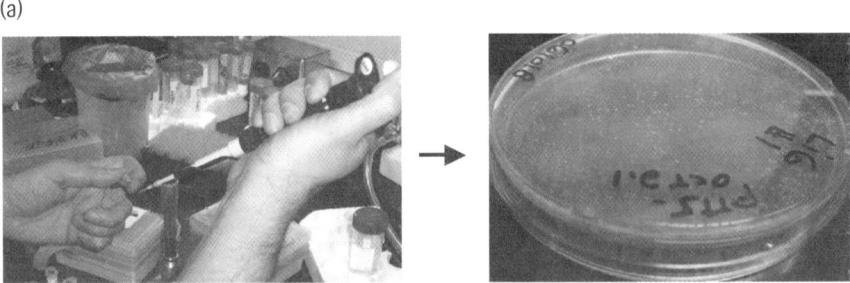

Days 1–4: cDNA cloning into pTT plasmid

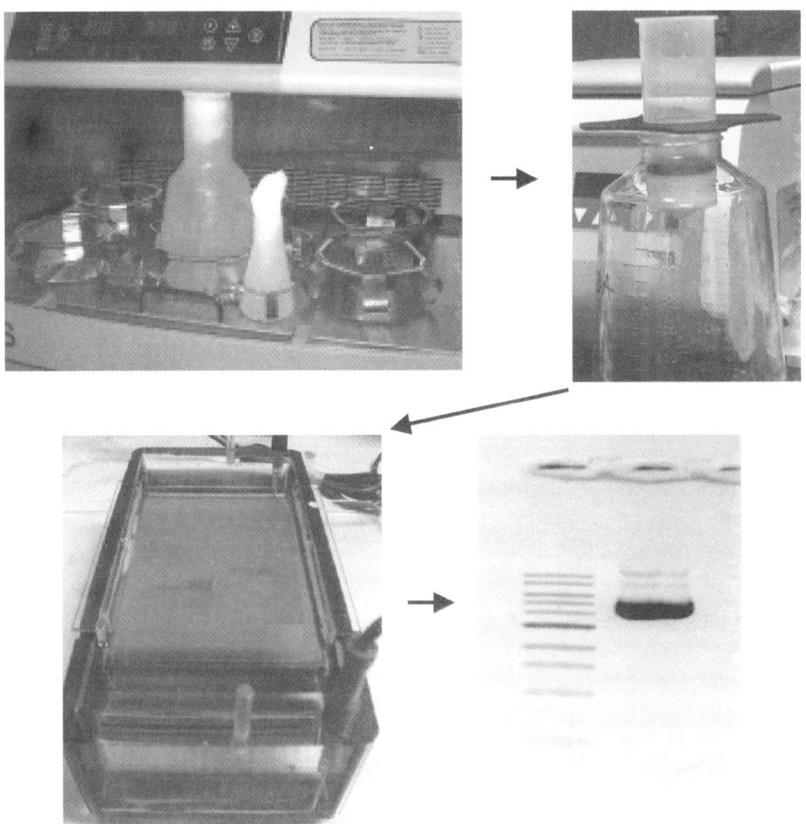

Days 3–5: plasmid production, purification, and verification

Figure 1. Transient gene expression time line.
(*a*) Steps for cDNA cloning in an expression vector, propagation, purification, and verification. (*b*) Small-scale experiments to determine expression kinetics in six-well plates (shown here for human kallikrein-10) and 125 ml shake flasks. (*c*) Transfection in a 10 l bioreactor, concentration of cleared culture medium, and purification in an AKTA system (GE Healthcare Life Sciences).

INTRODUCTION 205

(b)

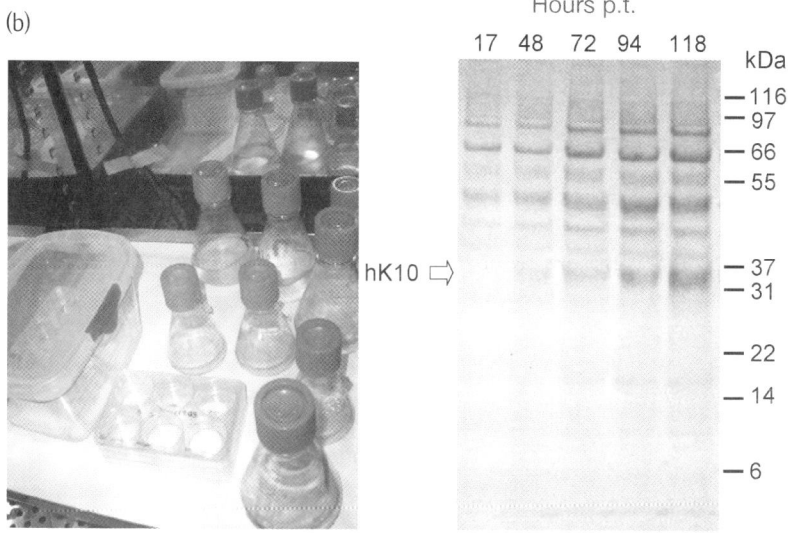

Days 4–11: small-scale transfections for expression kinetics

(c)

Days 10–17: bioreactor inoculation, transfection, and harvest

Day 17: concentration by ultrafiltration

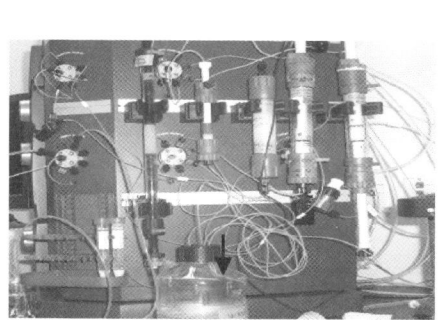

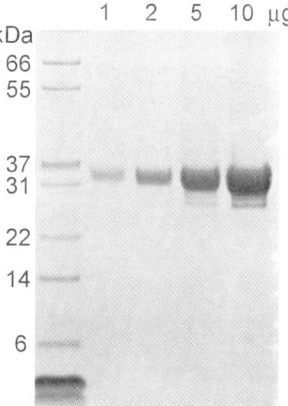

Day 18: protein purification and analysis

effective way: calcium phosphate and polyethylenimine (PEI). Calcium phosphate precipitate-mediated transfection was first reported in the 1970s when it was used to transform HEK cells with sheared adenovirus DNA (9). Whilst calcium-mediated large-scale transfection is very effective, it is usually achieved in serum-containing medium under tightly controlled conditions that are difficult to achieve on a large scale (10). Nevertheless, extensive optimization has allowed this method to be scaled up to 100 l cultures (11). By contrast, PEI is much easier to use. PEI was shown to bind and precipitate DNA efficiently (12) and these DNA–PEI complexes were later exploited for efficient transfection of mammalian cells (13). PEI has been used successfully on a large scale in serum-containing (4) and serum-free (6, 7, 14–17) cultures. This chapter describes the steps needed for successful transfection of HEK293 cells adapted to serum-supplemented or serum-free medium in adherent or suspension culture.

2. METHODS AND APPROACHES

2.1 Cell culture

The HEK293 cell line stably expressing the Epstein–Barr virus nuclear antigen-1 (HEK293-EBNA1, or 293E) is the most commonly used cell line for large-scale transfection. When using expression vectors bearing the Epstein–Barr virus origin of replication, *oriP* (such as the pTT vector), a threefold improvement in r-protein yield is generally obtained over a similar non-*oriP* vector (4). The protocols described in this chapter were developed in our laboratory using our cell lines (293E and 293-6E) and pTT expression vectors (available upon request). Other HEK293 cell lines (e.g. 293T, 293S, and 293F) and expression vectors can also be used, but with somewhat lower r-protein yields. The protocols can be done in any well-appointed cell culture laboratory with access to current equipment such as an inverted microscope with fluorescence detection capability, autoclaves, centrifuges (1–1000 ml capacity), high-purity water system (e.g. Milli-Q; Millipore), freezers (–20 and –80°C), and a liquid nitrogen freezer. It is important to be aware that cells growing in serum-free media are highly sensitive to chemical contamination; thorough rinsing of all equipment that contacts the cells with highly pure water is compulsory.

Protocol 1

Thawing and freezing of 293E and 293-6E cells[a]

Equipment and Reagents
- 293E and 293-6E working cell banks (NRC-BRI)
- Heat-inactivated cosmic calf serum (HyClone)[b]
- Pluronic F68 (10% (w/v) stock solution; Invitrogen)
- G418 (50 mg/ml stock solution; Invitrogen)
- LC-SFM medium (custom low-calcium HSFM formulation; Invitrogen) supplemented with 1% (v/v) serum for 293E cells (50 µg/ml G418) or FreeStyle medium (Invitrogen)[c,d] or HyQ SFM4TRANSFX-293 (HyClone) for 293-6E cells (25 µg/ml G418)
- Erythrosin B (25 mg/ml in phosphate-buffered saline (PBS); Sigma)
- DMSO, cell culture grade (Sigma)
- Disposable Erlenmeyer plastic shake flasks (Corning)
- Reusable Erlenmeyer glass shake flasks[e,f]
- Humidified incubator at 37°C with 5% CO_2
- Orbital shaker[g,h]
- Cryovials
- Water bath at 37°C
- Styrofoam box to contain cryovial boxes

Method
Cell thawing and maintenance
1. Prepare a 15 ml conical tube containing 10 ml of pre-warmed medium without G418.
2. Rapidly thaw the cells in a 37°C water bath.
3. Add the cells to the 15 ml tube and invert to mix.
4. Remove a 100 µl aliquot and dilute with 100 µl of erythrosin B to determine the cell density and viability.
5. Centrifuge the cells at 200 *g* for 5 min at room temperature.
6. Decant the supernatant and loosen the pellet by gently tapping the tube. Dilute the cells to 2.5×10^5 cells/ml in a 125 ml disposable shake flask using medium without G418. Cells can be diluted to 2×10^5 cells/ml if thawed on a Friday.
7. Place the shake flask on the orbital shaker (~120 r.p.m.) in the incubator (see *Fig. 2*).
8. At 48 h post-thawing, count the cells and dilute (if necessary) to 2.5×10^5 cells/ml using medium without G418.
9. When the doubling time is 24 h for two consecutive passages, subculture every 2 or 3 days to maintain cell densities between 2.5×10^5 and 1.2×10^6 cells/ml using medium supplemented with G418.
10. Dilute the cultures to 1.5×10^5 or 7.5×10^4 cells/ml for weekends or long weekends (3 or 4 days, respectively).

Cell freezing[i]
1. Prepare the freezing mixture by adding DMSO to fresh medium (10:90, v/v).
2. Label the number of cryogenic vials needed.
3. Count the cells and determine the volume needed for cryopreservation. Cells must be in exponential phase (between 8×10^5 and 1.2×10^6 cells/ml).

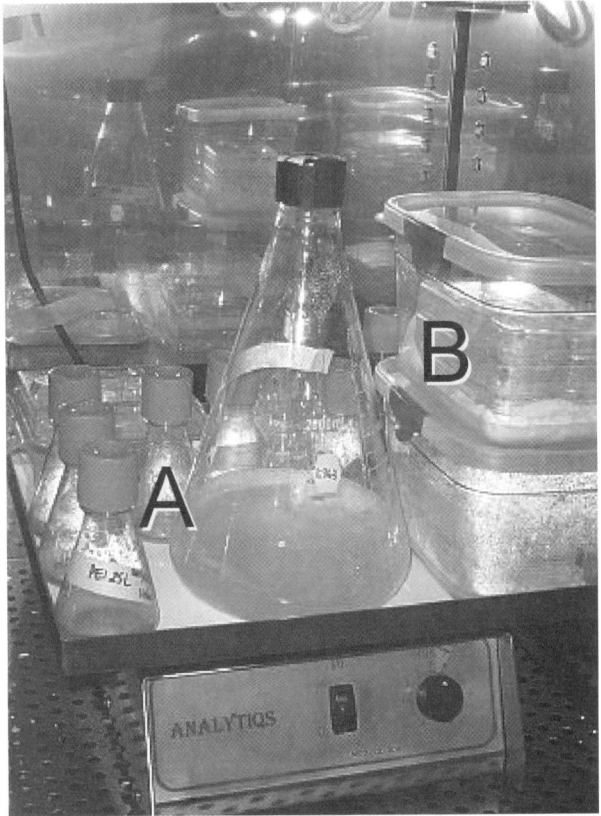

Figure 2. Growth of 293-6E cells in suspension.
Cells are grown in a humidified 37°C incubator with 5% CO_2. An orbital shaker with an anti-slip mat can accommodate various culture vessels at 120 r.p.m., such as shake flasks ranging in size from 125 to 2000 ml (*A*), and multi-well culture plates placed in a plastic sandwich box with a wet towel to help minimize medium evaporation (*B*). The box should be acclimatized with the wet towel a few minutes in the incubator before putting the plate(s) inside (see *Protocol 4*).

4. Centrifuge the cells at 200 *g* for 5 min, decant the supernatant, and dissociate the pellet by gently tapping the tube.
5. Add the freezing medium (dropwise) to the cells while swirling the tube to obtain the desired density.
6. Quickly aliquot into vials and immediately transfer to a –80°C freezer. Do not open the freezer door for at least 2 h.
7. Transfer the vials to liquid nitrogen freezer storage (vapor phase) the following day, up to a maximum of 4 days.

Notes

[a]Conduct all operations aseptically in a laminar flow hood; all reagents added to cultures and media must be sterile.

^{b}Thaw the frozen serum thoroughly at 37°C, then heat it for 30 min in a 56°C bath. Swirl the bottle occasionally for thorough heat distribution. Aliquots can be frozen at −20°C.
^{c}Add 10 ml of Pluronic F68 stock solution per liter of culture medium (0.1%, w/v, final).
^{d}Since the writing of this chapter, Freestyle medium has been replaced by F17 culture medium (Invitrogen).
^{e}Rinse glassware three times using Milli-Q water (or equivalent), followed by steam sterilization.
^{f}Operating culture volumes must be in the range of 16–33% of the nominal shake flask volume for sufficient oxygenation.
^{g}An orbital shaker must be used with a platform that contains holders for shake flask sizes ranging from 50 to 2000 ml. If no holders are supplied, the platform should contain an anti-slip adhesive mat. Shake flasks will stay on the platform up to 120 r.p.m. (see *Fig. 2A*).
^{h}Center the orbital shaker in a humidified incubator and make sure that the walls, door, and wire are not contacted during agitation. Check with the manufacturer that the orbital shaker can operate continuously under humidified conditions. Do not shut off the orbital shaker in the humidified incubator or it will seize upon subsequent use. If no cultures are in operation, leave the orbital shaker agitating or remove it from the incubator until needed. Do not use orbital shakers with digital agitation controls as their electronics are susceptible to humidity and they fail to start up after power failures.
^{i}Cells can be frozen at 5×10^6–5×10^7 cells/ml/vial, frozen at 1°C/min. This can be achieved by using a programmable cooler (if available) or by placing the vials in a small 20-position styrofoam box, which is put in a larger styrofoam box in a −80°C freezer.

2.2 Plasmid DNA preparation

A clean (RNA-free and protein-free) plasmid preparation containing mostly supercoiled DNA will facilitate quantification and ensure efficient transfection. If endotoxins need to be kept to a minimum in the final r-protein preparation (e.g. for *in vivo* animal studies), use endotoxin-free kits. Endotoxins do not interfere significantly with PEI-mediated transfection of HEK293 cells.

Protocol 2

Plasmid DNA preparation (maxiprep)

Equipment and Reagents
- Maxiprep DNA kit[a]
- CircleGro (Q-Biogen) or Terrific Broth medium[b]
- Orbital shaker plate in a non-humidified incubator at 37°C
- TE (10 mM Tris/HCl, 1 mM EDTA, pH 8.0)
- 50 mM Tris/HCl (pH 8.0)
- Centrifuge for 50 ml tubes (running at 10 000 g)
- Competent *E. coli* (DH5α strain)
- Equipment for agarose gel electrophoresis and UV spectrophotometer

Method
1. Pick a fresh colony and inoculate 50 ml of growth medium containing an appropriate antibiotic (e.g. 50 µg/ml ampicillin) in a 250 ml glass or polypropylene shake flask[c,d] (see *Fig. 1a*).
2. Incubate for 16–20 h with vigorous shaking (~300 r.p.m.).
3. Extract the DNA as recommended by the manufacturer.
4. Resuspend the DNA in 1 ml of sterile TE.
5. Measure the absorbance at 260 and 280 nm. DNA should be diluted in 50 mM Tris/HCl (pH 8.0) to obtain an accurate measurement. The 260/280 ratio of purified plasmid DNA should be between 1.80 and 1.95.
6. Verify the DNA integrity by 0.8% agarose gel electrophoresis (see *Fig. 1a*).

Notes
[a]We recommend using the Qiagen HiSpeed MAXI kit with high-copy-number plasmids. Other commercial kits using anion exchange chromatography may also be used.
[b]These rich media should generate about 1 mg of plasmid DNA when using a Qiagen HiSpeed kit and high-copy-number plasmids.
[c]Do not use a culture medium volume greater than 20% of the nominal flask volume to ensure efficient agitation and oxygenation.
[d]The use of baffled flasks such as Ultra Yield flasks (Thompson Instrument Co.) will result in higher biomass and plasmid yields.

2.3 Preparation of PEI

Many types of PEI are available on the market. The two most efficient and cost-effective ones for transfection are 25 kDa linear and 25 kDa branched PEI. The linear isoform is most effective for transfecting cells in suspension, whilst the branched isoform is most effective for adherent cells. Transfection with branched PEI will also promote attachment of cells to the plastic surface. It is thus very useful when the transfected cells are to be used in experiments requiring multiple medium exchanges or washing steps following transfection.

Protocol 3

Preparation of linear and branched PEI solutions

Equipment and Reagents
- Milli-Q water
- Glassware: 500 ml beaker and 500 ml graduated cylinder[a]
- 50 ml Conical tube
- 0.22 µm Vacuum filtration unit
- PEI (25 kDa linear, powder; Polysciences Inc.)
- PEI (25 kDa branched, liquid; Aldrich)
- 12 M HCl
- 10 M NaOH
- Tube heater

Method

Preparation of linear PEI (1 mg/ml solution)

1. Pour ~450 ml of Milli-Q water into a 500 ml glass beaker.
2. Weigh out 500 mg of PEI and add to the beaker with stirring.
3. Adjust the pH to <2.0 using concentrated HCl (dropwise).
4. Stir until dissolved (2–3 h) while maintaining the pH at <2.0[b].
5. Adjust the pH to 7.0 using concentrated NaOH (dropwise)[c].
6. Pour the solution into a 500 ml glass cylinder and adjust the final volume to 500 ml with Milli-Q water.
7. Filter-sterilize the solution through a 0.22 µm membrane.
8. Aliquot into the desired volumes and store at −20°C.

Preparation of branched PEI (1 mg/ml solution)

1. Weigh 500 mg of PEI[d] directly into a 50 ml conical tube.
2. Add 30 ml of Milli-Q water and heat to ~50°C to help it to dissolve.
3. Mix by shaking until completely dissolved.
4. Make up to 50 ml with Milli-Q water and mix again.
5. Pour 50 ml of the solution into a 500 ml graduated cylinder and adjust the final volume to 500 ml with Milli-Q water.
6. Filter sterilize the solution through a 0.22 µm membrane.
7. Aliquot into the desired volumes and store at −20°C.

Notes

[a] Rinse all glassware three times with high-purity water.
[b] Approximately 800 µl of 12 M HCl will be required for full PEI dissolution. There may still be some small fiber particles that will not dissolve.
[c] Approximately 500 µl of 10 M NaOH will be required to neutralize the PEI solution.
[d] This is a highly viscous liquid. Dip a 5 ml graduated serological pipette into the branched PEI bottle and then let the liquid fall dropwise into the 50 ml tube.

2.4 Transfection of 293E and 293-6E cells

The following transfection protocols were developed in our laboratory using our cell lines. The procedures describe scaleable protocols for transfection of cultures using 1 µg of DNA/ml of culture complexed with 2–3 µg/ml PEI. The DNA:PEI ratio selected for production is determined by testing various ratios in six-well plate experiments (see *Figs 1c* and *3*). These small-scale tests must include regular samplings to determine the optimal harvest periods (see *Figs 1b* and *3*). Secreted protein production is enhanced by supplementation with the casein hydrolysate TN1 (see *Protocol 4*, note d) at 24 or 48 h post-transfection (p.t.). TN1 supplementation is not necessary for intracellular protein production.

We highly recommend monitoring cell transfection using a plasmid encoding a fluorescent protein, such as green fluorescent protein (GFP). Adding a GFP plasmid to 5% in the transfection mixture does not significantly alter expression of the gene of interest and allows visual (or quantitative, if using flow cytometry) confirmation of transfection efficiency. A few GFP-positive cells can be detected as early as 3–4 h p.t. using a fluorescence microscope. A high transfection efficiency is when more than 40% of cells express GFP at 48 h p.t. when using 5% GFP plasmid in the transfection mixture. We have found that conditioned medium does not interfere with transfection (under the conditions described in *Protocols 4–7*), greatly simplifying the process. We usually seed cultures 1 or 2 days before transfection so that optimal cell density is reached on the day of transfection. This avoids having to centrifuge the cells, which may have a detrimental effect on transfection efficiency if performed shortly prior to transfection.

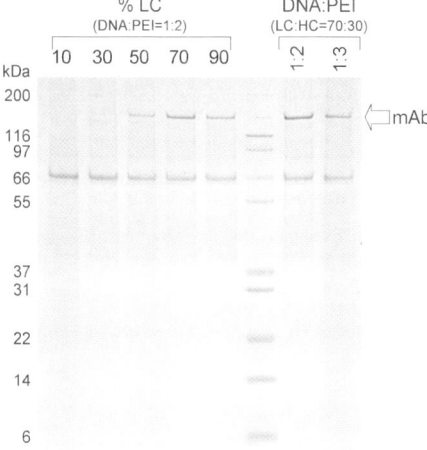

Figure 3. Expression optimization for a monoclonal antibody.
Small-scale tests in six-well plates allow multiple tests to be carried out to confirm the DNA:PEI ratios required for optimal protein expression. Antibody production calls for optimization of the light chain (LC):heavy chain (HC) ratio. Expression is evaluated between two DNA:PEI ratios by comparing 1 µg/ml DNA with 2 or 3 µg/ml PEI. The gel was run under nonreducing conditions.

Protocol 4

Transfection of 293E and 293-6E cells in suspension with linear PEI (DNA : PEI ratio of 1 : 2)[a,b]

Equipment and Reagents
- 293E or 293-6E maintenance cells in shake flask[c]
- PBS
- 1 mg/ml Linear PEI solution (see *Protocol 3*)
- 20% (w/v) TN1, sterile stock solution[d]
- Purified plasmid DNA of interest (see *Protocol 2*)
- Disposable six-well plates (Nunc)
- Plastic (Tupperware-type) containers
- Culture medium for 293E or 293-6E cells (see *Protocol 1*)
- Humidified incubator at 37°C with 5% CO_2

Method
Small-scale transfection in a six-well plate (2 ml) or a 125 ml shake flask (20 ml)[e]

1. Warm the PBS (transfection buffer) to 25–37°C and take out the DNA and PEI from the −20°C freezer to thaw.
2. From a maintenance flask, determine the cell density and viability[f,g].
3. Add 1.8 ml of cells to each of the six wells (or 18 ml to a new 125 ml shake flask) and transfer to an incubator. To minimize medium evaporation, put the plates in a plastic container (lined with a wet paper towel to preserve humidity) making sure you 'trap' some CO_2 before closing the lid (see *Fig. 2b*).
4. Add 200 µl of PBS to six 1.5 ml tubes (or 2.0 ml to a 15 ml tube for the 125 ml shake flask).
5. Add 2 µg of DNA[b] to each tube (or 20 µg to the 15 ml tube for the 125 ml shake flask) and vortex gently.
6. Add 4 µl of PEI to each tube (or 40 µl to the 15 ml tube for the shake flask) containing DNA solution and immediately vortex (3 × 3 s) after PEI addition[a].
7. Incubate the mixture for 15 min at room temperature.
8. Remove the culture (step 3) from the incubator, add the DNA/PEI mixture(s) and swirl.
9. Return the culture to the incubator.
10. For secreted protein production, add 50 µl of pre-warmed 20% TN1 at 24–48 h p.t. to each of the six wells (or 500 µl for the 125 ml shaker) and return the culture to the incubator. The final TN1 concentration in the culture(s) should be 0.5% (w/v).

Harvesting r-proteins[h,i]

1. Harvest the cells at 48–72 h p.t. for production of intracellular r-proteins.
2. Harvest the supernatants between 96 and 168 h p.t. for secreted r-proteins, as long as the culture viability is high and the product titer is increasing (see *Figs 1b* and *4*)[j].

Notes
[a]Compare different DNA : PEI ratios (1 : 2 vs 1 : 3, etc.) to see whether different lots and/or production batches affect transfection efficiency and productivity (see *Fig. 3*). Each new PEI preparation must be tested.

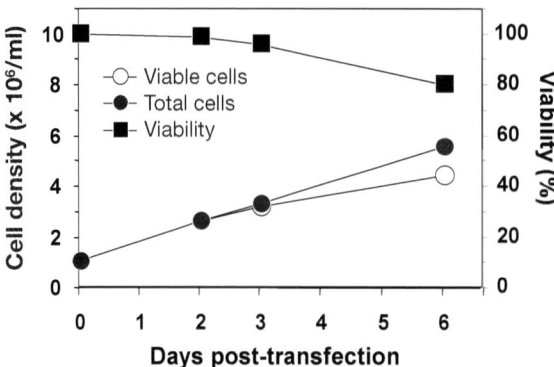

Figure 4. Cell viability curve in a 500 ml culture for hClusterin production.
293-6E cells were transfected at a density of 1×10^6 cells/ml and TN1 was added at 24 h p.t. to a final concentration of 0.5% (w/v). Cell viability was monitored over a 6 day period.

[b]To evaluate transfection efficiency, a GFP expression plasmid can be added at 5% of the final DNA content used. A reasonable indicator of successful transfection is when the percentage of GFP-positive cells is between 30 and 50% by 48–72 h p.t.
[c]The presence of G418 does not interfere with transfection.
[d]Tryptone N1 (OrganoTechnie) can be dissolved in Freestyle or another basal culture medium such as low-calcium HSFM (LC-SFM). The final TN1 stock solution must be supplemented with Pluronic F68 at 0.1% (w/v) and filter-sterilized.
[e]This is scaleable to 650 ml in a 2 l shake flask.
[f]Cell density at transfection can range from 8×10^5 to 1.2×10^6 cells/ml, provided the doubling time is 24 h and viability is greater than 95%.
[g]Cells can be diluted to 2.5×10^5 cells/ml at 48 h prior to transfection or to 5×10^5 cells/ml at 24 h prior to transfection.
[h]Establish the best harvesting time by performing an expression kinetics study (see *Fig. 1b*).
[i]As culture viability decreases with time, it is important to harvest when the viability is relatively high (over 75%) to protect the r-protein from proteolytic degradation (see *Fig. 4*).

Protocol 5

Transfection of adherent 293E cells in a six-well plate with branched PEI[a]

Equipment and Reagents
- 293E maintenance cells in T-flask or shake flask
- PBS
- LC-SFM medium (custom-made low-calcium HSFM formulation; Invitrogen) supplemented with 1% (v/v) serum and 50 µg/ml G418
- Branched PEI solution (1.0 mg/ml; see *Protocol 3*)
- 20% TN1 solution (refer to *Protocol 4*, note d, for TN1 preparation)
- Purified plasmid DNA of interest (see *Protocol 2*)
- Disposable CellBIND six-well plates (Corning)[a]
- Humidified incubator at 37°C with 5% CO_2

Method
Plating the cells 24 h before transfection
1. Warm the culture medium to 37°C.
2. Remove the cell maintenance T-flask from the incubator and dislodge the cells by gently tapping the side of the flask (this is not necessary if the maintenance cultures are in shake flasks). Remove a cell aliquot to determine cell density and viability.
3. In a 15 ml tube, dilute the cells to 2.5×10^5 cells/ml in a volume of 12 ml of medium[b].
4. Prepare the plates as described in *Protocol 4*, step 3 (see *Fig. 2b*).

Transfection (DNA : PEI ratio of 1.5 : 2)[c]
1. Add 100 µl of PBS to each of six 1.5 ml tubes.
2. Add 3 µg of DNA to each tube and vortex.
3. Add 672 µl of PBS to another 1.5 ml tube. Add 28 µl of branched PEI solution and vortex.
4. Add 100 µl of PEI solution to each DNA-containing tube and immediately vortex each mixture three times for 3 s each.
5. Incubate for 15 min at room temperature.
6. Remove the six-well plate from the incubator, add the DNA/PEI mixture to each well, and swirl the plate. Return the plate to the incubator.
7. For secreted proteins, at 24–48 h p.t., add 50 µl of pre-warmed 20% TN1 solution to each of the six wells and return the plate to the incubator.
8. Refer to *Protocol 4* for details of r-protein harvesting.

Notes
[a]Branched PEI works well for adherent cells and can be used to adhere suspension-growing cells to the plate during transfection; linear PEI works less efficiently for adherent cultures. In a different application, for the establishment of stable cell lines, branched PEI can be used to help adhere transfected cells, combined with the use of CellBIND six-well plates.
[b]Can be scaled down to 1 ml in 12-well plates.
[c]Refer to the Notes in *Protocol 4*.

Protocol 6

Transfection of 293-6E cells in suspension with X-tremeGENE Ro-1539 (DNA:Ro-1539 ratio of 0.1:1)[a]

Equipment and Reagents
- 293-6E maintenance cells in shake flask
- PBS
- X-tremeGene Ro-1539 (Roche)
- 20% TN1 solution (see *Protocol 4*, note d, for preparation)
- Purified plasmid DNA of interest (see *Protocol 2*)

Method

Small-scale transfection in a six-well plate (2 ml) or in a 125 ml shake flask (20 ml)[b]

1. Warm the PBS (transfection buffer) to 25–37°C. Remove the DNA from the freezer to thaw and the Ro-1539 from the refrigerator.
2. From a maintenance flask, determine the cell density and viability.
3. Prepare plates as described in *Protocol 4*, step 3 (see *Fig. 2b*).
4. Add 200 µl of PBS to each of six 1.5 ml tubes (or 2.0 ml to a 15 ml tube for a 125 ml shake flask).
5. Add the required DNA volume corresponding to 0.2 µg to each tube (or 2 µg to the 15 ml tube) and vortex.
6. Add 2 µl of Ro-1539 to each tube (or 20 µl to the 15 ml tube) containing the DNA solution and immediately vortex.
7. Incubate the DNA/Ro-1539 mixture(s) for 15 min at room temperature.
8. Remove the culture (step 3) from the incubator, add the DNA/Ro-1539 complexes to the wells (or the 15 ml tube contents to the shake flask) and swirl.
9. Return the culture to the incubator.
10. For secreted proteins, add 50 µl of pre-warmed 20% TN1 at 24–48 h p.t. to each of the six wells (or 500 µl for the 125 ml shaker) and return the cultures to the incubator.
11. Refer to *Protocol 4* for details of r-protein harvesting.

Notes

[a] Evaluate different volumes of Ro-1539 per µg of DNA to determine the best transfection efficiency and production.

[b] Refer to *Protocol 4*, notes b–i.

Protocol 7

Transfection of 293-6E cells in suspension with 293fectin (DNA : 293fectin ratio of 1.5 : 1.3)[a]

Equipment and Reagents
- 293-6E maintenance cells in shake flask
- 293fectin (Invitrogen)
- Opti-MEM I medium (Invitrogen)
- 20% TN1 solution (see *Protocol 4*, note d, for preparation)
- Purified plasmid DNA of interest (see *Protocol 2*)

Method
Small-scale transfection in a six-well plate (2 ml) or in a 125 ml shake flask (20 ml)[b]

1. Warm the Opti-MEM I medium (transfection medium) to 25–37°C. Remove the DNA from the freezer and the 293fectin from the refrigerator.
2. From a maintenance flask, determine the cell density and viability.
3. Prepare plates as described in *Protocol 4*, step 3 (see *Fig. 2b*).
4. Add 100 µl of Opti-MEM I to each of six 1.5 ml tubes (or 1.0 ml to a 15 ml tube for a 125 ml shake flask).
5. Add the required DNA volume corresponding to 3 µg to each tube (or 30 µg to the 15 ml tube) and vortex.
6. To another 1.5 ml tube, add 682 µl of Opti-MEM I and 18 µl of 293fectin (or to another 15 ml tube, add 1.0 ml Opti-MEM I and 26 µl 293fectin). Vortex gently.
7. Incubate for 5 min at room temperature.
8. Add 100 µl of diluted 293fectin to each DNA-containing tube (or the entire 1.0 ml of the 293fectin dilution to the 15 ml DNA tube) and immediately vortex after addition.
9. Incubate the mixtures for 20 min at room temperature.
10. Remove the culture (step 3) from the incubator, add the DNA/293fectin mixture(s), and swirl the transfected culture.
11. Return the culture to the incubator.
12. For secreted proteins, add 50 µl of pre-warmed 20% TN1 at 24–48 h p.t. to each of the six wells (or 500 µl for the 125 ml shake flask) and return the culture(s) to the incubator.
13. Refer to *Protocol 4* for r-protein harvesting periods.

Notes
[a]Evaluate different volumes of 293fectin per µg of DNA to determine the best transfection efficiency and production.
[b]Refer to *Protocol 4*, notes b–i.

2.5 Purification of His-tagged r-proteins

Recombinant proteins often need to be as pure as possible before any characterization study can begin. Although many types of protein tag are available, histidine is the most popular (18). The following protocol describes the immobilized metal-affinity column (IMAC) purification technique, thus completing the 'gene-to-protein' process. It also determines the production scale that will be needed to provide enough pure material for a given study. We have successfully expressed and purified many secreted proteins using the following protocol. Most of our His-tagged proteins were expressed from the pTTSH8Q1 (or similar) vectors (19). Whilst small-scale IMAC purification (e.g. <500 ml of culture medium) can easily be achieved using gravity chromatography columns, larger volumes could be processed with the aid of automated chromatography systems such as the AKTA system (see *Fig. 1c*).

When performing small-scale trials ranging from 20 to 500 ml, a gravity column can be packed with 1-5 ml of resin, with a bed height of at least 1 cm (see *Fig. 5*). Refer to the manufacturer's binding capacity specifications in order to use an adequate volume of resin. If a large volume of culture medium is to be purified, the clarified culture supernatant can be concentrated 10-20-fold by tangential flow filtration (see *Fig. 1c*), followed by r-protein capture on the IMAC column.

(a)

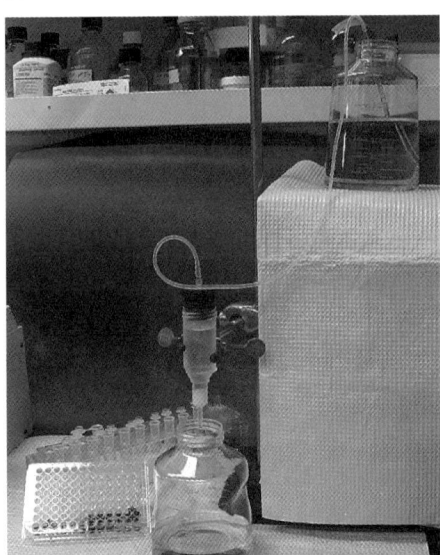

(b)

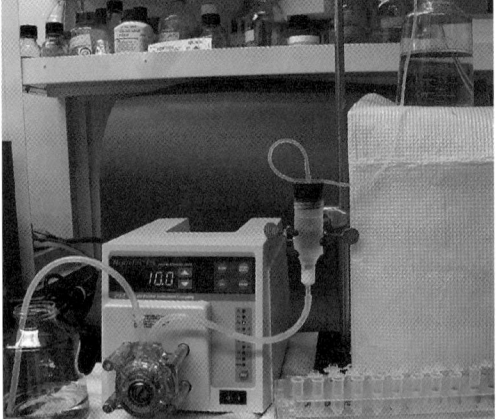

Figure 5. IMAC purification of a 1 cm bed height Fractogel column (1 ml total volume).
Small sample volumes can be loaded into the reservoir and passed through by gravity (*a*). For volumes greater than 100 ml, a peristaltic pump can be used to pull the sample through the column faster (*b*). To prevent the column from drying out, stop the pump early enough and allow the last few milliliters to pass through by gravity. The flow-through and two washes must be collected in separate receptacles.

Protocol 8

Purification of His-tagged proteins using Fractogel–cobalt[a]

Equipment and Reagents
- Fractogel EMD Chelate (M) resin (EMD Biosciences)
- Empty column to load the resin[b]
- Peristaltic pump (optional)
- Charging buffers:
  - 0.5 M NaCl
  - 200 mM $CoCl_2$
  - 0.5 M NaCl (pH 5.0±0.2)
- PBS
- Wash buffer 1 (50 mM sodium phosphate, pH 7.0, 300 mM NaCl)
- Wash buffer 2 (wash buffer 1 mixed with 8.3%, v/v, elution buffer to obtain a final imidazole concentration of 25 mM)
- Elution buffer (wash buffer 1 + 300 mM imidazole, followed by pH adjustment to 7.0)
- Column regeneration buffers:
  - Cobalt stripping buffer (50 mM EDTA, 0.5 M NaCl)
  - Milli-Q water
  - Cleaning buffer (0.5 M NaOH, 1 M NaCl)
- Storage buffer (150 mM NaCl, 20%, v/v, ethanol)
- Bio-Rad protein assay reagent, 5× concentrate, diluted to 1× in Milli-Q water
- Spectrophotometer
- Culture supernatant[c]

Method

Charging the 5 ml Fractogel gravity flow column with cobalt

1. Add 5 column volumes of 0.5 M NaCl.
2. Add 4 column volumes of 200 mM $CoCl_2$.
3. Add 2 column volumes 0.5 M NaCl (pH 5±0.2).

Sample loading and r-protein elution

1. Add 10 column volumes of PBS to equilibrate the column.
2. Load the supernatant sample (feed)[c,d]. Keep the flow-through and store at 4°C. Retain a 500 µl aliquot of both feed and flow-through for subsequent gel analysis[e].
3. Add 10 column volumes of wash buffer 1. Retain a 500 µl aliquot[e].
4. Add 10 column volumes of wash buffer 2 to remove impurities bound to the column. Retain a 500 µl aliquot[e].
5. Add elution buffer in 1 ml increments. Harvest individual 1 ml fractions in Eppendorf tubes (pre-labeled 1, 2, 3...) until the equivalent of 4 column volumes of elution buffer have passed through.
6. Add 5–20 µl of each eluted fraction to a 96-well plate and add 200 µl of diluted protein assay reagent to each well. Make sure that each sample is well mixed as the reagent is being added.
7. Wait 5–10 min for color development. The wells with the darkest blue coloration correspond to the fractions containing the r-protein[f] (see *Fig. 5a*).

8. Pool the 1 ml fractions containing r-protein.
9. Desalt the purified r-protein using a desalting column[g] equilibrated in PBS or in your preferred buffer.
10. Based on the molar extinction coefficient[h] of the protein, determine its concentration after measuring its absorbance at 280 nm (use an appropriate buffer as the blank).

Column washing and storage
1. Add 2 column volumes of cobalt stripping buffer.
2. Add 2 column volumes of Milli-Q water.
3. Add 15 column volumes of cleaning buffer.
4. Add 5 column volumes of Milli-Q water, followed by 5 column volumes of PBS, or more if necessary, to neutralize the column to pH 7.0–7.2.
5. Add 5 column volumes of storage buffer.
6. Add an additional 1–2 ml of storage buffer on top of the column, cover the top and bottom, and store at 4°C.

Notes

[a]All solutions must be filtered (0.45 μm) and degassed to minimize air bubble formation in the column.
[b]We suggest using a clean empty Qiagen-tip 500 column (Maxi kit; 5 ml minimum bed volume) or HiSpeed Maxi Tip column (10 ml minimum bed volume) to prepare the IMAC column.
[c]The supernatant must be filtered (0.45 μm) before loading it onto the column.
[d]A peristaltic pump can be attached to the column outlet to pull the sample feed through the column more quickly for volumes greater than 100 ml. We recommend a flow rate up to a maximum of 10 ml/min for a 5 ml Qiagen-tip 500 column (2 cm diameter or 190 cm/h) and 20 ml/min for a 10 ml HiSpeed Maxi Tip column (3.5 cm diameter or 125 cm/h). Stop and disconnect the pump 1–2 min before the entire sample empties from its reservoir. Add the last few milliliters of sample to the reservoir with a pipette and allow it to flow through by gravity. This will prevent the column from drying out (see *Fig. 5b*).
[e]Do not discard the flow-through or wash fractions until it has been determined that the r-protein has been entirely captured and eluted from the column. Subsequent analysis of the original sample feed, flow-through, washes, and pooled protein on a gel determines whether the purification process was sufficient or whether the column resin was overloaded beyond its protein-binding capacity. Sometimes the r-protein can be found eluted in wash buffer 2. A wash buffer with a lower imidazole concentration should then be used.
[f]To get a better qualitative reading of the blue coloration, absorbance at 595 nm can be read using a microplate reader.
[g]We recommend commercial disposable pre-packed columns such as Econo-Pac 10DG desalting columns (Bio-Rad).
[h]The molar extinction coefficient of your protein can be obtained by uploading its amino acid sequence at http://ca.expasy.org/tools/protparam.html[12.1].

2.6 Results

Some of the r-proteins produced in our laboratory using the pTT expression vector and our scaleable transient gene expression technology are illustrated in *Fig. 1(b)* and *(c)* and *Figs 3, 4,* and *6*.

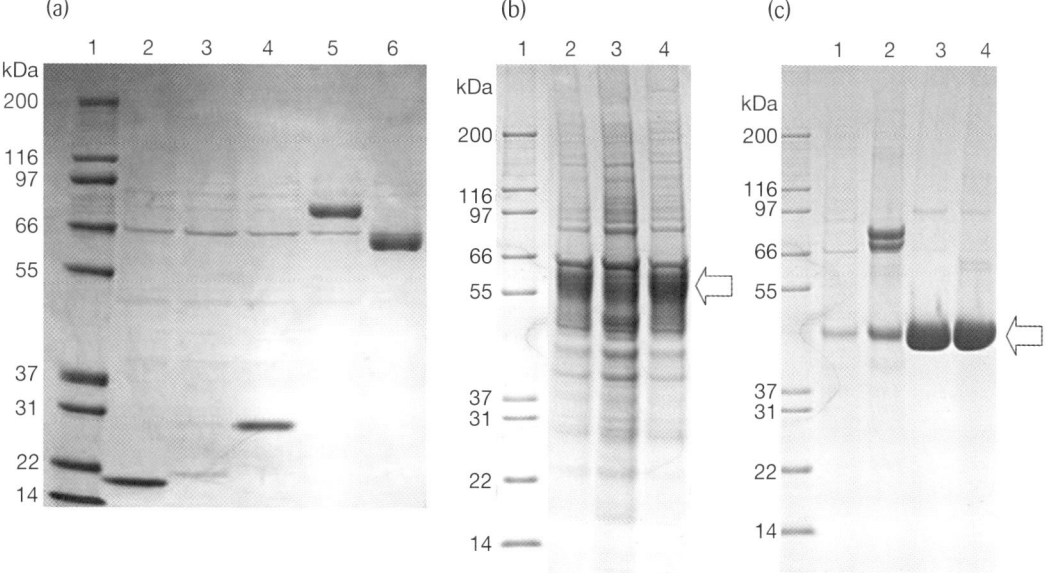

Figure 6. Examples of secreted proteins produced by transient gene expression in 293-6E cells. All proteins were expressed using the pTT vector and culture medium was harvested at day 5 p.t. (a) SDS-PAGE of culture supernatants (20 µl per lane). Lanes: 1, molecular mass markers; 2, human growth hormone, hGH1v2 (GenBank no. BC062475); 3, human growth hormone, hGH2v1 (GenBank no. BC020760); 4, rat CD4 ectodomains 3 and 4 (20); 5, human transferrin (GenBank no. M12530); 6, secreted alkaline phosphatase (19). (b) SDS-PAGE of Fc-tagged protein (~64 kDa) produced at three different scales. Lanes: 1, molecular mass markers; 2, a 2 ml culture from a six-well plate; 3, a 20 ml culture from a 125 ml shake flask; 4, a 530 ml culture from a 1 l Wavebag. (c) Comparison of two culture media for 500 ml Fc-tagged protein production and purification. Lanes: 1, Freestyle (Invitrogen) supernatant (20 µl); 2, HyQ SFM4TRANSFX-293 (Hyclone) supernatant (20 µl); 3, purified protein from Freestyle culture medium (total yield 7 mg); 4, purified protein from HyQ SFM4TRANSFX-293 culture medium (total yield 21 mg). Note that HyQ SFM4TRANSFX-293 culture medium is supplemented with a protein (see *lane 2*).

3. TROUBLESHOOTING

- **The cells do not grow well**
  Cell doubling times and viability, especially in serum-free media, are affected by many parameters that are sometimes within the control of the operator. If a rigorous cell culture routine is practiced, one can tell immediately when the cultures are compromised.
  - If cells do not grow with the expected doubling time, verify the following causes: cultureware cleanliness, source of ultrapure water, and the filtration system (if applicable) used to make it. Glassware for all culture manipulations and reagent preparations must be rinsed at least three times with high-purity water.

- Cell aggregation can give a false impression of slower growth rates. Check regularly that the orbital shaker is at the right speed and that the incubator is at the correct temperature and is supplying the prescribed amount of CO_2.
- The number of cell passages indicates the age of the cells in service. A new cryovial can be thawed every 3 months, on average.
- Verify expiration dates, culture medium lots, and batches of additives used: glutamine, Pluronic F68, and antibiotics.
- Check for mycoplasma contamination; if positive, thaw new cells.

- **The transfection does not work**
  Low transfection efficiencies are readily indicated when the percentage of GFP-positive cells is below 30% by 48 h p.t. The previously mentioned parameters can be considered when trying to debug transfection problems.
    - Make sure the cells are in exponential growth phase for transfections. If the culture grows to higher-than-prescribed transfection densities, medium nutrient depletion and metabolic by-products will compromise the transfection.
    - DNA plasmid quality can be verified for contaminants by A_{260}/A_{280} ratio readings; acceptable values range between 1.80 and 1.95. In addition, the DNA plasmid can be digested and run on an agarose gel to verify its integrity (it should be mostly in a supercoiled form, see *Fig. 1a*) and/or to detect RNA contamination.
    - Cell line productivity can be monitored occasionally by transfecting the cells with an easily measured protein such as secreted alkaline phosphatase or GFP. For each new medium or reagent lot, transfection efficiencies can be monitored using six-well plate assays (see *Fig. 1b*).
    - Anti-clumping agents such as dextran sulfate can inhibit transfection.

- **Purification from IMAC does not work**
    - Perform SDS-PAGE analysis of the feed, flow-through, washes, and pooled fractions to see whether the r-protein was captured by the resin or not. If the His-tagged protein is in the flow-through, perhaps the His tag on the protein itself may be sterically hindered and inaccessible for capture by the column.
    - The r-protein may also be aggregated, thus hampering its capture onto the column (this may be checked by gel filtration chromatography).
    - Try increasing the pH of your feed to 8.0 or higher to enhance r-protein binding to the column.
    - Try charging the IMAC column with nickel instead of cobalt to increase binding affinity.

4. REFERENCES

1. **Wurm FM** (2004) *Nat. Biotechnol.* **22**, 1393–1398.
2. **Pham PL, Kamen A & Durocher Y** (2006) *Mol. Biotechnol.* **34**, 225–238.
3. **Wurm F & Bernard A** (1999) *Curr. Opin. Biotechnol.* **10**, 156–159.
★★ 4. **Durocher Y, Perret S & Kamen A** (2002) *Nucleic Acids Res.* **30**, E9. – *Development of a simple, robust, and scaleable transfection process using linear 25 kDa PEI that does not require a medium exchange prior to transfection.*
5. **Jordan M, Köhne C & Wurm FM** (1998) *Cytotechnology*, **26**, 39–47.
★ 6. **Schlaeger E-J & Christensen K** (1999) *Cytotechnology*, **30**, 71–83. – *First demonstration that PEI can be used for the production of r-proteins by large-scale transfection of 293E cells.*
7. **Baldi L, Muller N, Picasso S, et al.** (2005) *Biotechnol. Prog.* **21**, 148–153.
8. **Farrell P & Iatrou K** (2004) *Protein Expr. Purif.* **36**, 177–185.
★★ 9. **Graham FL & van der Eb AJ** (1973) *Virology*, **52**, 456–467. – *First description of the use of calcium phosphate precipitates as a vehicle for transfection of animal cells.*
10. **Jordan M & Wurm F** (2004) *Methods*, **33**, 136–143.
★★ 11. **Girard P, Derouazi M, Baumgartner G, et al.** (2002) *Cytotechnology*, **38**, 15–21. *First report of 100 l transfection using calcium phosphate.*
12. **Atkinson A & Jack GW** (1973) *Biochim. Biophys. Acta*, **308**, 41–52.
★★★ 13. **Boussif O, Lezoualc'h F, Zanta MA, et al.** (1995) *Proc. Natl. Acad. Sci. U.S.A.* **92**, 7297–7301. – *First paper showing that PEI is an efficient nonviral DNA carrier for mammalian cells.*
14. **Pham PL, Perret S, Doan HC, et al.** (2003) *Biotechnol. Bioeng.* **84**, 332–342.
★ 15. **Pham PL, Perret S, Cass B, et al.** (2005) *Biotechnol. Bioeng.* **90**, 332–344. – *Paper showing that peptone addition post-transfection significantly improves r-protein yields.*
16. **Derouazi M, Girard P, Van Tilborgh F, et al.** (2004) *Biotechnol. Bioeng.* **87**, 537–545.
17. **Geisse S & Henke M** (2005) *J. Struct. Funct. Genomics*, **6**, 165–170.
18. **Arnau J, Lauritzen C, Petersen GE, et al.** (2006) *Protein Expr. Purif.* **48**, 1–13.
19. **Cass B, Pham PL, Kamen A, et al.** (2005) *Protein Expr. Purif.* **40**, 77–85.
20. **Chapple SD, Crofts AM, Shadbolt SP, et al.** (2006) *BMC Biotechnol.* **6**, 49.

CHAPTER 13

Nisin- and subtilin-controlled gene expression systems for Gram-positive bacteria

Oscar P. Kuipers and Jan Kok

1. INTRODUCTION

Following the discovery that nisin, an antimicrobial peptide of the lantibiotic family, can act as autoinducer of its own biosynthesis (1), this knowledge was used to construct the first tightly controlled gene expression systems employing nisin as the inducer (2). Some background on nisin biosynthesis, mode of action, immunity, and regulation can be found in the review of van Kraaij and co-workers (3) and references therein. In short, nisin is a post-translationally modified peptide containing three dehydrated amino acid residues and five thioether bridges called lanthionines. The enzymes catalyzing the dehydration of serine and threonine residues and the thioether formation are called NisB and NisC, respectively. The nisin gene cluster consists of 11 genes, i.e. *nisABTCIPRKFEG*, encompassing the two inducible promoters upstream of *nisA* (the structural gene) and *nisF* (4). There also is a constitutive promoter upstream of *nisR*. The NisT protein is involved in secretion of the precursor peptide, NisP is the protease that cleaves off the leader peptide, and NisIFEG are involved in conferring immunity to the producing cells. Important for the development of the nisin-inducible gene expression system are the two proteins belonging to the class of two-component regulatory systems, NisR and NisK, which are involved in the autoregulation of nisin biosynthesis. Nisin is used in the food industry as a natural preservative, as it efficiently kills other Gram-positive bacteria via a combined mechanism involving binding to lipid II and pore formation (5, 6).

For the construction of the inducible system, there were only three essential requirements: an inducible promoter (P_{nisA}, the most widely used, or P_{nisF}), an inducer (nisin), and a signal transduction system involving NisR, a response regulator (7), and NisK, a sensor histidine kinase. NisK is able to sense the extracellular presence of nisin at very low, subinhibitory concentrations and will subsequently

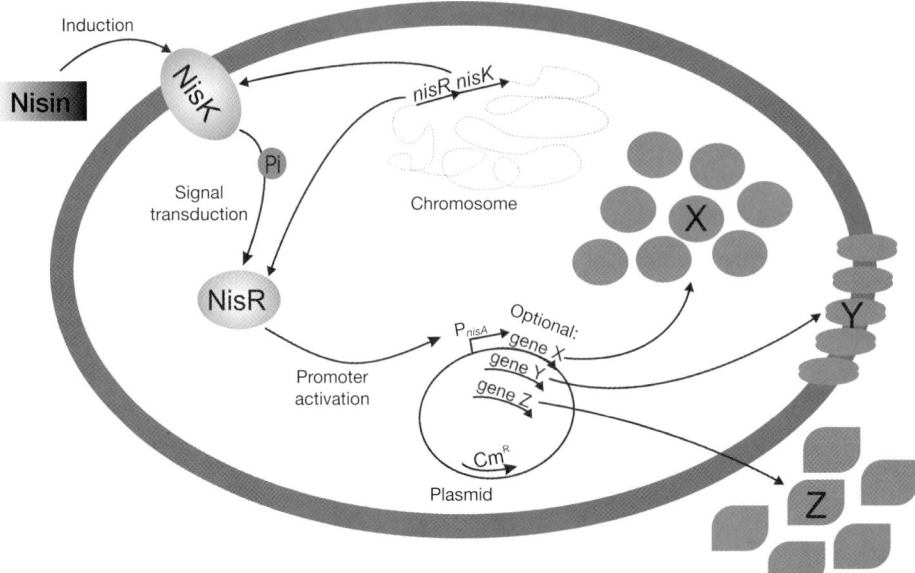

Figure 1. Principle of the NICE system.

autophosphorylate and donate the phosphate moiety to NisR. Phosphorylated NisR, in turn, can bind to P_{nisA}, thereby activating transcription (1).

Lactococcus lactis harboring the two-component regulatory system encoded by *nisRK* was used as host and plasmids carrying either P_{nisA} or P_{nisF} and suitable multiple cloning sites were used for expressing the gene(s) of interest. The system was designated nisin-controlled expression, yielding the acronym NICE (2, 8, 9). The principle of the induction process is outlined in *Fig. 1*.

L. lactis expressing chromosomally located *nisRK* genes at a relatively low constitutive level harbors a multi-copy expression plasmid with a suitable selection marker, e.g. chloramphenicol resistance (Cm^R), and P_{nisA} followed by a multiple cloning site. Any gene of interest can be cloned downstream of P_{nisA}, either as a transcriptional or a translational fusion. Genes *X*, *Y*, and *Z* encode an intracellular, an integral membrane and a secreted protein, respectively. The system is induced by extracellular addition of NisA or NisZ at subinhibitory levels, typically ranging between 0.02 and 5 µg/l, preferably at mid- or late-exponential phase. Optimal expression is usually achieved at approximately 2 h after induction.

It soon became clear that the system developed had several major advantages over other existing systems for lactic acid bacteria (for reviews, see 8–11):

1. The inducer nisin is food-grade and safe to use in all applications (NisA and NisZ function equally well).
2. The inducer is easily obtained from culture supernatants of nisin-producing strains (e.g. *L. lactis* strain NZ9700) or is commercially available as a powder (e.g. Nisaplin).
3. The inducer can be used at very low concentrations that are well below the minimal inhibitory concentration (ranging between 10 and 2000 µg/l for var-

ious Gram-positive bacteria). In *L. lactis*, concentrations between 0.02 and 5 µg/l can be used for induction.
4. Various host strains can be used, with the minimal requirement that they contain the *nisRK* genes either integrated on the chromosome (this is preferred, as in *L. lactis* strains NZ3900, NZ9000, and NZ9800) or on a second plasmid (usually a pIL-derived plasmid, e.g. pNZ9530; 12).
5. Host strains are easily modified because of the abundant number of genetic tools available.
6. Various expression plasmids with multiple cloning sites downstream of P_{nisA} or P_{nisF} (2) can be used, either for transcriptional or translational fusions. Specific plasmids for cloning genes of interest with tags or signal sequences are now also available.
7. The plasmids from the pNZ8xxx series are shuttle vectors, e.g. for *Escherichia coli*, as they are based on the pSH71 replicon (2).
8. The induction by extracellularly added nisin shows a linear dose response over a large dynamic range (2).
9. The levels of protein produced can be very high, up to 50% of the total soluble intracellular protein (2).
10. The system is tightly controlled: without inducer added, there is hardly any leakage from P_{nisA}, enabling the cloning and expression of 'difficult' or even lethal genes.
11. *L. lactis* is a host with low protease activity, preventing undesirable degradation of the proteins produced. It also cannot sporulate.
12. *L. lactis* is easy to grow and industrial-scale fermentation applications have been achieved (13, 14).

The NICE system is especially helpful in those situations where the protein to be overexpressed is detrimental or even lethal to the cell. This is often the case with (integral) membrane proteins and is one of the reasons for the limited number of three-dimensional structures of (eukaryotic) membrane proteins. Nisin-induced functional overexpression of these types of protein to levels that are useful for protein structural analysis is possible in *L. lactis* (15). In a subsequent study, it was shown that the level of expression of a number of eukaryotic transporters is negatively influenced by the N-terminal domains of the proteins. Expression levels could be increased more than tenfold by truncating or removing these regions, or by replacing them with lactococcal signal sequences (16). Moreover, lytic cassettes, such as the holin–lysin cassette from phage ΦUS3, could successfully be used for food-grade controlled lysis, by sole use of lactococcal DNA (self-cloning) (17).

Quickly after the implementation of the NICE system in *L. lactis*, the system was successfully extended for use in other Gram-positive bacteria, e.g. *Leuconostoc lactis* and *Lactobacillus helveticus* (12), by the development of low- and high-copy-number vectors for heterologous expression of *nisRK* in these species together with P_{nisA}-containing plasmids. A similar approach was used for developing NICE systems for other Gram-positive organisms including several pathogens: *Streptococcus pyogenes*, *Streptococcus pneumoniae*, *Enterococcus faecalis*, and *Bacillus subtilis* (18). However, because of the use of plasmids for expression of *nisRK*, the optimal expression level is hard to achieve, due to the fact that too high expression of NisRK leads to premature activation of the system without induction and, thus, to considerable leakage.

Therefore, systems making use of integrated single copies of *nisRK* were developed, which performed better than the multi-copy systems, e.g. for *Lactobacillus plantarum*, *B. subtilis*, and *S. pneumoniae* (19–21). Also the use of a single plasmid harboring P_{nisA} and *nisRK* was shown to work reasonably well in the case of *Lactobacillus reuteri* (22), although considerable leakage was observed.

The major paradigm Gram-positive bacterium is *B. subtilis*, being very well characterized at the genetic, physiological, and biochemical level. The subtilin-producing *B. subtilis* ATCC 6633 carries all of the genes for the density-dependent production of subtilin (*spaBTCSIFEGRK*). The genes *spaRK* are involved in signal transduction and the promoter upstream of the structural subtilin gene *spaS* is inducible by subtilin. This promoter typically contains two pentanucleotide direct repeats close to the predicted −35 region, where phosphorylated SpaR presumably binds (23). An optimized inducible system was developed for this organism in analogy to the NICE system, using its own native lantibiotic subtilin as inducer and the two-component signal transduction system encoded by *spaRK*. These genes were integrated in the chromosomal *amyE* locus, whilst the P_{nisA}-containing vectors were changed into P_{spaS}-containing vectors. This resulted in the <u>s</u>ubtilin-<u>r</u>egulated <u>e</u>xpression (SURE) system, which was only slightly leaky and could reach very high levels of protein production (20).

During the last decade, many more antimicrobial peptides have been identified that show autoinducing properties or are embedded in gene clusters that encode special inducing peptides. Some of these elements (inducing peptides, inducible promoters, and signal transduction systems) have been employed to generate novel expression systems that could have high value for specific applications in desired species. A successful example is the system using sakacin for induction in *Lactobacillus sakei* and *L. plantarum* (24).

2. METHODS AND APPROACHES

2.1 The NICE system

The NICE system consists of three elements: a strain (we provide an extensive protocol for *L. lactis*, which can be used also for other nisin-inducible systems when taking into account the original literature on that species) that expresses the *nisRK* genes to the appropriate level, the inducer molecule nisin (or one of its inducing analogs or mutants), and a vector that contains a multiple cloning site preceded by a nisin-responsive promoter. The latter could either be P_{nisA} or P_{nisF}, the promoter of the *nisABTCIP* or *nisFEG* genes, respectively.

2.1.1 Sources of nisin

For induction, one can use either a commercial preparation of nisin (usually NisA), such as Nisaplin from Danisco, or diluted supernatants of a nisin-producing

L. lactis strain, e.g. strain NZ9700, which produces ~10 mg NisA/I in a standing culture grown in GM17 or SPYS medium. Dilutions should always be made in weak organic acid solutions, e.g. 0.05% acetic acid, as nisin has a low stability at pH values >6 and is especially unstable at alkaline pH. This also helps to prevent sticking of the nisin molecules to the tube material or at the water–air interface, which negatively influences the actual available concentration. It is advisable always to prepare the dilutions fresh on ice and immediately use them for induction. Nisin is very stable at pH 2–5 and can even be boiled for several hours at pH 4 without significant loss of activity.

2.1.2 Host strains and plasmids for use of NICE in *L. lactis*

The copy number and strength of expression of the *nisRK* genes is of eminent importance for the proper functioning of the NICE system. The plasmid-free *L. lactis* MG1363 derivative NZ9000 is the most sensitive and most used host strain for nisin-induced expression in this species (25). This strain carries a single-copy chromosomal insertion of the *nisRK* genes driven by their native constitutive promoter in the *pepN* locus and, consequently, lacks peptidase N activity (see *Table 1*).

In cases where a fully active proteolytic system is required this would pose a problem, but for most applications the strain is very useful. If a specific *L. lactis* strain is required other than NZ9000, the original integration plasmid pNZ9573 could be used to integrate *nisRK* in *pepN* of this strain, as most if not all *L. lactis* strains will carry a homologous *pepN* in their chromosomes. Alternatively, for example if the homology between the *pepN* genes proves to be too limited, the genes may be positioned elsewhere in the chromosome using appropriate plasmid integration methods (30). If need be, the *nisRK* genes can be introduced in the strain to be studied on a plasmid, thus making use of a two-plasmid approach for nisin-induced gene expression (for details, see below).

The *L. lactis* MG1363 derivative NZ3900 (*pepN::nisRK*; Lac⁻) carries a chromosomal insertion of the *lac* operon of *L. lactis* NCDO712 in which part of the *lacF* gene is deleted (2, 17) (see *Table 1*). The strain is able to grow on lactose when *lacF* is provided *in trans* on a plasmid, thus presenting a stable plasmid selection and maintenance system that has been used for the food-grade use of the NICE system. A good example is the development of a system for controlled lysis of *Lactococcus* by *in situ* induction in cheese using only self-cloning and food-grade components. This system can conveniently be used to achieve accelerated cheese ripening (17). An improved derivative of *L. lactis* NZ9000 for the analysis of protein secretion was made by introducing a deletion in the chromosomal *htrA* gene encoding an extracellular housekeeping proteinase (see *Table 1*; 27).

A variety of vectors is available for use of the NICE system in *L. lactis* (see *Table 1*). Two principally different vectors should be distinguished: (i) the expression plasmid, which carries the nisin-inducible P_{nisA} (or P_{nisF}) upstream of the gene of interest to be (over)expressed, and (ii) in a two-plasmid system only, the vector that carries the *nisRK* genes. The most commonly used nisin-inducible gene expression plasmid is pNZ8048 (see *Table 1*; 25). It carries the nisin-inducible P_{nisA}

Table 1. Strains and plasmids used for NICE and SURE in Gram-positive bacteria

Component	Relevant properties[a]	Reference
NICE system		
Strain		
L. lactis[b]		
NZ9700	Nisin producer; Rif[R] Strp[R]; carries nisin–sucrose transposon Tn*5276*	26
NZ9000	NICE system host; *pepN::nisRK*	25
NZ3900	Host for food-grade use of NICE; *pepN::nisRK*; chromosomally integrated *lac* operon but Lac⁻ due to Δ*lacF*	2, 17
NZ9000 *htrA*	Clean deletion of *htrA* gene; host for protein secretion	27
Plasmid(s)		
pNZ9573	*pepN::nisRK* integration vector; Cm[R], Em[R], *E. coli* p15A *ori*	2
pNZ9520	*nisRK* under control of P_{rep} of pIL253 (pAMβ1 replicon; high copy number); Em[R]	12
pNZ9521	*nisRK* driven by P_{nisR} and P_{rep} read-through; pIL253 (pAMβ1 replicon; high copy number); Em[R]	12
pNZ9530	*nisRK* under control of P_{rep} of pIL252 (pAMβ1 replicon; low copy number); Em[R]	12
pNZ9531	*nisRK* driven by P_{nisR} and P_{rep} read-through; pIL252 (pAMβ1 replicon; low copy number); Em[R]	12
pNZ8048	Most used NICE expression vector; *Nco*I site used for translational fusions; Cm[R]	25
pNG8048E	pNZ8048 with Em[R] in MCS for easy cloning; Em[R], Cm[R]	Leenhouts; Molecular Genetics lab collection
pNZ8008, pNZ8020, pNZ8030, pNZ8032, pNZ8037	Early vectors for transcriptional fusions; same make-up as pNZ8048 including MCS	2
pNZ8045	Vector for food-grade applications with the *lacF* selectable marker	2
pNZ8150	Standard NICE gene expression vector; *Sca*I site used for translational fusions; Cm[R]	11
pNZ8021	Vector for transcriptional fusions; Cm[R]	11
pNZ8110	Protein secretion using signal sequence of *L. lactis* Usp45; translational fusion via *Nae*I site; Cm[R]	11
pMSP3535	pAMβ1 (low copy number) and ColE1 replicons; carries *nisRK* and P_{nisA}; Em[R]	28
pMSP3545	pMSP3535 derivative; *Nco*I site for translational fusions at ATG start of *nisA*	28
pNICE	*E. coli*–*L. reuteri* shuttle vector; nisin-controlled expression vector containing P_{nisA} and *nisRK*; Em[R]	22
pNIES	Expression-secretion vector; pNICE with sequence encoding signal peptide of *B. licheniformis amyl*; Em[R]	22

Component	Relevant properties[a]	Reference
SURE system		
Strain		
B. subtilis		
ATCC 6633	Subtilin producer	ATCC collection
NZ8900	*B. subtilis* 168; *amyE::spaRK*; Km[R]	20
Plasmid		
pNZ8900	*amyE::spaRK* insertion vector; Amp[R] Km[R]	20
pNZ8901	SURE expression vector; $P_{spaSmut}$; Cm[R]	20
pNZ8902	SURE expression vector; $P_{spaSmut}$; Em[R]	20
pNZ8910	SURE expression vector; P_{spaS}; Em[R]	20
pNZ8911	SURE expression vector; P_{spaS}; Cm[R]	20

[a]$P_{spaSmut}$, mutant *spaS* promoter: *spa* box changed to a perfect pentanucleotide repeat; P_{rep}, promoter of plasmid replication gene; Amp[R], ampicillin resistance; Km[R], kanamycin resistance; Em[R], erythromycin resistance; Cm[R], chloramphenicol resistance; Rif[R], rifampicin resistance; Strp[R], streptomycin resistance; MCS, mulitple cloning site.
[b]All *L. lactis* strains mentioned are derivatives of the plasmid-free strain *L. lactis* subsp. *cremoris* MG1363 (29).

and a Cm[R] gene for selection. Also available is the same vector but with an additional erythromycin resistance (Em[R]) marker, i.e. pNG8048E (see *Table 1*).

Translational fusions to the ATG start codon of *nisA* can be made by cloning a properly PCR-amplified gene of interest in the *NcoI* site in pNZ8048. The disadvantage that, in some cases, it is necessary to change the first base of the second codon of the gene of interest is alleviated in pNZ8150. In this plasmid, a *ScaI* site immediately upstream of the *nisA* start codon replaces the *NcoI* site. Genes of interest that are amplified starting with their start codon are correctly translationally fused to *nisA* by blunt-end ligation (11). The vector pNZ8021 can be used for transcriptional fusions, whilst pNZ8110 carries the nucleotide sequence encoding the signal peptide of the major secreted protein of *L. lactis*, Usp45, and can be employed for protein secretion purposes (see *Table 1*; 11).

All materials can be obtained from NIZO Food Research, who own a patent (*Method for Controlling the Gene Expression in Lactic Acid Bacteria*; inventors Oscar Paul Kuipers and Willem Meindert de Vos; US5914248 and EP1162271) on the application of the NICE system and related systems. Signing a Material Transfer Agreement and paying a modest fee for handling, administration, and shipping is required for academic groups to obtain the requested materials. Commercial use of the system should be negotiated with NIZO Food Research.

Protocol 1

Nisin-controlled expression in *L. lactis* (2)

Equipment and Reagents
- NICE system expression vectors, e.g. pNZ8048, pNZ8030, pNZ8032, or pNZ8037 (see *Table 1*)
- Expression strain: *L. lactis* NZ9000 or NZ3900 (allows use of *lacF* marker for food-grade constructs)
- Either purified nisin, a stock solution of Nisaplin, or nisin-producing strain *L. lactis* NZ9700 (or any other nisin producer that produces similar amounts, e.g. R5 or 22186)
- M17 medium with 0.5–1% glucose and 5 µg/ml chloramphenicol
- Glycerol
- 0.05% Acetic acid
- Glass beads or French press

Method
1. Amplify the desired gene to be expressed by PCR and, after restriction, ligate in a suitable NICE expression vector using standard cloning techniques. Vector pNZ8048, which allows cloning directly behind the nisin ribosome-binding site using an *Nco*I site, is often used and generally gives good results.

2. Transform the construct into *L. lactis* (e.g. strain NZ9000 or NZ3900) or *E. coli* (e.g. DH5α) for subcloning[a]. For a transformation protocol for *L. lactis*, see *Protocol 2*.

3. Optional, when first cloned in *E. coli*: isolate the correct construct and transform the plasmid DNA into *L. lactis*. Check transformants by plasmid DNA isolation and restriction. Direct sequencing can be performed on purified plasmids.

4. Grow an overnight culture of *L. lactis* (e.g. NZ9000) with the overexpression construct in M17 with 0.5–1% glucose and 5 µg/ml chloramphenicol at 30°C. Do not shake – use a standing culture. The culture volume should be as large as possible (for example, an 8 ml culture in a 10 ml tube). Glycerol stocks (10% glycerol) can be prepared for *L. lactis* after transformation (fresh transformation is not necessary). Volumes can be chosen as desired (1–10 ml, for instance).

5. The following morning, dilute the culture 1:50 in fresh medium (typically in 10–100 ml total volume; doubling time approximately 20–30 min), and when the OD_{600} = 0.5–0.8, induce with nisin (10^{-6} or 10^{-7} dilution of Nisaplin (the powder contains 2.5%, w/w, NisA) from a stock solution of ~20 mg/ml)[b]. Alternatively, use 2 ng/ml of purified nisin for optimal induction or 1:1000-diluted supernatant of an overnight culture of strain NZ9700. Freshly prepare all dilutions in 0.05% acetic acid on ice and use immediately for induction. Typical induction time is 2 h[c].

6. Take samples at desired time points for further analysis, e.g. by SDS-PAGE and Coommassie staining or enzyme assay. Cells can be lysed with glass beads (for example, resuspend the cell pellet of a 1 ml culture in the desired buffer (with a little lysozyme), add 500 mg of glass beads (75–150 µm diameter) and shake two to five times for 1 min in a shaking apparatus, with cooling steps on ice in between) or with a French press.

7. Include controls (e.g. uninduced sample and a strain containing empty plasmid).

8. Also see the protocol for NICE in *S. pneumoniae* (21)[d].

Notes

[a] Note that cloning first in *E. coli* can cause expression of the construct due to the leakiness of P_{nisA} in the absence of *nisRK* and to unpredictable recombinations, especially when a detrimental gene is to be cloned. It is recommended to clone directly in *L. lactis*.

[b] Preparation of Nisaplin stock solution: add 1 : 1 (v/w) 50% ethanol to Nisaplin (e.g. 0.5 ml to 0.5 g); vortex well, spin down for 2 min at 20 000 ***g*** in an Eppendorf centrifuge, and remove the supernatant to new tube (this contains approximately 20 mg/ml nisin). This supernatant is used for nisin induction at an approx. 10^{-6} or 10^{-7} dilution.

[c] The desired level of expression can be reached by varying the amount of nisin added as inducer. There is a linear dose–expression relationship between 0.01 and 1 ng/ml.

[d] NICE usage for *S. pneumoniae* (21):
- Genetic background: use streptococcal strains that have *nisRK* integrated in the *bgaA* locus (using plasmid pTK3 from (21) e.g. D39*nisRK*).
- Appropriate plasmids are pNZ8048 or pNG8048E.
- Aliquots of strains should be grown in M17 containing 0.5% glucose or another medium with 2.5 µg/ml chloramphenicol until mid-exponential phase and can be stored in 10% glycerol at −80°C.
- For nisin-induced (over)expression, cells should be washed and inoculated 1 : 20 or 1 : 50 in fresh GM17 medium (antibiotic pressure is not normally necessary). Nisin-induced expression can be accomplished at the desired OD_{600} (mid-exponential) in the same way as described for *L. lactis*.

Protocol 2

Electrotransformation of *L. lactis*

Reagents
- Wash buffer: 0.5 M sucrose and 10% glycerol
- SMGG: M17 (Difco), 0.5 M sucrose, 0.5% glucose, and 1% glycine
- SMG17MC: M17, 0.5 M sucrose, 0.5% glucose, 20 mM $MgCl_2$, and 2 mM $CaCl_2$
- GSM17 agar: M17 agar with 0.5 M sucrose and 0.5% glucose

Method

Preparation of competent cells
1. Prepare a 10 ml overnight culture in 100 ml of SMGG.
2. Grow to an OD_{600} of 0.2–0.7.
3. Wash three times with 50 ml of ice-cold wash buffer.
4. Resuspend in 1 ml of wash buffer.
5. Aliquot as 10% glycerol stocks and store at −80°C or use directly.

Electroporation
1. Add 1 µl of DNA to 39 µl of cell suspension in an ice-cold cuvette.
2. Electroporate at 2.5 kV, 25 µF, 200 Ω.
3. Add the contents of the cuvette to 4 ml of SMG17MC in a test tube.
4. Incubate for 2 h at 30°C.
5. Concentrate the cells to a volume of 0.5 ml.
6. Plate on GSM17 agar.

2.1.3 Use of the NICE system in other bacteria

The use of broad-host-range vectors in the construction of the NICE system has allowed evaluation of the system via a two-plasmid approach in other bacterial species. In these bacteria, one plasmid provides the *nisRK* regulatory genes, whilst the other plasmid carries the nisin-inducible P_{nisA} driving expression of a reporter gene. The quantity of NisRK is an important parameter for proper functioning of the NICE system; for example, it has been observed that high expression of *nisR* results in constitutive expression from P_{nisA}, possibly due to autophosphorylation of NisR or nonspecific cross-talk leading to active (phosphorylated) NisR (7). The required amount of the two proteins *a priori* is not known for the various bacterial species tested. Thus, four plasmids were constructed in which *nisRK* expression was varied by using two plasmid copy number variants of pAMβ1, pIL252 (low copy number) and pIL253 (high copy number), and by cloning the *nisRK* genes in these vectors in such a way that they were either expressed by read-through from the promoter of the plasmid *rep* gene alone or by this promoter in combination with P_{nisR} (plasmids pNZ9520, pNZ9521, pNZ9530, and pNZ9531 in *Table 1*; 12). The other plasmid in the system was pNZ8008 (1; see *Table 1*), in which the gene for the enzyme β-glucuronidase is driven by P_{nisA}. The two-plasmid system allowed nisin-induced gene expression in *Leuconostoc lactis*, *L. helveticus*, *Streptococcus agalactiae*, *S. pneumoniae*, *B. subtilis*, and *E. faecalis* (12, 18). Unexpectedly, the absolute copy numbers and the copy number differences of the pIL derivatives did not hold in the different bacteria and, consequently, the amount of expressed NisRK could not easily be predicted (18). The lesson from these studies is that for each new bacterial species, the proper regulatory (*nisRK*-specifying) vector needs to be identified.

To circumvent various problems associated with the implementation of the two-plasmid NICE system in *L. plantarum* NCIMB8826, e.g. poor growth of the strains and low expression levels, the *nisRK* genes were stably integrated in the strain's chromosomal tRNA^{Ser} locus (19). After optimization of the induction conditions (nisin concentration, timing of addition, and length of induction), high-level production (10% of total soluble protein) of model antigen tetanus toxin fragment C was attained. An *S. pneumoniae* host strain for nisin-induced gene expression was made by integration of the *nisRK* genes in the *bgaA* locus for β-galactosidase. Nisin dose-dependent gene expression was observed (21).

A number of publications detail the use of a single plasmid carrying both the *nisRK* genes and the nisin-inducible P_{nisA} (see *Table 1*). Plasmids pMSP3535 and pMSP3545, which both carry the low-copy-number replicon-derivative of pAMβ1, have been employed successfully in *E. faecalis* for the nisin-inducible overproduction of cytosolic, integral-membrane, and cell-surface proteins (28). The vector pMSP3535 was also shown to lead to nisin-inducible expression in *L. lactis* and *Streptococcus gordonii*. The required nisin concentration and the optimal time of induction varied among the three species, as did the production of, in this case, the extracellular cell-wall-bound protein PrgB (aggregation substance Asc10) (31). The vectors pNICE (gene expression vector) and pNIES (secretion vector employing the signal sequence of *Bacillus licheniformis* amylase L) were constructed for use in

L. reuteri (22). Apart from the *Col*E1 replicon for replication in *E. coli*, they contain the replication region from an indigenous *L. reuteri* plasmid. Unfortunately, the system is leaky, as expression is observable in cells that have not been induced with nisin. Using α-amylase as the reporter enzyme, the maximum induction in this system was estimated to be approximately sevenfold. Most probably, this problem is caused by levels of the NisRK two-component regulator system that are too high (see above) and could be solved by integrating the *nisRK* genes in the chromosome of the organism, as has been done successfully for other *Lactobacillus* species, such as *L. plantarum*, *L. casei* and *L. gasseri* (19, 32, 33).

2.1.4 Industrial use of NICE

Recently, it has been shown that the NICE system can also be employed on an industrial scale (13, 14). Large amounts of the antibacterial protein lysostaphin from *Staphylococcus simulans* were produced in 1–3000 l fermentations. Induction with 10 ng/ml nisin at an OD_{600} of 1.0 resulted in approximately 100 mg/l of the protein, independent of the size of the fermentation. Thus, 300 g of lysostaphin was produced in independent 3000 l fermentations (14). The medium composition is extremely important for optimal protein production with the NICE system. For example, production of human and animal pharmaceuticals requires the use of bovine spongiform encephalitis-free media, and a medium based on yeast extract and plant protein was developed. Careful adjustment of key parameters such as culture temperature and pH (choice of neutralizing agent), concentration of zinc, phosphate and nisin, and point of addition led to a threefold increase in lysostaphin yield to 300 mg/l (13).

2.2 The SURE system

Although NICE, as detailed above, has been used in *B. subtilis*, the SURE system described below has several advantages over NICE in *B. subtilis*: (i) it is a single-plasmid system, whilst NICE employs two plasmids; (ii) maximum expression levels with SURE are significantly higher; and (iii) this high expression level is obtained with less of the (antimicrobial) inducer molecule (20).

Two elements are crucial to the proper functioning of the subtilin-induced regulation of gene expression in *B. subtilis*, namely the two-component system comprised the SpaR and SpaK proteins and a promoter carrying a specific *cis*-acting sequence called the *spa* box (23).

2.2.1 Source of subtilin

B. subtilis ATCC 6633 is used as the source for the inducer subtilin. The strain is grown aerobically at 37°C in standard TY medium. A fresh overnight culture of cells is spun down and the resulting supernatant is heated for 10 min at 80°C to kill residual producer cells. As the effective concentration of subtilin may differ with strain, a range of concentrations (generally from 0.1 to 1%, v/v) of the subtilin-containing culture supernatant is initially added to the culture to be induced to determine the optimum.

2.2.2 Host strains and plasmids for use of SURE in *B. subtilis*

The genes *spaRK* are present between the *amy*-front and *amy*-back fragments of the *B. subtilis amyE* locus in pNZ8900 (see *Table 1*). This plasmid, which is kanamycin resistant (Km^R), was used to insert *spaRK* in the corresponding locus in the chromosome of *B. subtilis* 168 (20). The resulting strain, *B. subtilis* NZ8900 (*amyE::spaRK*, Km^R), is the subtilin-responsive production host for the SURE expression plasmids described below. As *amyE* is well conserved in strains of *B. subtilis*, pNZ8900 could be used to integrate the two-component subtilin-triggered *spaRK* genes in these backgrounds as well, extending the potential of the SURE system. Although the *spaRK* genes are expressed in a growth-phase-dependent manner (P_{spaR} is under the additional control of sigma H; 34), the SURE system operates perfectly during early- and mid-exponential-phase induction with subtilin (20).

Three promoters in the *spa* operon, P_{spaB}, P_{spaS}, and P_{spaI}, contain a *spa* box and are subject to SpaRK-mediated quorum sensing control mediated by extracellular subtilin (23). Of these, P_{spaS} shows the highest strength at a given subtilin concentration. This promoter, and a mutant derivative of it in which the *spa* box has been changed into a perfect pentanucleotide repeat (TTGAT) ($P_{spaSmut}$), have been employed to construct subtilin-inducible gene expression of SURE plasmids for *B. subtilis* (20; see *Table 1*). Both promoter variants allowed the tightly subtilin-controlled overexpression of β-glucuronidase and green fluorescent protein (GFP). Although the mutant P_{spaS} led to higher levels of the reporter proteins and generated maximum levels of induction with lower levels of subtilin, it was also slightly more active under noninducing conditions. P_{spaS}- and $P_{spaSmut}$-driven expression was increased 110- and 80-fold, respectively, following induction with subtilin. The GFP reporter was used to show that culture heterogeneity, a well-known phenomenon in growth-phase-dependent regulation of competence development and sporulation in *B. subtilis* (35, 36), does not occur in the SURE system: all cells in a culture of *B. subtilis* responded equally and fully to subtilin induction.

Two types of antibiotic-resistance markers, effective against erythromycin and chloramphenicol, have been used in the SURE expression vectors (see *Table 1*), allowing the choice of a suitable marker for use in different antibiotic-resistant genetic backgrounds (20). In addition, the availability in each of these plasmid variants of either the wild-type or the mutated *spaS* promoter enables choice of the best conditions for any gene to be overexpressed, as these will depend on both the nature of the gene and that of the encoded protein.

Protocol 3

SURE in *B. subtilis* (20)

Equipment and Reagents
- SURE system expression vectors, e.g. pNZ8901, pNZ8902, pNZ8910, or pNZ8911[a]
- *L. lactis* (e.g. strain MG1363) or *E. coli* (e.g. DH5α)
- Expression strain *B. subtilis* NZ8900 (168 *amyE::spaRK*, Km^R)
- Shake flasks
- TY (or LB) medium
- Strain *B. subtilis* ATCC 6633 or another subtilin-producing strain

Method
1. Amplify the desired gene to be expressed by polymerase chain reaction (PCR) and, after restriction, ligate in a suitable SURE expression vector using standard cloning techniques[b].
2. Transform the construct into *L. lactis* (e.g. strain MG1363) or *E. coli* (e.g. DH5α) for subcloning.
3. Isolate the correct construct and transform the plasmid DNA into *B. subtilis* NZ8900 using natural competence as described in (37). Grow all *Bacillus* strains at 37°C with vigorous shaking in shake flasks with 10% culture volume/90% air. Use an empty vector as a control.
4. Check for transformants by plasmid DNA isolation and restriction.
5. Grow *B. subtilis* ATCC 6633 overnight in TY (or LB) medium. The expected level of subtilin production is ~10 mg/l.
6. Spin down the cells and isolate the supernatant.
7. Incubate the supernatant at 80°C to kill any remaining cells.
8. Aliquot into desired volumes and store at −80°C.
9. Grow *B. subtilis* NZ8900 containing the expression plasmid overnight in TY medium (also grow the control strain).
10. Dilute 20–50 times in fresh TY medium (to typical volumes of 10–50 ml).
11. Induce expression at an OD$_{600}$ of 0.5–1.0 by adding subtilin-containing supernatant (isolated as described above) to a concentration of 0.01–1% (v/v)[c].
12. Take samples at desired time points for further analysis, e.g. SDS-PAGE or enzyme activity measurement. Maximum expression is usually reached after 2 h. Also include an uninduced culture as a control as well as a strain with the empty plasmid (induced and uninduced).

Notes

[a]The different expression vectors contain either the chloramphenicol (pNZ8901 and pNZ8911) or erythromycin (pNZ8902 and pNZ8910) resistance marker. Plasmids pNZ8910 and pNZ8911 have the wild-type P$_{spaS}$, whereas pNZ8901 and pNZ8902 have a mutated, optimized derivative P$_{spaSmut}$ containing a perfect pentanucleotide direct repeat in the *spa* box. Wild-type P$_{spaS}$ shows hardly any background expression under noninducing conditions, whereas P$_{spaSmut}$ shows some leakage. The latter, however, gives a higher level of expression under maximal induction conditions (20). It appears from unpublished work that the expression from pNZ8902 is less well controlled, i.e. expression is either 'on' (maximal level) or 'off' (background level), when a range of subtilin-containing supernatant levels with small intervals is used. Although they have the same promoter, the expression level from pNZ8901 is well controlled using subtilin-containing supernatant concentrations ranging from 0.005 to 0.2% (unpublished results). The only difference between the two vectors is the antibiotic-resistance marker.

[b]Preferably construct translational fusions with P$_{spaS}$/P$_{spaSmut}$ by cloning the start codon of the gene at the *Nco*I site in the multiple cloning site of the expression vector.

[c]The maximal level of expression is usually reached using 0.5% (P$_{spaS}$) or 0.2% (P$_{spaSmut}$) subtilin-containing supernatant of *B. subtilis* ATCC 6633.

2.3 Future applications

Several new vectors are being developed based on the first-generation NICE vectors by various laboratories. GFP/yellow fluorescent protein/cyan fluorescent protein fusion vectors (N- or C-terminal) are being constructed in our laboratory to enable protein subcellular localization experiments. Histidine-tagged, streptavidin-tagged and maltose-binding protein fusion vectors are being constructed at several institutions to enable easy purification of the overproduced proteins, e.g. for structural characterization.

Currently, attempts are ongoing in our laboratory to combine NICE and SURE in one bacterium (R. Eijlander et al., unpublished), i.e. *L. lactis*, to be able to have two independent possibilities for producing proteins in different ratios. However, it remains to be established whether possible cross-talk between the systems would prevent proper separate expression.

In addition, alternative induction protocols are being introduced. For example, *L. reuteri* induction has been reported to be optimal by adding 50 ng/ml nisin in the early exponential growth phase (i.e. 1 h after the 1:50 dilution from an overnight culture) and proceeding for 3–5 h with induction, which is considerably different from the protocols described for other Gram-positive bacteria (22).

Finally, it is clear that NICE- and SURE-like systems show great promise for implementation in other Gram-positive organisms following the strategies outlined here. Considering the high relevance and ubiquity of industrial, probiotic, and pathogenic Gram-positive bacteria, this is certainly worth the effort.

3. TROUBLESHOOTING

- **The correct expression construct cannot be obtained or seems to be unstable**
  Try to avoid cloning via *E. coli*. Introduce the ligation mix directly into *L. lactis* NZ9000 or NZ3900. It is important to use *L. lactis* strains that harbor *nisRK*, as these proteins help to keep the promoter turned off when no inducer is present. Sometimes, cleaving PCR products with *Nco*I and another enzyme is inefficient, which hampers efficient cloning. In such cases, consider using blunt-end cloning of the PCR product first, or use the Topo kit for *E. coli*. Subsequently, the cloned fragment can be obtained by cutting it out of the plasmid. In exceptional cases, the problem can persist; in this case, trying another vector or even another host is advised.
- **Nisin or subtilin induction does not seem to work or no expression is seen**
  Check the concentration of nisin or subtilin in the stock solution, e.g. by tricine SDS-PAGE. Stocks should be stored at –20 or –80°C at slightly acidic pH (e.g. in 0.05% acetic acid). When in doubt, prepare a new stock. Make fresh dilutions on ice just before induction.
- **Following induction, the induction level is too high and the cells cease to grow**
  Try lower amounts of inducer and also use controls, e.g. strains with the empty plasmid, to assess whether the expressed protein is causing the problem. It is

quite common that cells will suffer as a result of high expression levels of certain proteins, so some growth delay is normal and indicates expression of the gene of interest. For each different protein, the optimal induction level needs to be determined. Harvesting 30 or 60 min after induction can be a solution. Membrane protein expression can cause various problems. Sometimes it helps to remove particular regions, e.g. the N-terminal region, or to fuse signal peptides to the protein of interest (16).

- **A high level of leakage is observed from the promoter on the expression plasmid when using the two-plasmid system in a host other than *Lactococcus***

  The level of NisRK is probably too high in this host, leading to premature activation. Try low-copy-number vectors and low-level constitutive promoters upstream of *nisRK* or *spaRK*. Even better, try to integrate the two-component system into the chromosome.

Acknowledgements

We thank Tomas Kloosterman and Hein Trip for assistance with protocol descriptions and Reindert Nijland for excellent artwork. O.P.K. thanks former colleagues at NIZO Food Research, who were involved in the early development of NICE and SURE, for a pleasant collaboration: Pascalle de Ruyter, Marke Beerthuyzen, Ingrid van Alen-Boerrigter, Roger Bongers, Roland Siezen, Willem de Vos, and Michiel Kleerebezem.

4. REFERENCES

★ 1. Kuipers OP, Beerthuyzen MM, de Ruyter PG, Luesink EJ & de Vos WM (1995). *J. Biol. Chem.* **270**, 27299-27304. – *The original publication describing autoinduction of nisin biosynthesis.*

★ 2. de Ruyter PG, Kuipers OP & de Vos WM (1996) *Appl. Environ. Microbiol.* **62**, 3662-3667. *The original publication of the nisin-inducible system in L. lactis.*

★★★ 3. van Kraaij C, de Vos WM, Siezen RJ & Kuipers OP (1999) *Nat. Prod. Rep.* **16**, 575-587. – *A recommended review on biosynthesis, mode of action, and regulation of nisin.*

4. de Ruyter PG, Kuipers OP, Beerthuyzen MM, van Alen-Boerrigter I & de Vos WM (1996) *J. Bacteriol.* **178**, 3434-3439.

5. Breukink E, Wiedemann I, van Kraaij C, Kuipers OP, Sahl H, de Kruijff B (1999) *Science*, **286**, 2361-2364.

★★★ 6. Hasper HE, Kramer NE, Smith JL, *et al.* (2006) *Science*, **313**, 1636-1637. – *The latest views on the mode of action of nisin.*

7. van der Meer JR, Polman J, Beerthuyzen MM, Siezen RJ, Kuipers OP & de Vos WM (1993) *J. Bacteriol.* **175**, 2578-2588.

8. de Vos WM, Kleerebezem M & Kuipers OP (1997) *Curr. Opin. Biotechnol.* **8**, 547-553.

9. Kuipers OP, de Ruyter PGGA., Kleerebezem M & de Vos WM (1997) *Trends Biotechnol.* **15**, 135-140.

★★★ 10. Kleerebezem M, Quadri LE, Kuipers OP & de Vos WM (1997) *Mol. Microbiol.* **24**, 895-904. – *An excellent review on two-component regulatory systems in Gram-positive bacteria.*

★★★ 11. Mierau I & Kleerebezem M (2005) *Appl. Microbiol. Biotechnol.* **68**, 705-717. – *A nice overview of the achievements of NICE.*

★★ 12. Kleerebezem M, Beerthuyzen MM, Vaughan EE, de Vos WM & Kuipers OP (1997) *Appl.*

Environ. Microbiol. **63**, 4581–4584. – *Implementation of nisin inducible systems in other Gram-positive bacteria.*

★★ 13. Mierau I, Olieman K, Mond J & Smid EJ (2005). *Microb. Cell Fact.* **4**, 16. – *Industrial-scale production using NICE.*

★★ 14. Mierau I, Leij P, van Swam I, et al. (2005) *Microb. Cell Fact.* **4**, 15. – *Industrial-scale production using NICE.*

15. Kunji ER, Slotboom DJ & Poolman B (2003) *Biochim. Biophys. Acta.* **1610**, 97–108.

★ 16. Monne M, Chan KW, Slotboom DJ & Kunji ER (2005) *Protein Sci.* **14**, 3048–3056. – *Improvements in eukaryotic membrane protein production.*

17. de Ruyter PG, Kuipers OP, Meijer WC & de Vos WM (1997). *Nat. Biotechnol.* **15**, 976–979.

★ 18. Eichenbaum Z, Federle MJ, Marra D, et al. (1998) *Appl. Environ. Microbiol.* **64**, 2763–2769. – *Implementation of nisin-inducible systems in other Gram-positive bacteria.*

★ 19. Pavan S, Hols P, Delcour J, et al. (2000) *Appl. Environ. Microbiol.* **66**, 4427–4432. – *Implementation of nisin-inducible systems in other Gram-positive bacteria.*

★ 20. Bongers RS, Veening JW, van Wieringen M, Kuipers OP & Kleerebezem M (2005) *Appl. Environ. Microbiol.* **71**, 8818–8824. – *The first description of the SURE system in* Bacillus.

★ 21. Kloosterman TG, Bijlsma JJ, Kok J & Kuipers OP (2006) *Microbiology*, **152**, 351–359. – *An efficient NICE system for several strains of* S. pneumoniae.

22. Wu CM, Lin CF, Chang YC & Chung TC (2006) *Biosci. Biotechnol. Biochem.* **70**, 757–767.

23. Kleerebezem M, Bongers R, Rutten G, de Vos WM & Kuipers OP (2004) *Peptides* **25**, 1415–1424.

24. Sorvig E, Gronqvist S, Naterstad K, Mathiesen G, Eijsink VGH & Axelsson L (2003) *FEMS Microbiol. Lett.* **229**, 119–126.

25. Kuipers OP, de Ruyter PGGA, Kleerebezem M & de Vos WM (1998) *J. Biotechnol.* **64**, 15–21.

26. Kuipers OP, Beerthuyzen MM, Siezen RJ & de Vos WM (1993) *Eur. J. Biochem.* **216**, 281–291.

27. Lindholm A, Smeds A & Palva A (2004) *Appl. Environ. Microbiol.* **70**, 2061–2071.

28. Bryan EM, Bae T, Kleerebezem M & Dunny GM (2000) *Plasmid*, **44**, 183–190.

29. Gasson MJ (1983) *J. Bacteriol.* **154**, 1–9.

30. Leenhouts K, Buist G, Bolhuis A, et al. (1996) *Mol. Gen. Genet.* **253**, 217–224.

31. Hirt H, Erlandsen SL & Dunny GM (2000) *J. Bacteriol.* **182**, 2299–2306.

32. Martin MC, Fernandez M, Martin-Alonso JM, Parra F, Boga JA & Alvarez MA (2004) *FEMS Microbiol. Lett.* **237**, 385–391.

33. Neu T & Henrich B (2003) *Appl. Environ. Microbiol.* **69**, 1377–1382.

34. Stein T, Borchert S, Kiesau P, et al. (2002) *Mol. Microbiol.* **44**, 403–416.

35. Smits WK, Eschevins CC, Susanna KA, Bron S, Kuipers OP & Hamoen LW (2005) *Mol. Microbiol.* **56**, 604–614.

36. Veening JW, Hamoen LW & Kuipers OP (2005) *Mol. Microbiol.* **56**, 1481–1494.

37. Leskela S, Kontinen VP & Sarvas M (1996) *Microbiology*, **142**, 71–77.

CHAPTER 14
Protein production using lentiviral vectors

Rénald Gilbert, Sophie Broussau, and Bernard Massie

1. INTRODUCTION

1.1 Properties of lentiviral vectors (LVs)

LVs are derived from lentiviruses, a family of complex retroviruses (for review, see 1, 2). The best-characterized lentivirus is human immunodeficiency virus (HIV), the causative agent of AIDS. For this reason, many LV systems used today are derived from this virus. Lentiviruses, as for other retroviruses, carry their genetic material in the form of two identical RNA molecules within a proteinaceous shell (capsid), which is surrounded by a lipid bilayer (envelope) studded with glycoproteins. An important property of all retroviruses is their capacity to integrate their genome efficiently into the chromosomes after infection. In contrast to other retroviruses, the integration complex of lentivirus can cross the cell nuclear envelope. This means that they can integrate their genome (and thus activate transcription of their genes) in dividing and nondividing cells such as muscle and neurons, whereas other retroviruses need the nuclear envelope to break down during mitosis to do so. A 'typical' retrovirus, as opposed to a lentivirus, can thus infect only actively dividing cells. The use of lentiviruses as gene transfer vehicles is relatively recent. The first experiment employing this vector was published in 1996 by the group of Inder Verma at the Salk Institute in California, who demonstrated that a replication-defective lentivirus could stably express a reporter gene in nondividing cells (neurons) of the mouse brain (3). Since this ground-breaking study, several modifications have been made to improve the efficacy and safety of LVs.

1.2 Essential components of LVs

In addition to the two long terminal repeats (LTRs) and the *pol*, *gag*, and *env* genes present in every retrovirus, the lentiviral genome contains genes for five additional accessory proteins (Vif, Nef, Tat, Rev, and Vpr). The *pol*, *gag* and *env* genes produce the viral enzyme, capsid protein, and envelope glycoprotein, respectively. The presence of Vif, Nef, and Vpr are not essential to produce LVs. Tat is a powerful transcription factor that stimulates transcription from the LTR. Consequently, the presence of Tat is essential if the lentivirus LTR is used to control transgene expression or to produce RNA from the viral genome. Rev is a protein that promotes efficient transport of RNA to the cytoplasm. Rev interacts with a sequence of the viral RNA known as the Rev-responsive element (RRE). Hence, efficient production of LVs requires the presence of Rev and RRE. In addition to the packaging signal, the two LTRs, and RRE, the viral RNA should contain the central polypurine track (cPPT) (4). This is a sequence of 118 nucleotides that facilitates nuclear translocation of the pre-integration complex. Inclusion of the woodchuck hepatitis virus post-transcriptional regulatory elements increases RNA stability and thus the production level of the gene of interest (5). The type of envelope protein determines the type of cell a lentivirus can infect, as this protein mediates attachment of the virion to the cell surface, as well as fusion with the plasma membrane. For safety reasons and to extend their natural tropism, an envelope protein different from that of the lentivirus envelope glycoprotein is normally used.

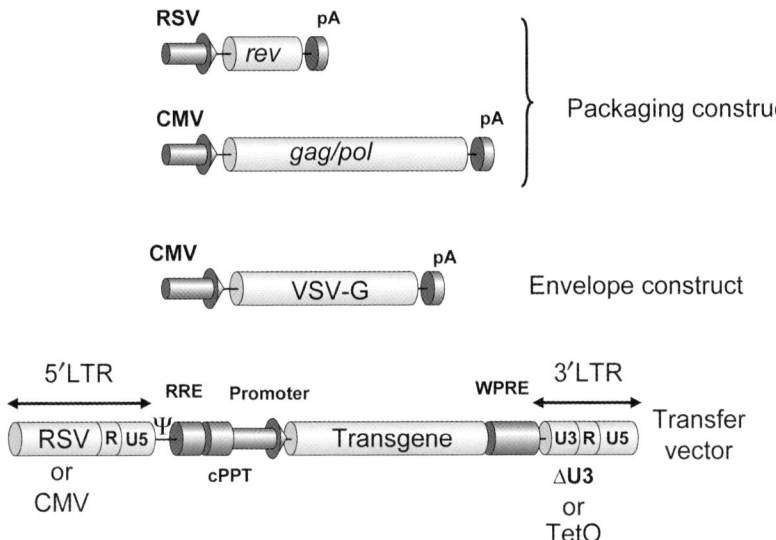

Figure 1. Components necessary for the production of LVs.
One of the most straightforward methods for producing LVs is to transfect cells with a mixture of plasmids: the packaging, envelope, and transfer vectors. The two packaging constructs provide the structural proteins and enzymes necessary for the assembly of the LV and processing of the RNA. The envelope construct encodes the gene for the envelope proteins, in this example vesicular stomatitis virus G protein (VSV-G). The transfer vector contains the transgene expression cassette and all of the necessary cis-acting DNA sequences (see text for details). In this example, the cytomegalovirus (CMV) or Rous sarcoma virus (RSV) promoter regulates elements of the packaging and envelope constructs. In the transfer vector, the U3 region of the 5'LTR is often replaced with the RSV or CMV promoter to confer efficient and Tat-independent transcription. To generate a self-inactivating LV (SIN-LV), the U3 region of the 3'LTR is deleted (ΔU3). A conditional SIN-LV can be generated by replacing the U3 region with an inducible promoter such as the tetracycline-regulated operator (TetO). U3, U5, and R, unique 3', unique 5', and repeat regions of the LTR, respectively; Ψ, packaging signal.

INTRODUCTION 243

Glycoprotein G of vesicular stomatitis virus (VSV-G) is often employed for pseudotyping LV and retroviral vectors. Vectors pseudotyped with VSV-G possess an extended tropism and are more stable. They can therefore be concentrated by ultracentrifugation without losing significant infectivity. The major components necessary to produce LVs are summarized in *Fig. 1*.

1.3 Strategies to control protein expression from LVs

LVs are often produced by simultaneously transfecting cells with plasmids that encodes the essential enzymes and structural proteins mentioned above in conjunction with a plasmid (transfer construct) that contains the necessary cis-acting genetic elements and the gene of interest (transgene) (see *Fig. 1*). The simplest form of transfer vector contains the transgene flanked by the two LTRs of HIV in addition to the packaging signal, RRE, and cPPT. Transcription from the 5'LTR, using the enhancer and promoter sequences located in the U3 region, will produce an RNA molecule that will be encapsidated and that possesses all of the cis-acting genetic elements essential for reverse transcription and integration. The enzymes necessary for reverse transcription and integration are produced by the cells or the packaging constructs (see *Fig. 1*) and are incorporated into the capsid,

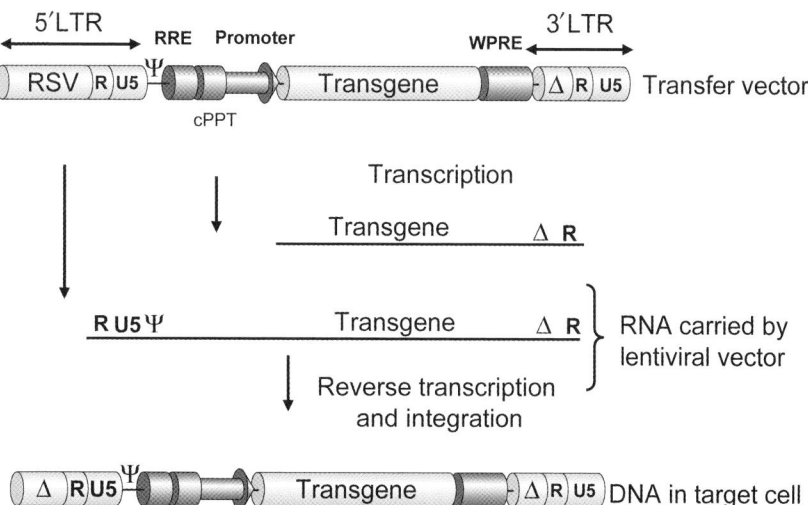

Figure 2. Transport and processing of LV RNAs.
Two types of RNA can be produced by the transfer vector. One originates from the 5'LTR and produces the complete genome of the LV including the transgene. This RNA will be encapsidated into viral particles. The second RNA originates from the internal promoter and contains only the transgene. After transduction, the LV RNA is reverse transcribed and is integrated into the genome of the targeted cells. During this process, the U3 region of the 3'LTR replaces the U3 region of the 5'LTR. In this example, because the U3 region of the 3'LTR is deleted (as in the case of SIN-LV), no LV genomic RNA can be produced after integration. Only the transgene RNA is synthesized, because it is controlled by an internal promoter. ΔU3, U5, and R, deletion of unique 3', unique 5', and repeat regions of the LTR respectively; Ψ, packaging signal.

which is assembled at the cell surface and released by budding into the extracellular environment. Normally, the cell is not lysed by the budding process and will release virions continuously into the extracellular medium. The culture medium can be used directly to infect cells or the virions can be concentrated further by ultracentrifugation (see below). The transgene delivered by the LV is integrated in the transduced cells (see *Fig. 2*). It becomes a permanent constituent of the chromosomes and is duplicated during mitosis. The transgene carried by the LV can be regulated by the enhancer/promoter elements of the 5'LTR or by a strong ubiquitous promoter such as that of human cytomegalovirus (CMV) flanking the transgene. The natural promoter of the U3 region of the 5'LTR is often replaced with a strong constitutive promoter such as CMV (6), or with the promoter of another retrovirus such as Rous sarcoma virus (RSV). This modification permits constitutive transcription of viral RNA without the need for the Tat protein. An additional improvement is the deletion of the enhancer/promoter elements within the U3 region of the 3'LTR, thus eliminating its transcriptional activity. The 5'LTR drives transcription of the full-length RNA. After reverse transcription and integration, the 5'LTR is replaced by the 3'LTR and will be transcriptionally inactive (see *Fig. 2*). Such vectors, referred to as self-inactivating LVs (SIN-LVs), are safer because they cannot produce viral RNA after integration and because the 3'LTR cannot drive transcription of the adjacent cellular gene, which could be disastrous if it was an oncogene (6, 7). The enhancer and promoter region or the 3'LTR can also be replaced with the regulatory elements of an inducible promoter, such as that for the tetracycline-inducible system (8). In this type of vector (conditional SIN-LV or cSIN-LV), transcription from the LTR remains silent until an active transactivator such as the tetracycline-regulated transactivator is provided.

2. METHODS AND APPROACHES

2.1 Production of LVs

A convenient method of producing LVs is to generate a cell line that will secrete the LV carrying the gene of interest. Such cell lines can be generated by infecting packaging cells with a cSIN-LV obtained by transient transfection. After integration, the cells (now called producers) assemble complete virions at their surface and release them into the extracellular medium when the promoter driving the LV RNA is activated. The culture medium can be used directly to infect cells, or the virions can be concentrated by ultracentrifugation (see below). Obviously, to generate good producers, one needs good packaging cells that express all of the necessary structural proteins and enzymes for assembly of the virion, except for the LV RNA. Although several packaging cell lines have been described (9–16), they are difficult to produce, as several essential protein components of the LV (such as VSV-G and the protease) are toxic. Therefore, expression of these proteins must be repressed during normal cell growth and turned on only at the time of LV production. For this reason, LVs are often produced by transfecting an immortalized cell line, such as human 293 cells, using a mixture of plasmids consisting of the transfer, packaging, and envelope constructs (see *Fig. 1*). Commercial systems currently available to produce LVs are based on this transient transfection method. Although this method works fine for the generation of small amounts of vector (10^6–10^7 transducing units (TU)/ml of culture medium), it is not readily amenable to scaling up.

In this section, we present two different methods for the production of LVs. In *Protocol 1*, a LV is generated by transfecting packaging cells with a mixture of plasmids encoding the necessary vector components. As it can be difficult to obtain a good plasmid preparation of the transfer construct, most likely because of the two LTRs it carries, which can facilitate recombination, we present in *Protocol 2* a simple procedure to prepare this DNA. The other plasmids (packaging and envelope constructs) are purified using conventional methods of plasmid preparation. *Protocol 3* describes a method to generate a producer cell line using a cSIN-LV and *Protocol 4* presents the procedure to produce the LV once the producer cells are available. With the development of improved packaging cell lines, the use of stable producers will most likely become increasingly popular, as they provide a continuous supply of vectors and are amenable to scaling up.

Protocol 1

Production of LVs by transient transfection

Equipment and Reagents
- 293 Cells (the cell line 293T is commonly utilized for lentivirus production) or a packaging cell line such as 293SF-PacLV (Broussau *et al.*, unpublished)
- Low-calcium, serum-free medium (LC-SFM; Invitrogen) or normal growth medium for the cell line used
- Tissue culture dishes (150 × 20 mm; Sarstedt)
- Transfer vector and the other complementary plasmids encoding the essential components of the LV (Env, Gag, Pol, etc.)[a]
- Polyethylenimine (PEI, 25 kDa, linear; Polyscience) (see Chapter 12, *Protocol 3*, for preparation of PEI stock solutions)
- Doxycycline (1 mg/ml in water; Sigma). Filter the solution using a 0.22 μm pore size Millex GP filter unit with a PES membrane (Millipore). Aliquot 500 μl into 2 ml tubes and store at –20°C[b]
- Cumate (99% 4-isopropylbenzoic acid; Aldrich) (50 mg/ml in 95% ethanol; store at room temperature)[b]
- HT Tuffryn membrane filter unit (0.45 μm pore size; Pall Life Science)

Method
1. Plate the packaging cells into a 150 mm culture dish at a concentration of 1.0×10^7 cells/dish in LC-SFM and incubate at 37°C overnight in a 5% CO_2 humidified atmosphere[c].
2. Change the medium 2 h before transfection with 17 ml of fresh LC-SFM medium.
3. Combine the following reagents in the following order:
   - 45 μg of DNA[d]
   - 1.5 ml of LC-SFM
   - 90 μg of PEI[e]

   Mix well, incubate at room temperature for 10 min, and then slowly apply the solution to the cells.

4. At 4–6 h post-transfection, refresh the medium with 18 ml of fresh LC-SFM supplemented with doxycycline to a concentration of 1 µg/ml and cumate to 50 µg/ml[f].

5. Collect the LV suspension at 48 h post-transfection, clarify the suspension by low-speed centrifugation (1000 r.p.m. for 5 min) and store at 4°C for 24 h[g].

6. Continue LV production for 24 h by refreshing the cell medium with 18 ml of fresh LC-SFM medium supplemented with doxyxycline (1 µg/ml) and cumate (50 µg/ml).

7. Collect the LV as described in step 5.

8. Pool the two suspensions of LV and filter the solution using a 0.45 µm filter.

9. Keep the LV suspension at −80°C or concentrate it immediately by ultracentrifugation (see *Protocol 5*)[h].

Notes

[a]The number of plasmids required depends on the type of packaging cells used. For example, our packaging cells (293SF-PacLV) express all of the essential LV components. Consequently, only one plasmid (the transfer vector) is needed for transfection.

[b]Doxycycline and cumate are used to induce expression of the LV components by our packaging cells (293SF-PacLV). The type and quantity of inducers needed depend on the packaging cells used.

[c]Transfection efficiency can be improved if the LC-SFM is supplemented with 1% fetal bovin serum (FBS) or with 0.1% (v/v) Lipid Mixture (Sigma) and 0.5% (w/v) Gelatin Peptone N3 (Organotechnie) (17).

[d]If more than one plasmid is used for transfection, use 22.5 µg of the transfer vector and a total amount of 22.5 µg for the other plasmids (equimolar amounts of each).

[e]Transfection using PEI allows at best 60% transduction efficiency. We have found Lipofectamine 2000 (Invitrogen) to be more efficient in 293 cells.

[f]Cumate will precipitate when added to the medium. We recommend adding cumate and doxycycline to the LC-SFM in a separate bottle using medium pre-warmed to 37°C. The stock solution of cumate should be vortexed several times before use.

[g]Do not keep the LV suspension for more than 24 h at 4°C, as it is very unstable.

[h]LV suspensions should be frozen in a solution containing at least 1% FBS or 5% sucrose. The LV titer decreases after each freeze–thaw cycle.

Protocol 2

Preparation of transfer vector

Equipment and Reagents
- Competent TOP10 (Invitrogen) or competent HB101 *Escherichia coli* strains (American Type Culture Collection)
- Luria–Bertani (LB) medium (10 g/l bacto-tryptone, 5 g/l bacto-yeast extract, 10 g/l NaCl)
- Qiagen Plasmid Purification Maxi kit
- Equipment and reagents for DNA analysis using agarose gel electrophoresis
- Spectrophotometer

Method
1. Transform competent TOP10 or HB101 *E. coli*[a] with 10–100 ng of transfer vector following the manufacturer's instructions.
2. Pick a single colony from a freshly transformed selective plate (less than 24 h old) and inoculate a starter culture of 2 ml of LB medium containing the appropriate selective antibiotic[b].
3. Incubate for 8 h at 37°C with vigorous shaking (300 r.p.m.).
4. Transfer the entire starter culture into 250–500 ml of selective LB medium and grow at 37°C for 16 h[c]. Do not overgrow.
5. Extract the DNA using a Qiagen Plasmid Purification Maxi kit following the manufacturer's instructions.
6. Determine the DNA concentration by spectrophotometry and by quantitative analysis on an agarose gel.

Notes
[a]We recommend TOP10 and HB101 when cloning unstable inserts such as lentiviral DNA containing direct repeat sequences.
[b]Use a freshly transformed colony when working with the transfer construct. Colonies obtained from an old plate or from a glycerol stock may result in low DNA recovery.
[c]*E. coli* containing the transfer vector may grow more slowly.

Protocol 3

Transduction of packaging cells with LV to generate stable producers

Equipment and Reagents
- cSIN-LV (the minimal titer should be 2.5×10^6 TU/ml to allow a multiplicity of infection of 5 with 1×10^5 cells in 200 µl). The LV is prepared as described in *Protocols 1* and *4*
- A packaging cell line such as 293SF-PacLV (Broussau *et al.*, unpublished), which produces all of the essential LV elements
- LC-SFM (see *Protocol 1*)
- Polybrene (hexadimethrine bromide; Sigma). Prepare a stock solution in water at 1 mg/ml
- 24-Well cell culture dishes

Method
1. Mix 200 µl of LV suspension with 8 µg/ml of Polybrene and incubate at 37°C for 30 min.
2. Add the LV preparation directly on 1×10^5 293SF-PacLV cells[a] (or another packaging cell line) that have been plated the same day in a 24-well dish and incubate overnight at 37°C.
3. Replace the culture medium with fresh medium.
4. The cells can be transduced a second time using the same protocol[b], or they can be used to produce the LV as described in *Protocol 4*.

Notes
[a]Make sure that the cells are healthy before transduction.
[b]293SF-PacLV cells can be transduced several times to increase the percentage of transduced cells as well as the number of integrated LVs per cell. LV and protein production are higher after several cycles of transduction.

Protocol 4

LV production using producer cells

Equipment and Reagents
- A culture of LV producer cells (see *Protocol 3*)
- Sterile cell culture Erlenmeyer flasks (250 ml to 2 l flasks, depending on the amount of LV needed; the ratio of culture medium volume to flask volume should be 1:5)
- LC-SFM (see *Protocol 1*)
- Shaker at 37°C in a humidified atmosphere of 5% CO_2
- Doxycycline (see *Protocol 1*)
- Cumate (see *Protocol 1*)
- FBS

Method
1. Seed the culture of LV producer cells at a concentration of 3.0×10^5 cells/ml in a sterile culture Erlenmeyer flask in LC-SFM and incubate at 37°C at 120 r.p.m.[a]
2. When the cell density reaches $1.0–1.5 \times 10^6$ cells/ml, induce LV production by adding doxycycline[b] to a concentration of 1 µg/ml and cumate[b] to 25 µg/ml and incubate at 37°C at 120 r.p.m. for 24 h.
3. Collect the LV suspension daily by low-speed centrifugation (1000 r.p.m.) for 5 min and resuspend the cells in fresh LC-SFM supplemented with doxycycline (1 µg/ml) and cumate (50 µg/ml) to continue LV production[c].
4. Freeze the LV suspension at −80°C in the presence of 1% FBS. When production is completed, pool the different LV suspensions. The LV can then be purified by ultracentrifugation as described in *Protocol 5*.

Notes
[a] If the cells do not grow in suspension, plate them in a 150 mm dish at a density of 2.0×10^7 cells/dish and collect the medium daily as described in *Protocol 1* until complete cell mortality.
[b] Doxycycline and cumate are used to induce expression of the LV components by the packaging cells (293SF-PacLV). The type and quantity of inducers needed depend on the packaging cells used.
[c] The LV suspension can be collected for 6–8 days (until complete cell mortality).

2.2 Concentration of LVs

Unless the LV preparation contains toxic molecules or unwanted components (serum for example), LVs do not have to be purified further to perform most cell culture experiments. Cell transduction is normally achieved by adding the medium containing the vector to the cells in the presence of Polybrene. However, for *in vivo* experiments, or when high concentrations of vector are required, the vector has to be purified and concentrated. Density gradient ultracentrifugation and chromatographic methods, which can be complex and may require special equipment, have been developed to purify LVs and retroviral vectors (18, 19) (for review, see 20). User-friendly commercial purification kits are also available to purify small amounts of vector. The following two protocols describe a simple and popular procedure that our laboratory employs to concentrate LVs by centrifugation on a sucrose cushion. One should keep in mind that this is a crude purification procedure, as other components such as cellular membranes and intracellular organelles are also pulled down during centrifugation.

Protocol 5

Small-scale concentration of LV by ultracentrifugation

Equipment and Reagents
- Ultraclear ultracentrifuge tubes (25 × 89 mm; Beckman)
- SW28 rotor (Beckman)
- Beckman Ultracentrifuge L8-55M (Beckman)
- 20% Sucrose (prepared in phosphate-buffered saline (PBS) and filtered using a 0.22 μm filter)
- FBS
- Sterile PBS

Method
1. Transfer the LV suspension (from *Protocol 1* or *4*) in ultraclear ultracentrifuge tubes and add 2 ml of 20% sucrose directly at the bottom of the tube (very slowly). Balance the tubes using sterile PBS.
2. Centrifuge at 24 000 r.p.m. (100 000 *g*) for 2 h at 4°C.
3. Remove the supernatant and resuspend the LV pellet in 0.5–1.0 ml of culture medium containing a minimum of 1% FBS. Incubate overnight at 4°C to solubilize the LV[a,b].
4. Transfer the LV preparation to a 2 ml sterile tube and store at −80°C.

Notes
[a] The pellet is very difficult to see and sometimes is not visible.
[b] The pellet should be resuspended in the same medium used to culture the cells to be transduced.

Protocol 6

Large-scale concentration of LV by centrifugation

Equipment and Reagents
- 250 ml Polycarbonate bottles (Sorvall)
- Filtration and storage Stericup system (0.45 µm; Millipore)
- Ultracentrifuge Sorvall Discovery 100SE, A621 fixed-angle rotor (Sorvall)
- 20% Sucrose (prepared in PBS and filtered using a 0.22 µm filter)
- Sterile PBS
- Resuspension buffer: 5% sucrose in 20 mM Tris/HCl (pH 7.5)
- 15 ml Sterile tubes
- 0.45 µm pore size HT Tuffryn membrane filter unit (Pall Life Sciences)

Method
1. Filter the LV suspension (produced in *Protocol 4*) using the 0.45 µm Stericup system.
2. Distribute 200 ml of LV supernatant into 250 ml polycarbonate bottles (Sorvall).
3. Slowly add 20 ml of 20% sucrose to the bottom of the bottle. Balance the bottles using PBS.
4. Centrifuge at 15 000 r.p.m. (37 000 *g*) at 4°C for 3 h.
5. Remove the supernatant[a] and resuspend the pellet in culture medium or in resuspension buffer by pipetting up and down[b].
6. Transfer the LV suspension to a 15 ml sterile tube and let the pellet dissolve overnight at 4°C[c].
7. Optional step: filter using a 0.45 µm pore size filter[d].
8. Aliquot and store at −80°C.

Notes

[a]The pellet is very fragile, so particular attention should be taken not to lose it when removing the medium.
[b]Use a minimal volume to resuspend the LV. For example, the concentration of LV will be increased 100-fold if 2 ml of solution is used. However, such a small volume may be difficult to filter.
[c]The solution can be vortexed gently to break up the aggregates of LV.
[d]It may be difficult to filtrate the LV as aggregates may block the filter. If this happens, remove the filter and use a new one. Filtration may reduce the LV titer. The LV preparation should be tested for potential bacterial contamination if it is not filtered.

2.3 LV titration

After production, the preparation of LV should be titrated. The most accurate method is to determine the number of TU (sometimes called infectious particles) per milliliter of solution. If the LV carries the gene for a fluorescent protein, such as green fluorescent protein (GFP), the titer can be readily obtained by flow cytometry. LV titers can also be determined by flow cytometry if the transgene encodes a protein for which an antibody is available. The titer can also be estimated by measuring the total number of physical viral particles in the solution. This can be done by quantifying the amount of a structural component in the preparation, such as p24, by ELISA using commercial systems. It is important to note that measuring a structural protein such as p24 does not allow the distinction between functional and defective virions. The presence of soluble p24 in the preparation may also artificially increase the titer. For this reason, when titers obtained by the two methods (TU vs p24) are compared, the titer obtained with p24 is usually higher by a factor of about 100. Other titration methods that measure the activity of reverse transcriptase or the amount of viral RNA have been described (21-24), but these methods, as with p24, cannot discriminate between functional and defective virions. Polymerase chain reaction-based methods to quantify the genome of the vector after integration, and thus functional vectors, have also been developed (22, 24). In the next two protocols, we describe two different methods for quantification of LVs. In *Protocol 7*, a procedure is described that measures the amount of functional vector using GFP as a gene reporter. *Protocol 8* presents a method, first described by Bounou *et al.* (25), based on p24 quantification. This method is therefore used to measure the total amount of physical viral particles in the preparation.

Protocol 7

Titration of LVs by flow cytometry

Equipment and Reagents
- 293A cells
- Dulbecco's modified Eagle's medium (DMEM) supplemented with 5% FBS
- Trypsin
- 24-Well dishes
- 1 mg/ml Polybrene (Sigma) (see *Protocol 3*)
- Polypropylene tubes (12 × 75 mm)
- 10% Formaldehyde (ultrapure EM grade; Polyscience)
- EPICS XL flow cytofluorometer (Beckman Coulter)

Method
1. Seed 293A cells at a concentration of 47 000 cells/well in a 24-well dish 24 h prior to transduction[a].
2. Dilute the LV in DMEM supplemented with 5% FBS and 8 µg/ml Polybrene and incubate for 30 min at 37°C.
3. Count the number of cells in a few extra wells[b].
4. Remove the medium from the cells. Apply 200 µl of the LV/Polybrene complex directly onto the cells and incubate overnight at 37°C.
5. At 16 h post-transduction, replace the culture medium with fresh medium and incubate the cells at 37°C for at least 2 more days.
6. Trypsinize and transfer the cells into 12 × 75 mm polypropylene tubes.
7. Fix the cells while gently vortexing by adding 10% formaldehyde dropwise until the final formaldehyde concentration reaches 2%.
8. Determine the percentage of transduced cells (GFP$^+$) by flow cytometry.
9. The infectious titer (TU/ml) is obtained by multiplying the percentage of GFP-positive cells by the cell number at the time of infection, by the dilution factor, and by 5 (if 200 µl of LV was used initially).

Notes

[a]293A cells can be seeded at a concentration of 9×10^4 cells/well and transduced the same day after the cells have attached to the dish.
[b]A precise cell count is essential at the time of infection to obtain an accurate infectious titer.

Protocol 8

LV titration by ELISA (p24) (25)

Equipment and Reagents
- HIV-1 p24 antibody 183 H12-5C (McKesson Bioservices Corporation)[a]
- HIV-1 p24 antibody 31-90-25 (ATCC HB-9725)[a]
- Recombinant purified p25^{Gag}/SF2 (Chiron Corporation)
- BD Falcon microplate
- Plastic wrap (Fisher)
- EZ-LINK sulfo-NHS-biotin reagents (Pierce)[b]
- Horseradish peroxidase (poly-HRP40) conjugated to streptavidin (Research Diagnostics)
- TMB-S substrate (3,3′,5,5′-tetramethylbenzidine; Research Diagnostics)
- Microtiter plate reader
- Coating buffer (0.1 M NaHCO$_3$ buffer, pH 9.6–9.8): 8.4 g of NaHCO$_3$ and 4.0 g of Na$_2$CO$_3$ in a final volume of 1 l water supplemented with 0.02% thimerosal; store at 4°C)
- PBS (pH 7.2)
- Tween 20 (Sigma)
- BSA (albumin bovine fraction V; MP Biomedicals)
- Disruption buffer: 2.5% Triton X-100 (Bio-Rad), 0.05% Tween 20 in PBS supplemented with 0.02% thimerosal; store at 4°C
- 1 M H$_3$PO$_4$

Method

1. Using a multi-channel pipette, dispense 100 µl of purified p24 antibody 183 H12-5C[a] at a concentration of 2.5 µg/ml in coating buffer into each well of a BD Falcon microplate. Seal the plates using plastic wrap and incubate overnight at room temperature. The plates containing the antibody solution can be stored at 4°C for several weeks.

2. Before using the plates, rinse them three times with PBS supplemented with 0.05% Tween 20 (PBS-T).

3. Block unoccupied sites with 200 µl of PBS-T supplemented with 1% BSA and incubated for 30 min at 37°C.

4. Rinse the plates three times with PBS-T.

5. Prepare several dilutions in duplicate (2000, 1000, 500, 250, 125, 62.5, 31, 25, and 0 pg/ml) of the p24 protein standard in PBS-T supplemented with 1% BSA.

6. Prepare dilutions of the samples to be tested in PBS-T supplemented with 1% BSA.

7. Add 100 µl of samples and standards to the antibody-coated wells. Add 25 µl of disruption buffer. Mix gently by tapping the plate and incubate for 60 min at 37°C.

8. Rinse the plates three times with PBS-T.

9. Add 100 µl of antibody 31-90-25 previously conjugated to biotin[a,b] diluted 1:2000 in PBS-T supplemented with 1% BSA and incubate for 60 min at 37°C.

10. Rinse the plates three times with PBS-T.

11. Add 100 µl of HRP40–streptavidin diluted between 1:4000 and 1:15 000 in PBS-T supplemented with 1% BSA and incubate for 30 min at room temperature.

12. Rinse the plates three times with PBS-T.

13. Add 100 µl of TMB-S substrate and incubate for 30 min at room temperature.
14. Stop the reaction by adding 50 µl of 1 M H_3PO_4.
15. Measure the signal intensity using a microtiter plate reader at 450 nm.

> **Notes**
>
> [a]These are the hybridoma cells that secrete the monoclonal antibodies against p24. The cells are cultured using standard protocols for hybridoma and the antibody is purified using commercially available purification kits (Montage Antibody Purification kit; Millipore).
> [b]This kit is used to biotinylate antibody 31-90-25 according to the manufacturer's recommendations.

2.4 Cell marking and protein production using LVs

Two important LV components are responsible for determining their efficacy at producing recombinant proteins. One is the envelope protein, which influences the efficacy by which the LV penetrates the cells. The other is the promoter, which controls the transcription efficacy. In our hands, LVs pseudotyped with VSV-G infect 293 cells very well and Chinese hamster ovary (CHO) cells reasonably well. The CMV promoter is also relatively strong in these two cell types, although much stronger in 293 cells, and consequently excellent protein production levels can be obtained with this promoter. In CHO cells, however, it is possible to increase the protein production level by using, instead of CMV, an artificial inducible promoter (the cumate gene-switch) derived from the *cym* operon of bacteria (26, 27). Protein production using the cumate gene-switch is accomplished by transducting CHO cells that stably express the cumate transactivator (cTA) with a LV carrying a transgene regulated by the CR5 promoter. The cTA binds specifically to the CR5 promoter and efficiently activates transcription from this promoter. Protein production or cell marking in 293 or CHO cells can readily be performed using a cSIN-LV pseudotyped with VSV-G. The protocol to transduce 293 cells using such a vector is identical to that used to generate a producer cell line (see *Protocol 3*). The reader should thus refer to this protocol to transduce 293 cells. In *Protocol 9*, we present a simple procedure to transduce a suspension of CHO cells adapted to serum-free and protein-free medium. Once the cells have been transduced, they can be expanded and used for protein production. Examples of 293 and CHO cells transduced with LVs using the protocols described in this chapter are shown in *Fig. 3*.

(a)

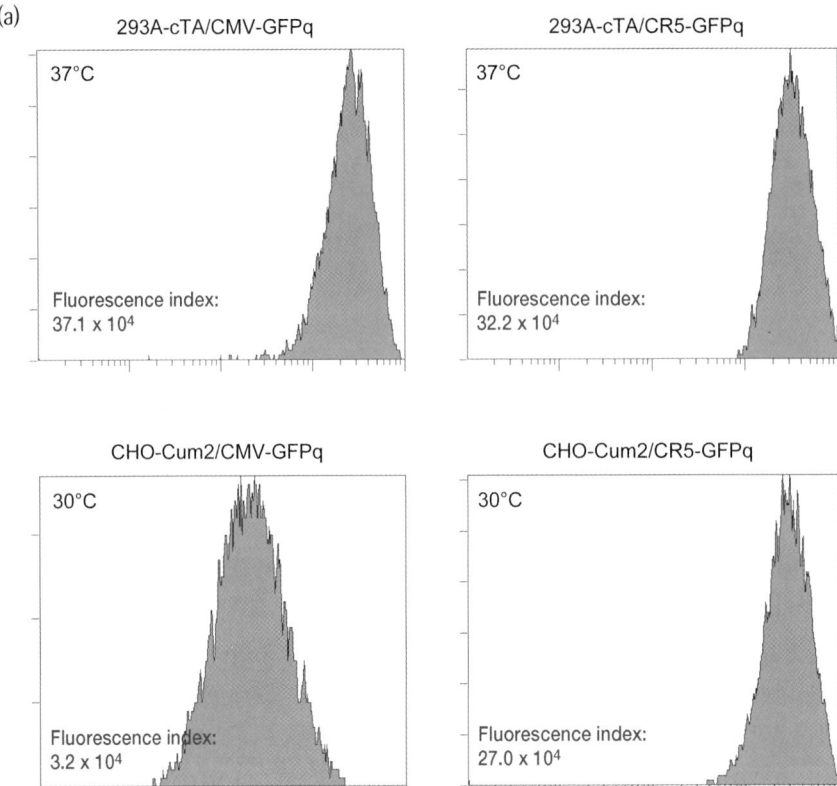

(b)

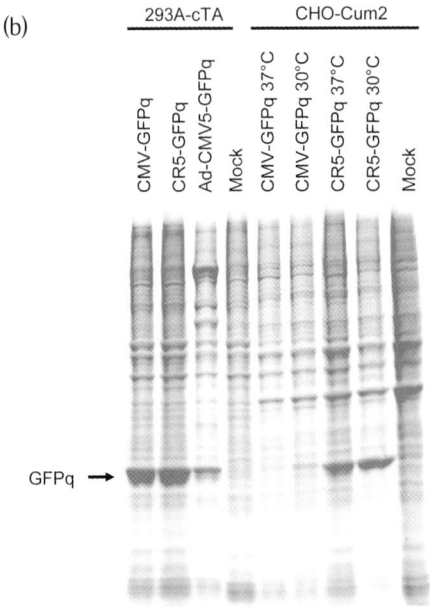

Figure 3. Efficient protein production following transduction of 293 and CHO cells using LV expressing GFP.
A LV carrying the gene for GFP regulated by the CMV or CR5 promoter was used to transduce 293-cTA cells and CHO-Cum2 cells expressing the cumate transactivator. The cells were infected using a multiplicity of infection of 100 and analyzed 3 weeks later by flow cytometry (a), whilst protein production was quantified using SDS-PAGE and a Typhoon Trio+ scanner (Amersham Biosciences) (b). The fluorescence indices (% of GFP-positive cells × relative mean fluorescence intensity of the cell population) are indicated. As shown, the level of GFP is in the range of 25% of total cell protein in 293-cTA cells LV-transduced with either the CMV or CR5 promoter, and in CHO-Cum2 cells with the CR5 promoter. This is twofold higher than 293 cells infected with adenovirus vector expressing GFP using the CMV promoter (Ad-CMV5-GFPq). It should be noted that the CR5 promoter is much stronger than the CMV promoter in CHO-Cum2 cells.

Protocol 9

Transduction of CHO cells using LV

Equipment and Reagents
- Preparation of LV (see *Protocols 1* and *4*)
- CHOSF cells (Invitrogen) or a different type of CHO cell adapted to serum-free culture
- Culture medium: CHO medium (BD), supplemented with 4 mM glutamine and 0.05 mg/ml dextran sulfate[a] (Amersham Pharmacia Biotech AB). A 100× stock solution of dextran sulfate is prepared by dissolving 5 mg/ml of dextran sulfate sodium salt in water followed by filtration using a 0.22 μm filter
- FBS
- Polybrene (see *Protocol 4*)
- 24 Well-dishes

Method
1. Mix 200 μl of LV (use a titer of at least 5×10^6 TU/ml) with 8 μg/ml Polybrene and incubate at 37°C for 30 min[b].
2. Wash 1.0×10^5 CHO cells in PBS to remove the dextran sulfate.
3. Resuspend the cells in 200 μl of LV solution and transfer them into a well of a 24-well dish[c].
4. The next day, or after 6–8 h, replace the medium with fresh medium supplemented with 1% FBS[d].
5. After cell recovery (approximately 3–7 days), they can be transduced again using the same procedure.
6. The cells are then expanded to produce the protein of interest.

Notes
[a]Dextran sulfate prevents the formation of cell aggregates. However, it interferes with the transduction and therefore has to be removed before adding the LV.
[b]The presence of 1% FBS helps the cells to recover from the transduction.
[c]Use an exponentially growing cell culture, as healthy cells recover more easily.
[d]If the cells are sufficiently attached to the well, remove as much medium as possible before adding fresh medium. As the cells are very fragile at this point, they should be centrifuged and manipulated with great care.

3. TROUBLESHOOTING

- If the LV titer produced using the transient transfection approach (see *Protocol 1*) is low, we suggest increasing the titer by generating a producer cell line (see *Protocol 3*), as the amount of LV generated by producer cells is usually higher.
- The use of a promoter derived from RSV to drive expression of the LV RNA might produce suboptimal quantities of LV. If this is the case, we suggest replacing the RSV promoter by the CMV promoter, as the latter is stronger in the cells (293 or derivatives of 293) that are normally used to produce the LV.
- If the titers obtained using the transient transfection procedure (see *Protocol 1*) are low because of poor transfection efficiency, the efficiency can be improved by using liposome–DNA complexes, such Lipofectamine 2000 (Invitrogen), instead of PEI.
- It can be difficult to obtain good producer cells if the titer of the LV used for transduction (see *Protocol 3*) is too low. If this is the case, we suggest concentrating the LV as described in *Protocol 5* before transducing the cells. Several batches of concentrated LV can be pooled and concentrated a second time by ultracentrifugation to increase the titer further. In addition, the cells can be transduced several times with the concentrated LV.
- The preparation of the transfer vector may result in poor DNA recovery. If this is the case, we highly recommend preparing this DNA as described in *Protocol 2*.
- LVs are relatively labile. To increase the stability of the LV during long-term storage at −80°C, we suggest freezing the preparation in the presence of a higher concentration of FBS (5%) or in the presence of 5% sucrose.
- Precipitates may form after freezing and thawing of the culture medium containing the LV. Such precipitates will block the filter and will be pelleted down with the LV during concentration by ultracentrifugation. To prevent this problem, we suggest filtering and concentrating the LV before freezing it.

4. REFERENCES

1. Lever AM, Strappe PM & Zhao J (2004) *J. Biomed. Sci.* **11**, 439–449.
2. Sinn PL, Sauter SL & McCray PB Jr (2005) *Gene Ther.* **12**, 1089–1098.
★★★ 3. Naldini L, Blomer U, Gallay P, *et al.* (1996) *Science* **272**, 263–267. – *First report demonstrating that LV can be used for stable transduction of nondividing cells* in vitro *and* in vivo.
4. Zennou V, Serguera C, Sarkis C, *et al.* (2001) *Nat. Biotechnol.* **19**, 446–450.
5. Zufferey R, Donello JE, Trono D & Hope TJ (1999) *J. Virol.* **73**, 2886–2892.
★★ 6. Miyoshi H, Blomer U, Takahashi M, Gage FH & Verma IM (1998) *J. Virol.* **72**, 8150–8157. – One of the first descriptions of *SIN-LV*.
★★ 7. Zufferey R, Dull T, Mandel RJ, *et al.* (1998) *J. Virol.* **72**, 9873–9880. – One of the first descriptions of *SIN-LV*.
★ 8. Xu K, Ma H, McCown TJ, Verma IM & Kafri T (2001) *Mol. Ther.* **3**, 97–104. – *The first description of cSIN-LV*.
★★ 9. Kafri T, van PH, Ouyang L, Gage FH & Verma IM (1999) *J. Virol.* **73**, 576–584. – *A description of one of the first packaging cell line for LVs.*
10. Klages N, Zufferey R & Trono D (2000) *Mol. Ther.* **2**, 170–176.
11. Farson D, Witt R, McGuinness R, *et al.* (2001) *Hum. Gene Ther.* **12**, 981–997.
12. Pacchia AL, Adelson ME, Kaul M, Ron Y & Dougherty JP (2001) *Virology*, **282**, 77–86.
13. Sparacio S, Pfeiffer T, Schaal H & Bosch V (2001) *Mol. Ther.* **3**, 602–612.
★ 14. Ikeda Y, Takeuchi Y, Martin F, Cosset FL, Mitrophanous K & Collins M (2003) *Nat. Biotechnol.* **21**, 569–572. – *Construction of improved packaging cell lines for LVs using a retroviral vector expressing Gag-Pol.*
15. Ni Y, Sun S, Oparaocha I, *et al.* (2005) *J. Gene Med.* **7**, 818–834.
16. Cockrell AS, Ma H, Fu K, McCown TJ & Kafri T (2006) *Mol. Ther.* **14**, 276–284.
17. Pham PL, Perret S, Doan HC, *et al.* (2003) *Biotechnol. Bioeng.* **84**, 332–342.
18. Yamada K, McCarty DM, Madden VJ & Walsh CE (2003) *Biotechniques*, **34**, 1074–8, 1080.
19. Baekelandt V, Eggermont K, Michiels M, Nuttin B & Debyser Z (2003) *Gene Ther.* **10**, 1933–1940.
20. Segura MM, Kamen A & Garnier A (2006) *Biotechnol. Adv.* **24**, 321–337.
21. Scherr M, Battmer K, Blomer U, Ganser A & Grez M (2001) *Biotechniques*, **31**, 520, 522, 524.
22. Sastry L, Johnson T, Hobson MJ, Smucker B & Cornetta K (2002) *Gene Ther.* **9**, 1155–1162.
23. Lizee G, Aerts JL, Gonzales MI, Chinnasamy N, Morgan RA & Topalian SL (2003) *Hum. Gene Ther.* **14**, 497–507.
24. Delenda C & Gaillard C (2005) *Gene Ther.* **12** (Suppl. 1), S36–S50.
★ 25. Bounou S, Leclerc JE & Tremblay MJ (2002) *J. Virol.* **76**, 1004–1014. – *A description of our preferred ELISA method for LV titration using p24.*
26. Mullick A, Xu Y, Warren R, *et al.* (2006) *BMC Biotech.* **6**, 43.
27. Gaillet B, Gilbert R, Amziani R, *et al.* (2007) *Biotechnol. Prog.* **23**, 200–209.

CHAPTER 15
Expression in mammalian cells using BacMam viruses

Hsiao-Ping Lee and Yu-Chen Hu

1. INTRODUCTION

The baculovirus expression vector system has been commonly employed as a rapid and inexpensive approach for recombinant protein production in insect cells (1, 2). Despite the wide applications, baculovirus infection results in insect cell death and lysis in a few days post-infection, which can lead to suboptimal protein processing late in the infection phase. In addition, some proteins that are synthesized initially as large inactive precursor proteins, such as peptide hormones, are not efficiently processed in insect cells (3). Furthermore, glycosylation in insect cells differs in many aspects from that in mammalian cells. In particular, insect cells generally fail to sialylate recombinant N-glycoproteins due to the absence of sialyltransferase activities and CMP-sialic acids (4, 5). The resultant glycoproteins, lacking sialic acids, may thus have altered immunogenicity and extremely short half-lives *in vivo*.

In the mid-1990s, Hofmann *et al.* (6) and Boyce & Bucher (7) first reported that recombinant baculoviruses harboring a mammalian expression cassette could efficiently transduce mammalian cells. Their data indicated a strong preference of baculovirus to enter hepatocytes of different origins. Later, Shoji *et al.* (8) showed that cells that are not transduced by a baculovirus expressing β-galactosidase under the control of a cytomegalovirus (CMV) promoter can be transduced efficiently by a baculovirus expressing the same reporter protein under the control of the stronger CAG promoter, comprising the β-actin promoter and CMV immediate-early enhancer. These pioneering studies lead to the discovery of a growing list of permissive cells originating from human, rodent, porcine, bovine, fish (for reviews, see 3 and 9) and even chicken (10). Due to the high efficiency of baculovirus-mediated gene delivery to various nondividing and primary cells, increasing efforts have been directed towards employing baculoviruses for *in vitro* and *in vivo* gene delivery.

In recent years, baculoviruses have emerged as a new expression vector in mammalian cells, know as the BacMam system, thanks to the high gene delivery efficiency, its nonreplicative nature and low cytotoxicity inside mammalian cells, its large genome (~130 kb) capable of harboring multiple genes or large inserts, and the ease of recombinant virus construction and production. Compared with other mammalian cell expression systems (e.g. plasmid transfection, Sindbis virus, vaccinia virus, and Semliki Forest virus), baculoviruses offer a rapid, safe, and convenient alternative for protein expression in mammalian cells. Therefore, the BacMam system has been exploited for a growing number of *in vitro* applications, such as high-level production of proteins, virus-like particles, and virus vectors (1, 9). The expression of certain functional proteins such as transporters (11), ion channels (12), and nuclear receptors (13) has led to the development of various cell-based assays. The expression of other proteins also allows studies of gene function and virus replication (14), eukaryotic surface display (15), and delivery of small interfering RNA (16). The success of these *in vitro* applications hinges on efficient baculovirus transduction. Therefore, several methods for baculovirus transduction of mammalian cells in 2D and 3D cultures are reviewed in this chapter.

2. METHODS AND APPROACHES

2.1 Construction and production of BacMam virus

The BacMam virus can easily be constructed using commercially available baculovirus transfer vectors (e.g. pBacPAK8 and pFastBac1) and procedures (e.g. homologous recombination or bacmid-based site-specific recombination), provided that the foreign gene is cloned under the transcriptional control of a mammalian promoter functional in the target cells. The promoters that have been cloned into baculovirus include the Rous sarcoma virus long terminal repeat promoter, CMV immediate-early promoter, simian virus 40 promoter, CAG promoter, hepatitis B virus promoter/enhancer, human ubiquitin C promoter, hybrid neuronal promoter (for review, see 9) and pol III H1 promoter (16). The mammalian expression cassette can be cloned into the multiple cloning site offered by the transfer vector. The baculovirus promoters (e.g. polyhedrin or *p10*) in the transfer vector can be deleted (8) or left intact (17). The subsequent recombinant virus selection and purification are performed according to standard procedures depending on the expression kit used (e.g. Bac-To-Bac system, Invitrogen; Bac-N-Blue system, Invitrogen). After obtaining the pure BacMam virus stock, the baculovirus is amplified by infecting the insect cells in tissue culture flasks or spinner flasks, and harvested at 3–4 days post-infection by slow-speed centrifugation (1000 *g* for 5 min). It is notable, however, that the subsequent virus storage and titration protocols may vary depending on the transduction procedures.

2.2 Transduction of mammalian cells using culture medium as the surrounding solution

In the transduction protocol commonly adopted by many laboratories (see *Protocol 1*), BacMam virus is concentrated (or even purified) from the infected cell culture medium by ultracentrifugation and resuspended in phosphate-buffered saline (PBS). The mammalian cells are then transduced at 37°C for 1–2 h using culture medium as the surrounding solution to adjust the multiplicity of infection (m.o.i.) and liquid volume. In recent years, some laboratories have also used unconcentrated virus for transduction (see *Protocol 2*).

Protocol 1

Transduction in 2D culture using concentrated BacMam virus (6, 7, 18–20)

Equipment and Reagents
- Ultracentrifuge (e.g. L8-70M, Beckman) and rotor (e.g. Beckman SW28)
- Ultracentrifuge tubes (Beckman SW28 tubes or equivalent)
- PBS (pH 6.2): 1 mM $Na_2HPO_4 \cdot 7H_2O$, 10.5 mM KH_2PO_4, 140 mM NaCl, 40 mM KCl
- PBS (pH 7.4): 10 mM Na_2HPO_4, 1.8 mM KH_2PO_4, 137 mM NaCl, 2.7 mM KCl
- Sucrose cushion (25%, w/w, sucrose containing 5 mM NaCl and 10 mM EDTA)
- Media for mammalian cell (e.g. Dulbecco's modified Eagle's medium (DMEM); Sigma) and insect cell (e.g. TNM-FH; Sigma) culture
- Fetal bovine serum (FBS)
- Sodium butyrate
- Humidified 5% CO_2 incubator controlled at 37°C for mammalian cell culture

Method
Preparation of concentrated BacMam virus stock
1. The BacMam virus is capable of infecting insect cells. Amplify and harvest the BacMam virus-infected insect cell medium according to standard protocols (21).
2. Load 33 ml of virus stock into each of six 38 ml ultracentrifuge tubes.
3. Underlay with 3 ml of sucrose cushion solution.
4. Centrifuge at 80 000 *g* (24 000 r.p.m. in an SW28 rotor) for 75–90 min at 4°C.
5. Decant the supernatant and carefully remove traces of sucrose. The viral pellet appears translucent white.
6. Resuspend the pellets in 1–2 ml of PBS (pH 6.2).
7. Filter the concentrated virus through 0.22 μm filters and determine the virus titer by plaque assay or an end-point dilution method using Sf9 cells according to standard protocols (21)[a]. Store at 4°C.

Transduction with concentrated baculovirus
8. Plate the mammalian cells to be transduced (e.g. HeLa cells) on tissue culture plates or dishes at ~50–60% confluency[b] and culture overnight.
9. Estimate the number of cells on the plates. Dilute the concentrated baculovirus with culture medium (e.g. DMEM). The amount of virus required depends on the m.o.i. to be used[c].
10. Aspirate the medium and wash the cells with PBS (pH 7.4).
11. Add the virus-containing medium to the cells (with or without 10% heat-inactivated FBS[d]). The volume of medium should be sufficient to cover the cell monolayer. For example, 500–1000 μl of medium is used for each well in six-well plates.
12. Incubate the cells at 37°C for 1–2 h.
13. Replace the virus-containing medium with fresh sodium butyrate-containing medium[e] and culture the transduced cells at 37°C until analysis[f].

Notes

[a] The virus can be further purified by 25–60% (w/w) sucrose gradient ultracentrifugation according to standard procedures (21) if *in vivo* applications are desired (20, 22). For *in vitro* transduction, the concentrated virus stock is sufficient.

[b] The transduction efficiency is best when the cells are subconfluent. One of the reasons why transduction efficiency decreases upon overconfluency is the cellular junctions that become established (23). Use of EGTA is reported to transiently disrupt the cell junctions between liver cells and enhance gene delivery efficiency (23).

[c] The m.o.i. to be used for transduction ranges from 10 to 1000. An m.o.i. of up to 1000 does not cause significant cytotoxicity (6). It is common to use an m.o.i. of up to 100 (24) or 200 (17, 19) for satisfactory transduction.

[d] Transduction does not appear to be inhibited by serum (25). However, heat inactivation of serum is essential as the serum complement proteins can inactivate baculovirus (26, 27).

[e] Histone deacetylase inhibitors such as sodium butyrate and trichostatin A can decondense the baculovirus genome and enhance gene transcription (19). However, these drugs can cause cytotoxicity (17), and a butyrate concentration exceeding 5–10 mM is not recommended.

[f] Typically, gene expression peaks at 24–48 h post-transduction for genes driven by viral promoters (7) and then gradually decays as the cells divide and the viral DNA degrades (28, 29). The kinetics of expression need to be determined empirically. Analyze the gene expression at the desired time points.

Protocol 2

Transduction using unconcentrated BacMam virus (30)

Equipments and Reagents
- Medium for mammalian cell culture (e.g. DMEM)
- Mammalian cells
- Humidified CO_2 incubator at 37°C

Method
1. Produce and harvest the BacMam virus stock from the infected insect cell culture as in *Protocol 1*. Directly store the virus supernatant (in spent insect cell medium) at 4°C without concentration or purification.
2. Plate the mammalian cells in six-well plates (2×10^5 cells/well) and allow the cells to attach overnight at 37°C.
3. Remove the medium and add 1 ml of virus supernatant to the cells[a]. Incubate the cells at 37°C for 1–2 h.
4. Remove the virus inoculum and replace with 2 ml of medium with or without sodium butyrate. Incubate the cells at 37°C until analysis.

Note

[a] This protocol utilizes unconcentrated baculovirus; thus, the m.o.i. that can be achieved may be lower than the generally recommended 100 or 200. The relatively low m.o.i. resulting from this procedure may be used for procedures that require gene expression closer to physiological levels (11). Higher m.o.i. may require concentration of the virus by ultracentrifugation.

2.3 Improved protocol for BacMam transduction

Protocol 1 has been routinely utilized for baculovirus transduction. However, this protocol requires ultracentrifugation, which is time-consuming, labor-intensive, and causes a significant loss of virus infectivity. However, incubation of mammalian cells with virus in the insect cell medium, as in *Protocol 2*, may result in reduced transduction efficiency (25, 28, 29). To enhance transduction efficiency, we have developed a protocol (see *Protocol 3*) by which virus transduction is performed by incubating the cells with unconcentrated BacMam virus supernatant for 4–8 h at 25 or 27°C using Dulbecco's PBS as the surrounding solution to adjust the final liquid volume. This protocol has been employed successfully for enhanced transduction of HeLa (25), Huh-7 (29), HepG2 (29), BHK (31), A-549 (32) and BEAS-2B (32) cells, primary rat chondrocytes (28), human mesenchymal stem cells (hMSCs) (33), and hMSC-derived progenitor cells (34) for the expression of enhanced green fluorescent protein (EGFP), large hepatitis delta antigen (L-HDAg), hepatitis B virus surface antigen (HBsAg) and bone morphogenic protein 2 (BMP-2). Importantly, this protocol yields transduction efficiencies comparable or superior to those obtained using *Protocol 1* (29) and eliminates the need for virus ultracentrifugation, thus not only representing a simpler approach, but also considerably reducing possible virus inactivation during ultracentrifugation. The major differences between *Protocols 1* and *3* are illustrated in *Fig. 1*.

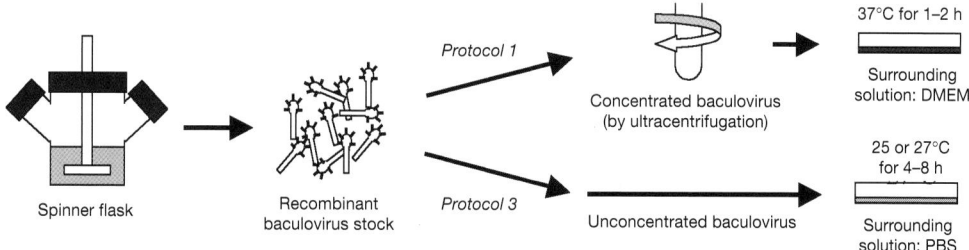

Fig. 1. Schematic illustration of *Protocols 1* and *3*.
In *Protocol 1*, the BacMam virus is concentrated by ultracentrifugation, diluted in DMEM, and incubated with cells at 37°C for 1–2 h. In *Protocol 3*, the unconcentrated virus is incubated directly with the cells using PBS as the surrounding solution at 25 or 27°C for 4–8 h.

Protocol 3

Transduction using PBS as the surrounding solution in 2D culture

Equipment and Reagents
- BacMam virus expressing EGFP
- Six-well culture plate
- PBS (pH 7.4) (see *Protocol 1*)
- Mammalian cell culture medium
- Humidified 5% CO_2 incubator at 37°C
- Rocking plate (Reliable Scientific Shaker; Midwest Scientific)

Method
1. Produce and harvest the BacMam virus from the infected insect cell culture as in *Protocol 1*. Directly store the virus supernatant (in insect cell medium) at 4°C without concentration or purification.
2. Plate the mammalian cells (e.g. HeLa cells) into six-well plates (5×10^5 cells/well[a]) and culture overnight.
3. Prepare the transduction solution by mixing PBS[b] with unconcentrated virus supernatant at a volumetric ratio of 4[c].
4. Prior to transduction, aspirate the spent medium and wash the cells with PBS.
5. Add 500–1000 μl[d] of transduction solution to each well.
6. Gently shake the plates on the rocking plate (ten rocking motions per min) for 4–8 h[e] at 25 or 27°C[f].
7. After transduction, replace the transduction solution with 2 ml of complete fresh medium[g] in each well. Continue to incubate at 37°C until analysis.

Notes
[a] As in *Protocols 1* and *2*, transduction efficiency is better when cells are subconfluent: 50–90% confluency is recommended.

[b] The use of PBS as the surrounding solution to adjust m.o.i. and liquid volume is critical in this protocol and allows better transduction than fresh TNM-FH or DMEM (25, 28).

[c] The volumetric ratio of PBS to virus supernatant influences the transduction efficiency and protein expression. A very low ratio (i.e. using a high proportion of virus supernatant and a low proportion of PBS) leads to decreased transduction efficiency. A volumetric ratio of 2–4 is recommended.

[d] The total volume of transduction solution in this protocol should be large enough to ensure uniform contact of cells with virus during shaking. The volume can be scaled up proportionally depending on the culture vessels (e.g. 10 cm dishes or T-150 flasks).

[e] Prolonging the incubation time enhances transduction by increasing virus uptake (25, 28). Some cell types that we have tested (e.g. HeLa, BHK, HepG2, and Huh-7) can be incubated for up to 8 h without appreciable cell detachment or negative effects on the cells. However, incubation for 12 h results in cell detachment. hMSCs can be incubated for up to 4 h.

[f] Transduction efficiency and protein expression are significantly higher at 25–27°C than at 37°C (25, 28).

[g] Sodium butyrate (2.5–10 mM) can also be added to the medium to enhance expression.

2.4 Protein production in a BelloCell-500 bioreactor

Protocol 3 has been used for the production of virus-like particles (VLPs) of hepatitis delta virus (HDV) in which hepatoma cells cultured in 10 cm dishes are co-transduced with two recombinant baculoviruses: Bac-GS2 expressing HBsAg and Bac-GD expressing L-HDAg. The co-transduction results in efficient self-assembly and secretion of HDV VLPs comprising HBsAg and L-HDAg (29). To scale up the production, we have developed a bioreactor process using a BelloCell-500 (Cesco Bioengineering Co.), a novel oscillating bioreactor composed of a packed bed in the upper chamber and a 'bellow' in the lower chamber (see *Fig. 2a*). The cells are seeded (see *Fig. 2b*) and immobilized onto the carriers (BioNOC II, which are nonwoven fabric strips made of 100% polyethylenephthalate) in the packed bed. The oscillating compression and relaxation of the bellow enables sequential exposure of the immobilized cells to medium and air for efficient nutrient and oxygen transfer, and allows the culture of mammalian cells to a very high density (35, 36). *Protocol 4* shows an example procedure for the co-transduction of BHK cells in the BelloCell-500 reactor using Bac-GS2 and Bac-GD (35). With this protocol, up to ~90% of the immobilized BHK cells can be transduced and a total yield of 427 µg based on HBsAg in the VLPs (harvested from 940 ml medium) can be obtained. The average volumetric yield (454 ng/ml) is approximately fivefold higher than that obtained by the commonly employed transfection method (80 ng/ml). Therefore, the combination of baculovirus for efficient transduction and the BelloCell-500 for high-density cell culture provides a new platform for transient protein production using mammalian cells on a larger scale.

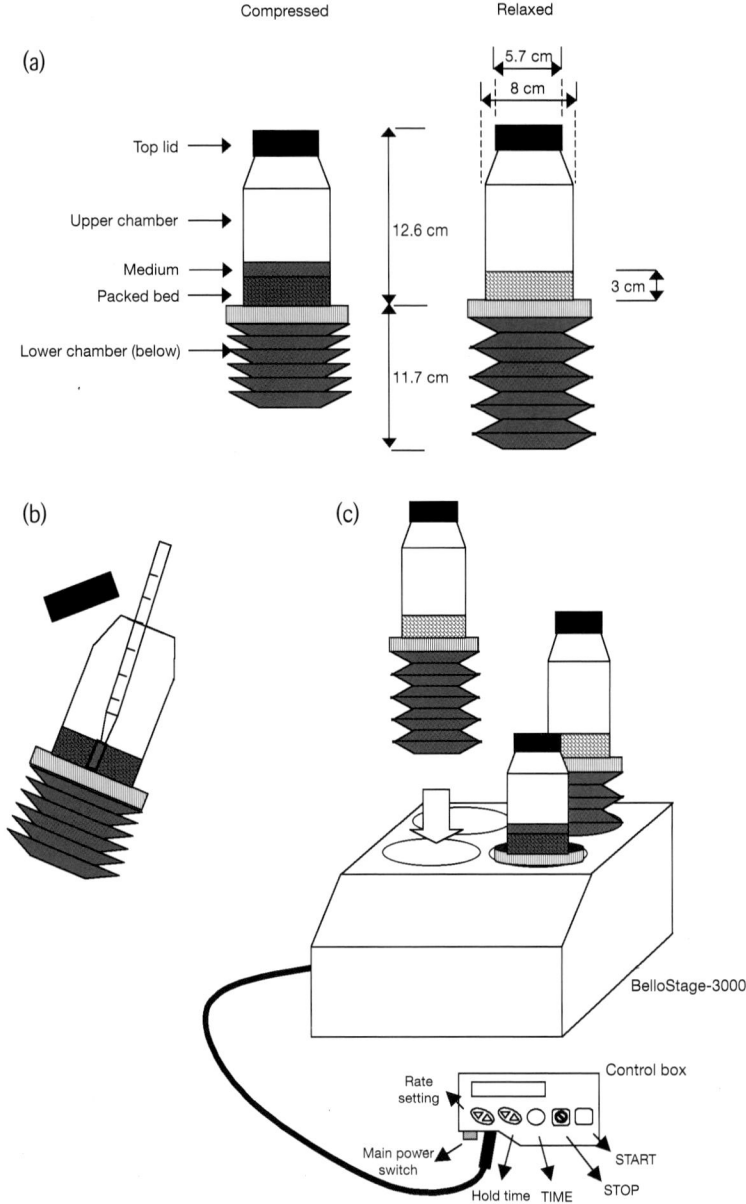

Fig. 2. Schematic illustration of the BelloCell and BelloStage.
(a) The BelloCell reactor consists of an upper chamber containing the packed bed for cell immobilization and a compressible bellow that can be alternately compressed and released by the BelloStage. The ascending movement of the holding plate inside the BelloStage (not shown) lifts the bellow and raises the medium level to submerge the carriers, thus allowing nutrient transfer. The descending movement of the holding plate drops the medium to the lower bellow, thus exposing the carriers to air for oxygen transfer. The top lid is equipped with a 0.22 μm PTFE filter for ventilation. (b) Inoculation procedure. The top lid can be opened and the cells are inoculated through the central duct. (c) The BelloStage platform can accommodate up to four BelloCell reactors. The linear moving rates of the holding plate and holding time can be changed using the control box.

Protocol 4

Baculovirus transduction of mammalian cells in a BelloCell-500 bioreactor

Equipment and Reagents
- PBS (pH 7.4) (see *Protocol 1*)
- Unconcentrated baculovirus stocks (e.g. Bac-GS2 and Bac-GD)
- Mammalian cells (e.g. BHK)
- Humidified CO_2 incubator controlled at 37°C
- Complete culture medium (DMEM containing 5% FBS)
- BelloCell-500 (working volume ~500 ml, containing ~860 pieces of BioNOC II carrier, with a bed volume of ~100 cm^3; the carriers are pre-sterilized by gamma irradiation)
- BelloStage-3000 (Cesco Bioengineering Co.)
- Sodium butyrate
- Lysis solution (0.1 % crystal violet in 0.1 M citric acid)
- YSI 2700 analyzer (Yellow Spring Instruments) and pH meter

Method

Cell seeding and culture

1. Pre-warm the DMEM in a 37°C water bath. Place the BelloCell-500 reactor in a laminar flow hood. Open the top lid aseptically and add 470 ml of medium to each reactor.

2. Detach the BHK cells cultured in T-flasks by a standard trypsinization protocol. Pellet the cells by centrifugation at 1000 *g* for 5 min.

3. Resuspend ~1.1×10^8 BHK cells in 30 ml of DMEM[a].

4. Inoculate the 30 ml cell suspension into the reactor through the central duct (see *Fig. 2b*). Gently shake the bottle to distribute the cells evenly in the medium. The cell suspension will appear cloudy.

5. Mount the BelloCell-500 reactor onto the BelloStage sitting in a 37°C incubator (see *Fig. 2c*).

6. For uniform cell attachment, set the linear moving rates (upward and downward) at 2 mm/s, the top holding time to 20 s and the bottom holding time to 0 s. Press START on the BelloStage controller to initiate the attachment phase.

7. After the cells have attached completely[b], lower the linear moving rates to 1.5 mm/s and set the top holding time to 0 s and the bottom holding time to 90 s for the rest of the culture phase.

8. Every 12–24 h during the culture phase, move the reactor vessel to the hood. Open the top lid to aspirate the medium aseptically for glucose (using the YSI 2700 analyzer) and pH measurements. When the pH falls to 6.8, or the glucose concentration falls to ~1–1.5 g/l, change the medium completely.

9. During the culture, take a total of six BioNOC II carrier strips from the top, middle, and bottom layers of the packed bed (the top screen covering the packed bed contains holes for sampling) using sterilized forceps. Distribute the strips evenly into three microfuge tubes.

10. Fill each tube with 1 ml of lysis solution (0.1 % crystal violet in 0.1 M citric acid) and incubate at 37°C for 1 h. Vortex the cells to release the nuclei and count the nuclei using a hemocytometer. Calculate the cell numbers from the three tubes and convert to total cell number in the reactor[c].

Baculovirus transduction and protein production phase

11. When the cell number reaches ~1–3 × 10^9 cells/reactor (mid-exponential phase for BHK cells[d]), pour off the spent medium and wash the immobilized cells with 400 ml of PBS with gentle swirling for 5 min. Discard the PBS after washing.

12. Add 125 ml of unconcentrated baculovirus containing Bac-GS2 and Bac-GD. Swirl the reactor several times to allow uniform contact between the cells and viruses, and then add 375 ml of PBS[e].

13. Mount the reactor onto the BelloStage and set the linear moving rate at 1.5 mm/s. Operate at 27°C for 6 h.

14. Decant the virus inoculum and PBS solution. Replenish with 500 ml of fresh medium containing 2.5 mM sodium butyrate[f].

15. Mount the reactor onto the BelloStage and operate under conditions identical to the culture phase[g].

16. Harvest the medium at 48 h post-transduction. Replenish with 500 ml of fresh medium without sodium butyrate.

17. Harvest the culture medium containing the product at 96 h post-transduction and terminate the reactor operation[h].

Notes

[a]The seeding cell number can be lowered to 5 × 10^7 cells to reduce the inoculum required. However, the cells may experience a longer lag phase.

[b]It takes ~2–3 h for BHK cells to attach completely to the carriers, as indicated by clearing of the medium.

[c]With this culture protocol, the lag phase is less than 1 day and BHK cells can grow to ~5.5 × 10^9 cells/reactor (~1 × 10^7 cells/ml of medium) in 192 h, which corresponds to a 49-fold cell growth and a doubling time of 22.8 h (35). The culture parameters, maximum cell density that can be reached, and optimal protocols for other mammalian cells (e.g. Vero, CHO, HEK293, Huh-7, and hybridoma cells) can be found in the manufacturer's web site (http://www.cescobio.com.tw/product.htm[15.1]).

[d]Transduction during the mid-exponential phase is recommended for BHK cells. Transduction at exceedingly high cell numbers may result in lower transduction efficiencies, whilst transduction at lower cell numbers may decrease the overall product yield. However, the optimal transduction timing for different cell types needs to be determined empirically.

[e]This transduction protocol uses PBS as the surrounding solution and a volumetric ratio (PBS to virus supernatant) of 2–4 is recommended for BHK cells. The total volume of virus supernatant (and total m.o.i.) can vary as long as the total working volume is 500 ml. For co-transduction, the relative amounts of both viruses are adjusted so that the total virus volume remains fixed.

[f]Sodium butyrate is somewhat toxic to cells. Due to the dynamic fluid environment, the sodium butyrate concentration is reduced to 2.5 mM from 0–2 days post-transduction and is not used late in the production phase.

[g]Because of the high cell density, the cells consume glucose and glutamine very quickly during the production phase. If the glucose concentration drops too quickly, add extra glucose (100 g/l) and glutamine (100 mM) to supplement the nutrients. Sodium bicarbonate (0.75 g/l) can be added to buffer the pH.

[h]Baculovirus-mediated expression is transient. The expression of HDV VLPs remains effective for 4 days, but decreases abruptly after day 4.

2.5 Determination of baculovirus transducing ability in mammalian cells

Protocols 2, 3 and *4* utilize unconcentrated baculovirus for transduction, yet the ability of different virus batches to transduce mammalian cells may vary from lot to lot, leading to inconsistent transduction results. Therefore, it is paramount to determine the baculovirus transducing ability for each batch. To date, the infectious titer and m.o.i. based on the ability of baculoviruses to infect insect cells are commonly adopted to indicate the virus dosage. However, the common titration methods such as plaque assays and end-point dilution methods are tedious and require 4–7 days to obtain results. Furthermore, it is difficult to obtain consistent, accurate infectious titers, as large variations tend to arise from experiment to experiment. As a result, transduction experiments based on *Protocol 1* using m.o.i. as the virus dosage indicator have shown inconsistent results. Therefore, we have developed a titration protocol to determine the ability of baculoviruses to transduce mammalian cells (see *Protocol 5* and *Fig. 3*). This protocol relies on expression of EGFP in the transduced cells and measurement of the percentage of GFP-emitting cells (% GFP$^+$ cells) by flow cytometry (37). Through this protocol, BacMam virus with a higher transducing titer can transduce cells at higher efficiencies and yield stronger and longer transgene expression, confirming that transducing titer is representative of the transducing ability of the BacMam virus (37). Compared with the infectious titer, the transducing titer can be determined for various mammalian cells (e.g. CHO, BHK, HeLa, and primary chondrocytes) in 24 h and the measurement is more reproducible as the standard deviations among measurements are smaller (37). Therefore, this titration protocol provides a simple, fast, and reliable measure to evaluate the quality of virus stocks during virus production and purification, and is helpful to predict the performance of vector supernatants and to ensure reproducible gene delivery experiments.

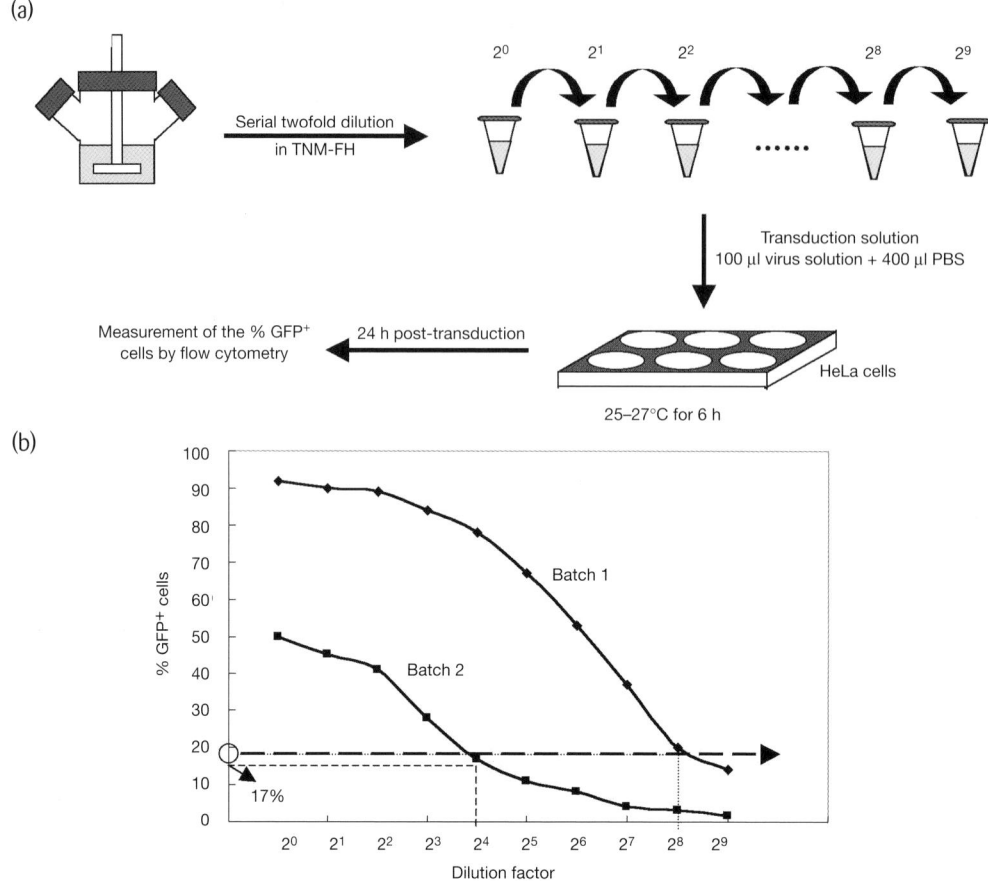

Fig. 3. Determination of BacMam virus transducing titer in mammalian cells.
(*a*) Experimental procedures. The unconcentrated virus is serially diluted twofold with insect cell medium (e.g. TNM-FH) and used for transduction following *Protocol 3*. The transduced cells are trypsinized at 24 h post-transduction for flow cytometric analysis. (*b*) Titration curves.

Protocol 5

Determination of baculovirus transducing titer in mammalian cells (37)

Equipment and Reagents
- BacMam virus expressing EGFP[a]
- Six-well plates
- Mammalian cells (e.g. HeLa)

- PBS (pH 7.4) (see *Protocol 1*)
- DMEM
- Trypsin/EDTA
- Humidified CO_2 incubator at 37°C
- Rocking plate
- Flow cytometer (FACSCalibur; BD Biosciences)

Method

1. Seed HeLa cells into six-well plates (5×10^5 cells/well) and incubate overnight.

2. Dilute the virus stock twofold serially (2^0, 2^1, 2^2...2^9) in fresh insect cell medium (the same medium as used for virus production, e.g. TNM-FH).

3. Prepare the transduction solution by mixing PBS with the virus at a volumetric ratio of 4.

4. Wash the cells with PBS.

5. For each well, add 500 µl of transduction solution containing the virus and PBS. For each virus dilution, conduct transduction in triplicate wells.

6. Shake the plates on a rocking plate at ~25–27°C for 6 h.

7. Aspirate the transduction solution and wash the cells with PBS.

8. For each well, add 2 ml of fresh DMEM and incubate at 37°C for 24 h.

9. For each well, detach the transduced cells using trypsin/EDTA, and wash and resuspend the cells in PBS (filtered through a 0.22 µm filter).

10 For the cells from each well, measure the % GFP⁺ cells using flow cytometry according to the manufacturer's instruction. Count 10 000 cells in each measurement.

11. Plot a titration curve as shown in *Fig. 3(b)*. Use the following equation to calculate the transducing titer, which is expressed as transducing units (TU)/ml: transducing titer = [(% GFP⁺ cells)/100] × dilution factor × cell number × reciprocal of volume (ml), where the dilution factor[b] is defined as the dilution when the % GFP⁺ cells is lower than 20%, the % GFP⁺ cells used in the equation is the corresponding value, the cell number is the number of cells in the well, and the reciprocal of the volume (1/0.1) converts the volume of virus (100 µl) to milliliters. For two batches of virus with titration curves as shown in *Fig. 3(b)*, calculate the transducing titer[c] as follows:

    Batch 1 = 20% × 2^8 × 500 000 × (1/0.1) = 2.56×10^8 TU/ml
    Batch 2 = 17% × 2^4 × 500 000 × (1/0.1) = 1.36×10^7 TU/ml

Notes

[a]This titration protocol is suitable for BacMam virus carrying target genes. The target genes and the *egfp* gene may be cloned into the same virus but transcribed independently under the control of separate promoters. Alternatively, the *egfp* and target genes may be transcribed together but translated separately via the use of an internal ribosomal entry site.

[b]Set the dilution factor to 1 if the virus transducing ability is so low that the transduction efficiency is below 20% without any dilution.

[c]In our hands, the transducing titers of BacMam virus expressing only EGFP under the control of the CMV promoter typically fall in the range of 1×10^7–3×10^8 TU/ml, depending on the production batch. Using HeLa cells as the host, the transduction efficiencies using *Protocol 3* are dependent on the transducing titers in the range of 1.4×10^7 (~50%) to 1.3×10^8 TU/ml (~92–95%). The transduction efficiency reaches a plateau at ~92–95% for viruses whose transducing titers exceed a threshold (~1.0–1.5×10^8 TU/ml for HeLa cells).

3. TROUBLESHOOTING

- In *Protocol 3*, the presence or absence of heat-inactivated serum and divalent cations (Ca^{2+} and Mg^{2+}) in the surrounding solution does not overtly affect transduction. However, $NaHCO_3$ in the medium profoundly suppresses transduction, partly due to the hindered baculovirus uptake (38). Therefore, we use PBS as the surrounding solution. If use of PBS for long-term transduction is of concern, use $NaHCO_3$-deficient medium (e.g. DMEM) instead. Omission of $NaHCO_3$ from DMEM during transduction significantly enhances transduction efficiency and gene expression compared with the use of DMEM with $NaHCO_3$.
- Proper production and storage of baculovirus is crucial for obtaining virus stocks of consistent quality for efficient and reproducible protein production. Here are some recommendations for virus production and storage:
  - BacMam virus can be produced using serum-free (e.g. Ex-Cell 420 or Sf-900 II) or serum-containing (e.g. TNM-FH) medium. In the case of serum-containing medium, the addition of 5% FBS results in equally high virus yield at 4 days post-infection compared with 10% FBS. However, further reducing the FBS concentration decreases the virus yield.
  - A suspension culture (e.g. spinner flask) produces more virus particles than a stationary culture (e.g. T-flask), but results in a higher fraction of virus particles devoid of transducing ability.
  - Store the virus at +4 or −80°C and protect from light. Baculoviruses are relatively stable under these conditions. Storage at −20°C causes a more rapid decrease in virus titer (39, 40).
  - Check the pH of the baculovirus stock. Baculovirus is more stable at pH 6.2 than at pH 5.7 because the virus can form aggregates under more acidic conditions (41, 42).

4. REFERENCES

★ 1. **Kost TA, Condreay JP & Jarvis DL** (2005) *Nat. Biotechnol.* **23**, 567–575. – *An excellent review of the BacMam system.*
★ 2. **Hu YC** (2005) *Acta Pharmacol. Sin.* **26**, 405–416. – *An excellent review of the BacMam system.*
★ 3. **Kost TA & Condreay JP** (2002) *Trends Biotechnol.* **20**, 173–180. – *An excellent review of the BacMam system.*
4. **Hill DR, Aumiller JJ, Shi XZ & Jarvis DL** (2006) *Biotechnol. Bioeng.* **95**, 37–47.
5. **Jarvis DL** (2003) *Virology*, **310**, 1–7.
★★ 6. **Hofmann C, Sandig V, Jennings G, et al.** (1995) *Proc. Natl. Acad. Sci. U.S.A.* **92**, 10099–10103. – *The original publication describing the BacMam system and* Protocol 1.
★★ 7. **Boyce FM & Bucher NLR** (1996) *Proc. Natl. Acad. Sci. U.S.A.* **93**, 2348–2352. – *The second publication describing the BacMam system and* Protocol 1.
8. **Shoji I, Aizaki H, Tani H, et al.** (1997) *J. Gen. Virol.* **78**, 2657–2664.
★ 9. **Hu YC** (2006) In: *Insect Viruses: Biotechnological Applications*, pp 287–320. Edited by Bonning BC. Elsevier, New York. – *An excellent review of use of the BacMam virus for gene therapy.*
10. **Ping WX, Ge JP, Li SX, et al.** (2006) *Avian Dis.* **50**, 59–63.

11. Hassan NJ, Pountney DJ, Ellis C & Mossakowska DE (2006) *Protein Expr. Purif.* **47**, 591–598.
12. Pfohl JL, Worley JF, Condreay JP, *et al.* (2002) *Receptors Channels*, **8**, 99–111.
13. Clay WC, Condreay JP, Moore LB, *et al.* (2003) *Assay Drug Develop. Technol.* **1**, 801–810.
14. Delaney WE & Isom HC (1998) *Hepatology*, **28**, 1134–1146.
15. Grabherr R, Ernst W, Oker-Blom C & Jones I (2001) *Trends Biotechnol.* **19**, 231–236.
16. Ong ST, Li F, Du J, Tan YW & Wang S (2005) *Hum. Gene Ther.* **16**, 1404–1412.
17. Hu YC, Tsai CT, Chang YJ & Huang JH (2003) *Biotechnol Prog.* **19**, 373–379.
18. Bilello JP, Cable EE & Isom HC (2001) *Hepatology*, **34**, 1299.
★ 19. Condreay JP, Witherspoon SM, Clay WC & Kost TA (1999) *Proc. Natl. Acad. Sci. U.S.A.* **96**, 127–132. – Description of the use of Protocol 1.
20. Tani H, Limn CK, Yap CC, *et al.* (2003) *J. Virol.* **77**, 9799–9808.
★★ 21. O'Reilly D, Miller L & Luckow V (1992) *Baculovirus Expression Vectors: a Laboratory Manual*. WH Freeman and Co, New York. – An excellent laboratory manual describing the baculovirus expression system.
22. Abe T, Takahashi H, Hamazaki H, *et al.* (2003) *J. Immunol.* **171**, 1133–1139.
23. Bilello JP, Delaney WE, Boyce FM & Isom HC (2001) *J. Virol.* **75**, 9857–9871.
24. Palombo F, Monciotti A, Recchia A, *et al.* (1998) *J. Virol.* **72**, 5025–5034.
★★ 25. Hsu CS, Ho YC, Wang KC & Hu YC (2004) *Biotechnol. Bioeng.* **88**, 42–51. – The first paper describing Protocol 3.
26. Hofmann C & Strauss M (1998) *Gene Ther.* **5**, 531–536.
27. Sandig V, Hofmann C, Steinert S, *et al.* (1996) *Hum. Gene Ther.* **7**, 1937–1945.
★★ 28. Ho YC, Chen HC, Wang KC & Hu YC (2004) *Biotechnol. Bioeng.* **88**, 643–651. – The second paper describing Protocol 3.
★★ 29. Wang KC, Wu JC, Chung YC, *et al.* (2005) *Biotechnol. Bioeng.* **89**, 464–473. – The first paper utilizing Protocol 3 for the production of HDV VLPs.
★★ 30. Merrihew RV, Kost TA & Condreay JP (2004) In: *Gene Delivery to Mammalian Cells:* Vol. 2. *Viral Gene Transfer Techniques*, pp. 355–365. Edited by WC Heiser. Humana Press Inc., Totowa, NJ. – This chapter describes the use of Protocol 2.
31. Chiang YW, Wu JC, Wang KC, *et al.* (2006) *World J. Gastroenterol.* **12**, 1551–1557.
32. Yang DG, Chung YC, Lai YK, *et al.* (2007) *Mol. Ther.* **15**, 989–996.
★ 33. Ho YC, Chung YC, Hwang SM, Wang KC & Hu YC (2005) *J. Gene. Med.* **7**, 860–868. – Description of the use of Protocol 3 for transduction of mesenchymal stem cells.
34. Ho YC, Lee HP, Hwang SM, *et al.* (2006) *Gene Ther.* **13**, 1471–1479.
★★ 35. Chen YH, Wu JC, Wang KC, *et al.* (2005) *J. Biotechnol.* **118**, 135–147. – The first paper to use Protocol 4 for the production of HDV VLPs in a bioreactor.
36. Ho L, Greene CL, Schmidt AW & Huang LH (2004) *Cytotechnology*, **45**, 117–123.
★★ 37. Chan ZR, Lai CW, Lee HP, Chen HC & Hu YC (2006) *Biotechnol. Bioeng.* **93**, 564–571. – The original paper developing the transducing titration method of Protocol 5.
38. Shen HC, Lee HP, Lo WH, Yang DG & Hu YC (2007) *J. Gene Med.* in press.
39. Jarvis DL & Garcia A Jr (1994) *Biotechniques*, **16**, 508–513.
40. Tsai CT, Chan ZR, Lu JT, *et al.* (2007) *Enzyme Microb. Technol.* **40**, 1345–1351.
41. Jorio H, Tran R & Kamen A (2006) *Biotechnol. Prog.* **22**, 319–325.
42. Jorio H, Tran R, Meghrous J, Bourget L & Kamen A (2006) *J. Virol. Methods*, **134**, 8–14.

APPENDIX 1
List of suppliers

ABgene – www.abgene.com
Alexis Corporation – www.alexis-corp.com
Amersham Biosciences – www.amershambiosciences.com
Amresco, Inc. – www.amresco-inc.com
Anachem Ltd – www.anachem.co.uk
Appleton Woods Ltd – www.appletonwoods.co.uk
Applied Biosystems – www.appliedbiosystems.com
Appropriate Technical Resources, Inc. – www.atrbiotech.com
AutoGen, Inc. – www.autogen.com
Axon Instruments – www.axon.com

Beckman Coulter, Inc. - www.beckman.com
Becton, Dickinson and Company - www.bd.com
Bio-Rad Laboratories, Inc. - www.bio-rad.com
BOC Group – www.boc.com
Brosch direct Ltd – www.broschdirect.com

Calbiochem – www.calbiochem.com
Cambrex Scientific – www.lonza.com
Cambridge Scientific Products – www.cambridgescientific.com
Carl Zeiss – www.zeiss.com
CellFree Sciences – www.cfsciences.com
Cesco Bioengineering Co. – www.cescobio.com.tw
Chemicon International, Inc. – www.chemicon.com
Chiron Corp. – www.chiron.com
Clontech – www.clontech.com
Corning, Inc. – www.corning.com

DakoCytomation – www.dako.com
Difco Laboratories - www.difco.com
Dionex Corporation – www1.dionex.com
DuPont – www2.dupont.com

E & K – www.eandk.scientific.com
Elliot Scientific – www.elliotscientific.com

EMD – www.emdbiosciences.com
Eppendorf – www.eppendorfna.com
European Collection of Animal Cell Culture – www.ecacc.org.uk

Findel Education Ltd – www.fipd.co.uk
Fisher Scientific International – www.fishersci.com
Fluka – www.sigmaaldrich.com
Fluorochem – www.fluorochem.co.uk

G-Biosciences – www.gbiosciences.com
GE Healthcare – www.gehealthcare.com
Gibco BRL – www.invitrogen.com
Goodfellow Cambridge Ltd – www.goodfellow.com
Greiner Bio-One International – www.gbo.com

Hampton Research – www.hamptonresearch.com
Harlan – www.harlan.com
Harvard Apparatus – www.harvardapparatus.com
Hybaid – www.hybaid.com
HyClone – www.hyclone.com

ICN Biomedicals, Inc. – www.icnbiomed.com
Insight Biotechnology – www.insightbio.com
Invitrogen Corporation – www.invitrogen.com

Jencons – www.jencons.co.uk

Kendro Laboratory Products – www.kendro.com
Kimble / Kontes – www.kimble-kontes.com
Kodak: Eastman Fine Chemicals – www.eastman.com

LabPlant Ltd – www.labplant.com
Lancaster – www.lancastersynthesis.com
Leica – www.leica.com
Life Technologies Inc. – www.lifetech.com
LOT-Oriel – www.lot-oriel.com

Matrix Impact – www.matrixtechcorp.com
Merck, Sharp and Dohme – www.msd.com
MetaChem – www.metachem.com
MicroFluidics – www.microfluidics.com
MIDSCI – www.midsci.com
Millipore Corporation – www.millipore.com
Miltenyi Biotec – www.miltenyibiotec.com
Misonix – www.misonix.com
MP Biomedicals – www.mpbio.com

MWG Biotech – www.mwg-biotech.com

Nalgene – www.nalgene.com
National Diagnostics – www.nationaldiagnostics.com
New England BioLabs, Inc. – www.neb.com
Nikon Corporation – www.nikon.com
Novagen – www.merckbiosciences.com

Olympus Corporation – www.olympus-global.com
Optivision Ltd – optivision.co.uk
Organotechnie – www.organotechnie.com
Oxford Expression Technologies – www.expressiontechnologies.com

Pall Life Sciences – www.pall.com
Perbio Science – www.perbio.com
PerkinElmer, Inc. – www.perkinelmer.com
Pharmacia Biotech Europe – www.biochrom.co.uk
Photonic Solutions plc – www.psplc.com
Pierce Biotechnology – www.piercenet.com
Polyscience – www.polyscience.com
Promega Corporation – www.promega.com

Qiagen – www.qiagen.com

R&D Systems – www.rndsystems.com
Research Diagnostics, Inc. – www.researchd.com
Research Organics – www.resorg.com
Roche Diagnostics Corporation – www.roche-applied-science.com

Sanyo Gallenkamp – www.sanyogallenkamp.com
Sarstedt – www.sarstedt.com
Schleicher and Schuell Bioscience, Inc. – www.schleicher-schuell.com
Scientifica – www.scientifica.uk.com
Serotec – www.serotec.com
Shandon Scientific Ltd – www.shandon.com
Sigma-Aldrich Company Ltd – www.sigmaaldrich.com
Sorvall – www.sorvall.com
Spectrum Labs – www.spectrapor.com
Stratagene Corporation – www.stratagene.com

Thames Restek – www.thamesrestek.co.uk
Thermo Scientific – www.thermo.com
Thistle Scientific – www.thistlescientific.co.uk
Topac, Inc. – www.topac.com

Vector Laboratories – www.vectorlabs.com

VWR International – www.vwr.com

Whatman – www.whatman.com
Wolf Laboratories – www.wolflabs.co.uk

Yellow Spring Instruments – www.ysi.com
York Glassware Services Ltd – www.ygs.net

Index

Actin, 103
Affinity tags
   c-Myc, 33, 125, 127
   FLAG, 33, 36, 37
   polyhistidine, 4, 13, 14, 24, 58–60, 63, 73, 113, 141, 218, 238
Albumin, 111, 144
alpha1-antitrypsin, 111
Antibody
   Fab, 29–30
   hybridoma, 29, 196, 255
   intrabodies, 32, 40
   monoclonal, 212
   ribosome display, 29
   scFv, 29–31, 33–37, 39–40, 44–45, 47–50
   stability engineering, 30–31, 49
   therapeutic, 3, 29
Antigenicity, 129
Arabidopsis thaliana, 102–103
Autographa californica multinucleopolyhedrovirus (AcMNPV), 147–149, 171

Bacillus subtilis
   expression screening, 237
   expression strains, 231, 236
   expression vectors, 231, 236
   subtilin, 226, 228, 231, 235–238
   subtilin regulated expression (SURE), 228, 231, 235–238
   transformation, 237
BacMam viruses
   concentration, 263
   multiplicity of infection (m.o.i.), 263
   production, 262
   storage, 274
   titer determination, 271–272

Baculovirus expression
   automation, 162–164
   BacMagic, 171, 176
   BacPAK, 171
   bacterial artificial chromosomes (BACs), 147
   Bac-to-Bac, 147
   BaculoDirect, 147
   BaculoGold, 171
   *baculo*Quant, 165–167
   BacVector, 171–172
   Cathepsin, 148
   Chitinase, 148–149
   co-transfection, 148–149, 155–156, 162–163
   expression screening, 160–162
   expression strains, 149
   flashBAC, 147–149, 155–157, 162–163, 171
   multiplicity of infection (m.o.i.), 156–157, 160, 167, 177
   ORF1629, 148, 171
   plaque assay, 157–159
   transfer vector, 147–148, 155–156, 163
   virus amplification, 149, 156–157
   virus production, 155–156
   virus quantitation by real time PCR, 164–167
Benzonase, 175, 178, 179
beta-galactosidase, 30, 144, 178, 261
Bleomycin (Zeocin), 125, 127, 131–133, 185, 192
Bone morphogenic protein 2 (BMP-2), 265
Brookhaven protein databank (PDB), 1, 4, 8–9
Butyric acid, 197, 263

Capped mRNA, 92
Carbohydrate group removal, 143
cdc2, 178
Cell-free expression
  E. coli, 87
  rabbit reticulocyte, 2, 87
  wheat-germ, 2, 87–107
Central polypurine track (cPPT), 242, 243
c-fos, 104
Chaperone, 2, 30, 32, 67, 129
Chlorophyll a/b-binding protein, 103
Chromatography
  Fractogel resin, 175
  Fractogel-cobalt, 219–220
  GST resin, 175
  ion-exchange, 60–61
  Ni-NTA resin, 175, 178, 179
  size-exclusion, 64
Circular dichroism, 63–64
c-jun, 104
Cloning
  cDNA, 6, 32, 79, 85, 89, 93–95, 97–98, 102–103, 113, 203
  clone collections, 5–6
  Cre-Lox recombination, 111
  GATEWAY, 6, 111
  gene synthesis, 6
  genomic DNA, 6, 133
  homologous recombination, 111, 130–131, 147–148, 155–156
  integration, 110, 131, 203
  ligation independent, 13, 15–17
  mutagenesis, 32–33, 37, 45, 48, 50, 71, 75, 77, 79, 85
  primer design, 7
  shuttle plasmids, 110
Codon
  optimization, 6
  rare, 31, 34, 56
  usage, 87, 102, 143
Constitutive
  expression, 111, 234
  promoter, 109, 111, 116, 125–126, 225–226, 229, 234, 239
Culture vessels, 187

Detergent, 60, 102, 172
Dextran sulfate, 222, 257
Dihydrofolate reductase (DHFR), 184–185
Dimethylsulfoxide, 114, 189, 207

Directed evolution
  DNA shuffling, 30, 68, 71–72, 76–77
  PCR mutagenesis, 30
Disordered regions, 3, 9
DNAse I, 24, 39, 57, 71, 77, 78, 85
Dynamic light scattering, 63–64

E. coli expression
  auto-induction, 19–20, 23–24
  BL21 (DE3), 22, 60, 72
  cell-lysis, 19–21, 24
  expression screening, 13, 20
  expression strain, 18, 30
  expression temperature, 27, 30, 76, 81
  expression vectors, 14
  general secretory pathway, 30
  periplasm, 30
  protease inhibitors, 23
  protein purification, 24–25
  transformation, 18, 72, 76, 79–80
  twin-arginine translocation pathway, 30
Enhancer
  CMV, 261
  hr5, 171, 180
  ie1, 171, 180
Enolase, 103
Ethylmethane sulfonate, 199
Expression vector
  inducible, 242, 244, 255
  flashback, 148
  NICE, 230
  pB1-7, 14
  pBacPAK8, 262
  pFastBac1, 262
  pIE1, 171
  pIEx, 172, 178, 179
  pIEx/Bac, 171–172, 175, 178
  pPIC, 125
  pTriEx, 178
  pTT, 206
  pYEXBX, 110
  SURE, 231
Extracellular, 8, 109, 112, 123–124, 135–140, 143, 225–227, 229, 234, 236

Factorial design, 61
Fermenters, 110, 124, 142, 204, 267, 269
Fructokinase, 103

Fusion protein
  glutathione-S-transferase (GST), 13–14, 102, 104
  maltose binding protein (MBP), 1, 13–14
  NusA, 14
  thioredoxin (Trx), 14, 103

Gene amplification, 184
Glyceraldehyde-3-phosphate dehydrogenase, 103
G-protein-coupled receptors
  pathways, 109
  refolding, 54
Gram-positive bacteria
  Bacillus subtilis, 227–228, 234–237
  Enterococcus faecalis, 227, 234
  Lactobacillus helveticus, 227, 234
  Lactobacillus plantarum, 228, 234
  Lactobacillus reuteri, 228, 235
  Lactococcus lactis, 226
  Leuconostoc lactis, 227, 234
  Staphylococcus simulans, 235
  Streptococcus agalactiae, 234
  Streptococcus pneumoniae, 227–228, 234
  Streptococcus pyogenes, 227
Green fluorescent protein (GFP)
  colony PCR, 71, 72, 73, 83, 83
  determination of BacMam titers using, 271
  DNA shuffling, 30, 68, 71–72, 76–77
  GFP insertion vectors, 69–70, 72, 74
  library construction, 79
  mutagenesis, 71, 75, 77, 79, 85
  reporter choice, 76, 252
  selection for correct folding, 81
  solubility screening, 73
  split-GFP, 68, 72–73, 84
  verification of optima fluorescence, 82
Guanidium hydrochloride, 57–60

Heat-shock proteins, 103
Hemocytometer, 157, 186, 188, 269
Hepatitis delta virus (HDV), 267
Histone, 102
Holoenzyme, 102
Hygromycin phosphotransferase, 185

Imidazole, 24–26, 58, 119–120, 142, 219–220

Immobilized metal-affinity chromatography (IMAC), 24, 26, 27, 170, 172, 175, 218, 220, 222
Importin, 103
Inclusion bodies (solubilization of), 54, 57–61
Induction, 22, 71–73, 76, 81–82, 111, 116, 118, 120, 135–137, 142, 144, 226–229, 232–239
Initiation factor, 92
Insect cells
  counting, 150
  High Five cells, 149, 151, 153, 160–161
  monolayer culture, 151, 153–154
  oxygen content, 156
  propagation, 151–155
  Sf21 cells, 149, 151, 153, 155, 157–159
  Sf9 cells, 149, 151, 153, 155, 157–161, 163–164
  suspension culture, 151–153
  viability, 149–150, 157–159, 163, 167
Interacting proteins, 4, 10
Intracellular, 30, 32, 49, 109, 113, 123–126, 135–138, 140–142, 144, 212–213, 226–227

Kinase, 2, 102–104, 106, 127, 179, 225
Kozak consensus sequence, 33

Lactic acid bacteria, 226, 231
Lactococcus lactis
  expression screening, 232
  expression strains, 229–230
  expression vectors, 229–231
  nisin controlled expression (NICE), 226–235, 238
  secretion, 225, 229–231, 234
  transformation, 233
Lectin
  Concanavalin A, 199
  Lens culinaris, 199
  Phaseolus vulgaris L, 199
Lentiviral vectors
  concentration, 250–251
  expression vectors, 241–244
  multiplicity of infection (m.o.i.), 248, 256
  self-inactivating (SIN-LV), 244
  storage, 258

INDEX

Liposome, 102, 156, 171, 191, 258
Low complexity, 3–5, 9
Luria-Bertani (LB) broth, 247
Lysostaphin, 235
Lysozyme, 21, 24, 55, 57–59, 232

Mammalian cells
   A-549, 265
   adherent culture, 186–187, 196–197
   African green monkey (CV-1, COS-1, COS-7), 203
   BEAS-2B, 265
   BHK, 265, 267, 271
   Chinese hamster ovary (CHO), 203
   CHO-DG44, 185
   CHO-DUKX, 184
   CHO-K1, 184–185
   CHO-S, 185
   CHO-SSF3, 184
   cloning, 193–196
   cryopreservation, 188, 207
   detachment, 186
   HEK293-6E, 207
   HEK293-EBNA1 or HEK293E, 206
   HEK293F, 206
   HEK293S, 206
   HEK293T, 206
   HeLa, 265, 271
   HepG2, 265
   Huh-7, 265
   human embryonic kidney (HEK) 293, 203
   lipofection, 191
   mutagenesis, 198–199
   primary chondrocytes, 265, 271
   serum-free, 203, 206
   stem cells, 265
   suspension culture, 188, 197, 203
   transduction by BacMam, 261–262, 265
   transfection, 190
Mammalian proteins, 109, 183, 203
Membrane proteins, 2, 8, 102, 109, 148–149, 226–227, 239, 262
Methotrexate (MTX), 184
Myelin basic protein, 102

NADPH thioredoxin reductase, 103
Neomycin phosphotransferase, 185
Neubauer counting chamber, 150

Nisin controlled expression (NICE), 2, 226–235, 238
Nuclear magnetic resonance (NMR), 9, 63, 105

p24, 252
p38 mapk, 178
P450 enzymes, 2
Peptide:N-glycosidase F, 143
Pichia pastoris
   cell lysis, 138
   expression screening, 135–137, 139–140
   expression strains, 127–129
   expression vectors, 124–126, 132
   glycosylation, 123, 129, 140, 143
   purification, 141–142
   scaling up, 142–142
   selectable markers, 124, 126
   temperature, 144
   transformant screening, 133–134
   transformation, 130
pIEX/Bac, 180
Plasmodium falciparum, 102
Poly(A) tail, 88, 92, 124
Polybrene, 248
Polyhedrin, 102, 147–148
Post-translational modification
   disulfide bond, 2, 30, 32, 39–40, 42, 49, 56, 60, 106, 123
   geranylgeranylation, 4
   glycosylation, 2, 109, 123, 129, 140, 143, 183, 198, 261
   glycosylphosphatidylinositol (GPI), 10
   insect cells, 261
   humanization, 129
   mammalian, 203
   methylation, 109
   N-myristoylation, 10
Phosphorylation, 109
   prenylation, 4, 10
   prosthetic group, 102
   proteolytic processing, 123
   sialylation, 261
Promoter
   actin, 171, 261
   ADH1, 111
   AOX1, 124
   CAG, 261, 262
   CMV, 180, 242, 245, 258, 261, 262

Promoter – *contd*
  CR5, 255
  CUP1, 111
  FLD1, 126
  GAL, 111
  GAP, 124
  H1, 262
  HBV, 262
  ie1, 155, 171, 180
  LTR, 242–245
  MFα, 111
  nisA, 225–230, 233–234
  nisF, 225–229
  OpIE2, 171
  p10, 155, 171, 178, 180, 262
  PEX8, 126
  polh, 147–148, 155–156, 167
  RSV, 244, 258, 262
  SP6, 89, 92–93, 95–96, 101, 103
  spas, 228, 236
  SV40, 262
  T7, 14, 22, 33, 36–40, 45, 48, 85, 180
  ubiquitin C, 262
  YPT1, 126
Protease
  enterokinase, 25, 127
  factor Xa, 25, 127
  thrombin, 4, 25
  tobacco etch virus (TEV), 13, 14, 25–27
Protein database, 4–5, 9
Protein disulfide isomerase, 40, 129
Protein microarrays, 105
Protein properties, 3
Protein
  aggregation, 60–61, 63–64
  refolding, 1, 53–54, 56, 67, 102
  therapeutic, 3, 54, 183, 203
  toxic, 2, 31, 49, 87, 126, 135, 244
  truncation, 3, 9
  turbidity, 62
Puromycin *N*-acetyl transferase, 185

Rev-responsive element (RRE), 242, 243
Ribosome display
  affinity selection, 44–46, 68
  display vector, 34–37
  E. coli extract preparation, 41–42
  in vitro transcription, 39
  in vitro translation, 40, 43
  in vivo analysis, 49

Ribosome display – *contd*
  library creation, 37
  RT-PCR, 44, 47
  ssrA oligonucleotide, 41
Ribosome inactivating proteins (RIPs), 88

Saccharomyces cerevisiae
  2 μm DNA, 110
  expression and analysis, 115–120
  expression strains, 112–113
  expression vectors, 110
  homologous recombination, 111
  promoters, 111
  secretion, 112
  shuttle plasmids, 110
  transformation, 114
S-adenosylmethionine synthetase, 103
Secreted alkaline phosphatase, 221–222
Secreted proteins, 2, 126, 133, 144,
    148–149, 212, 218, 221, 229–231
Secretion leader
  Bacillus licheniformis amylase L, 234
  L. lactis Usp45, 231
  P. pastoris PHO1, 126
  S. cerevisiae α mating type factor, 112
Secretion machinery, 111, 229
Selection agents, 192, 193
Shine-Dalgarno sequence, 33, 68
Signal peptide prediction, 10
Soluble expression, 9, 102, 124
Structural genomics, 1, 13, 105
Subtilin regulated expression (SURE), 228,
    230–231, 235–238
Supernatant concentration, 143, 218

Tangential flow filtration, 218
Transfection reagents
  293fectin, 217
  calcium phosphate, 206
  cationic lipid, 203
  CELLFECTIN, 156
  FuGENE 6, 156
  GeneJuice, 156, 181
  Lipofectamine 2000, 191, 246, 258
  Lipofectin, 155
  polyethylenimine (PEI), 206, 213, 215,
      245
  Tfx-20, 156
  X-tremeGENE Ro-1539, 216

Transformation
  Bacillus subtilis, 237
  E. coli, 18, 72, 76, 79, 80
  Lactococcus lactis, 233
  Pichia pastoris, 130
  Saccharomyces cerevisiae, 114
Transient expression in HEK293-EBNA1 cells
  culture medium concentration, 218
  maxiprep DNA preparation, 210
  media, 207
  PEI preparation, 211
  pluronic F68, 207
  protein purification, 218–220
Transmembrane, 3, 5, 8–10
Trichostatin A, 264
Tritin, 88
Trypsin, 154, 186, 253, 273

Untranslated regions (UTR), 6, 88, 124, 126, 143
Urea, 58–62

Vesicular stomatitis virus glycoprotein-G, 243, 245
Virus-like particle, 262, 267

WAVE technology, 151, 267
Wheat-germ cell-free system
  15N-labeling, 105
  bilayer reaction, 89, 97–100, 106
  continuous-flow cell-free (CFCF), 87, 95, 97–100, 102
  discontinuous batch reaction, 97–99
  embryo extract preparation, 89–91
  expression vector, 89, 92–93
  omega sequence, 92
  proteins expressed, 103–104
  transcription, 89, 92–96, 101, 103

X-ray crystallography
  selenomethionine labeling, 22, 105
  structure determination, 4, 9, 105, 183